Lecture Notes in Mathematics

Edited by A. Dold and B. Eckmann

477

Optimization and Optimal Control

Proceedings of a Conference Held at Oberwolfach, November 17–23, 1974

Edited by R. Bulirsch, W. Oettli, and J. Stoer

Springer-Verlag
Berlin · Heidelberg · New York 1975

Editors

Prof. Roland Bulirsch
Mathematisches Institut der
Technischen Universität München
8 München
Postfach 202420
BRD

Prof. Werner Oettli
Fakultät für Mathematik
und Informatik
Universität Mannheim
68 Mannheim
Schloß
BRD

Professor Josef Stoer
Institut für Angewandte Mathematik
und Statistik der Universität Würzburg
87 Würzburg
Am Hubland
BRD

Library of Congress Cataloging in Publication Data

Main entry under title:

Optimization and optimal control.

(Lecture notes in mathematics ; 477)
Bibliography: p.
Includes index.
1. Mathematical optimization--Congresses. 2. Control theory--Congresses. I. Bulirsch, Roland. II. Oettli, Werner. III. Stoer, Josef. IV. Series: Lecture notes in mathematics (Berlin) ; 477.
QA3.L28 no. 477 [QA402.5] 510'.8s [629.8'312]
75-23372

AMS Subject Classifications (1970): 41A20, 49AXX, 49BXX, 49CXX, 49DXX, 90AXX, 90CXX

ISBN 3-540-07393-0 Springer-Verlag Berlin · Heidelberg · New York
ISBN 0-387-07393-0 Springer-Verlag New York · Heidelberg · Berlin

Printed in Germany

Offsetdruck: Julius Beltz, Hemsbach/Bergstr.

TABLE OF CONTENTS

Rainer E. Burkard

Kombinatorische Optimierung in Halbgruppen 1

Richard W. Cottle

On Minkowski Matrices and the Linear Complementarity Problem 18

Hans Czap

Exact Penalty-Functions in Infinite Optimization 27

P.C. Das

Application of Dubovitskii-Milyutin Formalism to Optimal Settling Problem with Constraints 44

Peter Deuflhard

A Relaxation Strategy for the Modified Newton Method 59

E.D. Dickmanns

Optimale Steuerungen für Gleitflugbahnen maximaler Reichweite beim Eintritt in Planetenatmosphären 74

Ulrich Eckhardt

On an Optimization Problem Related to Minimal Surfaces with Obstacles 95

Klaus Glashoff

Optimal Control of One-Dimensional Linear Parabolic Differential Equations 102

Gene Golub

Nonlinear Least Squares and Matrix Differentiation+

Sven-Åke Gustafson

On the Numerical Treatment of a Multi-Dimensional Parabolic Boundary-Value Control Problem 121

Joachim Hartung

Penalty-Methoden für Kontrollprobleme und Open-Loop-Differential-Spiele 127

Gerhard Heindl

Ein Existenzsatz für konvexe Optimierungsprobleme 145

K.-H. Hoffmann, E. Jörn, E. Schäfer, H. Weber

Ein Approximationssatz und seine Anwendung in der Kontrolltheorie 150

P. Huard
Optimization Algorithms and Point-to-Set-Mapping+

Hansgeorg Jeggle
Zur Störungstheorie nichtlinearer Variationsungleichungen 158

H.W. Knobloch
On Optimal Control Problems with State Space Constraints 177

Michael Köhler
Explicit Approximation of Optimal Control Processes 188

P.J. Laurent
Un Algorithme Dual Pour le Calcul de la Distance Entre Deux Convexes 202

F.A. Lootsma
The Design of a Nonlinear Optimization Programme for Solving Technological Problems 229

H. Maurer, U. Heidemann
Optimale Steuerprozesse mit Zustandsbeschränkungen 244

Peter Spellucci
Algorithms for Rational Discrete Least Squares Approximation 261

Norbert Weck
Über das Prinzip der eindeutigen Fortsetzbarkeit in der Kontrolltheorie 276

Jochem Zowe
Der Sattelpunktsatz von Kuhn und Tucker in geordneten Vektorräumen 285

+ These papers, presented at the Conference, do not appear in this volume, and will be published elsewhere.

LIST OF PARTICIPANTS

Professor R.E. Burkard

Mathematisches Institut
der Universität zu Köln

5 Köln 41

Weyertal 86-90

Professor R.W. Cottle

Stanford University
Operation Research Department

Stanford, California 94305

U.S.A.

Dr. Hans Czap

Lehrstuhl für Mathematische
Verfahrensforschung der
Universität Göttingen

34 Göttingen

Nikolausbergerweg 9b

Dr. P.C. Das

I Mathematisches Institut
der Freien Universität Berlin

1 Berlin 33

Arnimallee 2-6

Dr. P. Deuflhard

Institut für Mathematik
der TU München

8 München 2

Arcisstr. 21

Dr. E.D. Dickmanns

DFVLR
Institut für Dynamik der Flugsysteme

8031 Oberpfaffenhofen

Post Weßling / Obb.

Dr. U. Eckhardt

Zentralinstitut für Angew. Mathematik
Kernforschungsanstalt Jülich

517 Jülich

Postfach 365

Dr. K. Glashoff
Fachbereich Mathematik
61 Darmstadt
Kantplatz 1

Dr. S.A. Gustafson
Royal Institute of Technology
Department for Computer Sciences
Numerical Analysis
S - 10044 Stockholm 70

Dr. J. Hartung
Institut für Angew. Mathematik
und Informatik
53 Bonn
Wegelerstr. 10

Dr. G. Heindl
Institut für Mathematik
der Technischen Universität
8 München 2
Arcisstr. 21

Dr. K.-H. Hoffmann
Mathematisches Institut
der Universität München
8 München 2
Theresienstr. 39

Professor H. Jeggle
Technische Universität Berlin
Fachbereich Mathematik
1 Berlin 12
Straße des 17. Juni 135

Professor H.W. Knobloch
Mathematisches Institut
87 Würzburg
Am Hubland

Dr. M. Köhler

Universität Zürich
Institut für Operations Research

CH - 8006 Zürich

Weinbergstr. 59

Professor P.J. Laurent

Mathematiques Appliquées
Université Scientifique et
Médicale de Grenoble

Cédex 53

F - 38041 - Grenoble

Dr. F.A. Lootsma

Afdeling Algemene Wetenschappen
Technische Hogeschool

Delft

Niederlande

Dr. H. Maurer

Mathematisches Institut
der Universität zu Köln

5 Köln 41

Weyertal 86-90

Dr. P. Spellucci

Fachbereich Mathematik

65 Mainz

Saarstr. 21

Professor N. Weck

Fachbereich Mathematik

61 Darmstadt

Kantplatz 1

Dr. J. Zowe

Institut für Angew. Mathematik
der Universität Würzburg

87 Würzburg

Am Hubland

Kombinatorische Optimierung in Halbgruppen

Rainer E. Burkard, Köln [1)]

Abstract

By an algebraisation of the objective function is achieved that combinatorial optimization problems with different kinds of objective functions (e.g. sums or bottleneck objective functions) now occur as special cases of one general problem. The algebraisation respects not only the structure of the underlying problems but also the structure of algorithms for solving these problems. Therefore the generalized problems belong to the same complexity class as the original problems.

Combinatorial optimization problems, which can be formulated without real variables (e.g. assignment problems), can be considered now in totally ordered semigroups, where $(S,*,\leq)$ obeys additionally a strong combatibility axiom and a divisibility axiom. For problems with real variables some additional combatibility axioms between the domain Ω of the variables and the semigroup S have to be fulfilled.

After the investigation of the structure of the systems $(S,*,\leq)$ and $(S,*,\leq;\Omega)$ general assignment problems and maximal flow problems in networks with generalized costs are considered and algorithms are given for solving these problems.

Zusammenfassung

In der vorliegenden Arbeit wird durch eine Algebraisierung der Zielfunktion erreicht, daß kombinatorische Optimierungsaufgaben mit verschiedenen Arten von Zielfunktionen (z.B. Summen- und Bottleneckzielfunktionen) nun als Spezialfälle eines verallgemeinerten Problems auftreten. Die Algebraisierung nimmt dabei nicht nur Rücksicht auf die Struktur der Probleme, sondern auch auf die Art der zugehörigen Lösungsverfahren. Dadurch wird erreicht, daß Algorithmen zur Lösung der allgemeinen Probleme angegeben werden können, die im wesentlichen denselben

---

1) Mathematisches Institut der Universität zu Köln,
D - 5 Köln 41, Weyertal 86 - 90

Aufwand haben wie Algorithmen für das zugehörige Standardproblem.

Variablenfrei formulierbare Probleme, wie etwa Zuordnungsprobleme, lassen sich in einer angeordneten Halbgruppe $(S,*,\leq)$ formulieren und lösen. Dabei muß $(S,*,\leq)$ zusätzlich ein starkes Verträglichkeitsaxiom und ein Teilbarkeitsaxiom erfüllen. Bei Optimierungsproblemen, in denen reelle Variable auftreten, wie etwa Transportprobleme, müssen für die äußere Verknüpfung zwischen dem Variablenbereich Ω und der Halbgruppe S zusätzliche Verträglichkeitsaxiome erfüllt sein. Nach einer Untersuchung der Struktur der Systeme $(S,*,\leq)$ und $(S,*,\leq;\Omega)$ werden als Beispiele verallgemeinerte Zuordnungsprobleme und maximale Flußprobleme in Netzwerken mit verallgemeinerten Kosten betrachtet und Lösungsmöglichkeiten für diese Probleme aufgezeigt.

1. Einleitung.

In der Praxis treten oft kombinatorische Optimierungsprobleme auf, die sich nur in der Form der Zielfunktion unterscheiden. Werden etwa Anzahlen minimiert, so sind es meist Summenprobleme, wie etwa das lineare Summenzuordnungsproblem:

"Finde eine Permutation φ von $N = \{1,2,...,n\}$, so daß

$\sum_{i \in N} c_{i\varphi(i)}$ minimal wird."

Handelt es sich hingegen um die Minimierung von Zeitfaktoren, so nimmt die Zielfunktion zum Beispiel Bottleneck Gestalt an. So läßt sich etwa ein lineares Bottleneckzuordnungsproblem folgenderweise formulieren:

"Finde eine Permutation φ von N, so daß $\max_{i \in N} c_{i\varphi(i)}$

minimal wird."

In der Literatur wurden bisher diese Probleme gesondert behandelt. Bei der Lösung quadratischer Zuordnungsprobleme

"Finde eine Permutation φ von N, die $\sum_{i \in N} \sum_{p \in N} d_{i\varphi(i)p\varphi(p)}$

minimiert"

fiel mir nun auf [1], daß sich ein Algorithmus zur Lösung des quadratischen Bottleneckproblems

$$\min_{\varphi} \max_{i \in N} \max_{p \in N} d_{i\varphi(i)p\varphi(p)}$$

direkt aus dem Algorithmus für quadratische Zuordnungsprobleme gewinnen läßt, wenn man jeweils die Operation "+" durch "max" ersetzt. Es zeigt sich nun generell, daß für kombinatorische Optimierungsprobleme die Struktur der Zielfunktion von jener der Restriktion weitgehend unabhängig ist und daß sich die Zielfunktion oftmals in einfacher Weise algebraisch beschreiben läßt.

In der kombinatorischen Optimierung hat man es mit zwei Problemtypen zu tun. Einerseits handelt es sich um Probleme, die variablenfrei geschrieben werden können (z.B. Zuordnungsprobleme, Rundreiseproblem,...) und andrerseits sind es Probleme, in deren Zielfunktion reelle Variable explizit auftreten (z.B. Transportprobleme, maximale Flüsse mit minimalen Kosten, Rucksackprobleme,...). Demnach muß man bei einer Algebraisierung der Zielfunktion zwei Fälle unterscheiden. Bei variablenfrei formulierbaren Problemen mit Summenzielfunktion wird das "+" durch eine innere Verknüpfung in einer geordneten Menge $(S,\leq)$ ersetzt. Bei Problemen, in denen reelle Variable explizit auftreten, tritt eine äußere Verknüpfung zwischen dem Variablenbereich Ω und der Halbgruppe S hinzu. In Abschnitt 2 wird zunächst die Struktur von $(S,*,\leq)$ bzw. $(S,*,\leq;\Omega)$ näher untersucht. Die Axiome für die Systeme $(S,*,\leq)$ und $(S,*,\leq;\Omega)$ orientieren sich nicht nur an der Lösbarkeit der in ihnen formulierten allgemeinen Probleme, sondern insbesondere auch daran, daß die üblichen Lösungsverfahren für Standardprobleme übertragen werden können zur Lösung der verallgemeinerten Probleme, die in $(S,*,\leq)$ und $(S,*,\leq;\Omega)$ formuliert sind. Dies ist ein wesentlicher Gesichtspunkt für kombinatorische Optimierungsaufgaben, die oftmals durch vollständige Enumeration auf triviale Weise lösbar sind. Die angegebenen Systeme garantieren, daß beim Übergang von Standardproblemen zu allgemeinen Problemen sich nicht viel am Rechenaufwand ändert. Präziser formuliert heißt dies: Sind die inneren und äußeren Verknüpfungen sowie die Reduktionen (vgl. Axiom IV, Abschnitt 2) in polynomial beschränkter Zeit auf einer deterministischen Turingmaschine ausführbar, so ist die Verallgemeinerung eines Problems der Komplexitätsklasse P wieder ein Problem der Komplexitätsklasse P (vgl. Karp [7]). Die Algebraisierungen anderer Autoren (z.B. Carré [4], Gondran [6], Roy [8]) sind darauf ausgerichtet, die algebraische Struktur der Restriktionen und Zielfunktion zu klären und zu axiomatisieren. Die vorliegende Arbeit möchte jedoch durch Berücksichtigung der Lösungsverfahren nicht nur einen Einblick in die Struktur der Probleme geben, sondern auch das Aufstellen von Lösungsverfahren für ganze Klassen von Problemen er-

Es wird sich zeigen, daß lediglich vier Axiome für $(S,*,\leq)$ bzw. einige weitere Verträglichkeitsaxiome für $(S,*,\leq;\Omega)$ genügen, um die üblichen Lösungsverfahren zu Lösungsverfahren für Probleme in $(S,*,\leq)$ bzw. $(S,*,\leq;\Omega)$ zu erweitern. Die entstehenden Strukturen sind so allgemein, daß eine Vielzahl von interessanten und für die Praxis relevanten Problemen dadurch erfaßt wird, wie etwa als Verknüpfungen Addition, Multiplikation, Maximumbildung, als Koeffizienten Zahlen, Vektoren und Elemente angeordneter Gruppen sowie als Ordnungen etwa $\geq$, $\leq$ und die lexikographische Ordnung. Für manche dieser Probleme läßt sich eine Transformation auf Summenprobleme in den reellen Zahlen angeben. Daß es oft nicht sinnvoll ist, derartige Transformationen durchzuführen, zeigt sich etwa an Bottleneckproblemen. So könen zum Beispiel lineare Bottleneck-Zuordnungsprobleme in einem Bruchteil der Rechenzeit von Summen-Zuordnungsproblemen gelöst werden [2]. Dieser Effekt, der ganz allgemein bei Bottleneckproblemen auftritt, findet in der nachfolgend entwickelten Theorie eine natürliche Erklärung.

2. Die Systeme $(S,*,\leq)$ und $(S,*,\leq;\Omega)$

Im Zusammenhang mit einer algebraischen Behandlung von Zuordnungsproblemen wurde das System $(S,*,\leq)$ von Burkard, Hahn und Zimmermann [3] näher untersucht. Es zeigte sich, daß folgende Forderungen an die nichtleere Menge S mit der inneren Verknüpfung * und einer Ordnungsrelation $\leq$ zu stellen sind:

(I) S ist totalgeordnet bezüglich "$\leq$"

(II) $(S,*)$ ist kommutative Halbgruppe

(III) Für alle $a,b,c \in S$ mit $a \leq b$ gelte $a * c \leq b * c$

(IV) Für alle $a, b \in S$ mit $a \leq b$ existiere ein $c \in S$ mit $a * c = b$

Beispiele für Systeme $(S,*,\leq)$, die das Axiomsystem erfüllen, sind etwa

a) $(S,*,\leq)$ mit $S=\mathbb{R}, \mathbb{R}^+, \mathbb{Q}, \mathbb{Q}^+, \mathbb{Z}, \mathbb{N}$ und $* =$ "+" oder "max"

b) $(\mathbb{R}^+\setminus\{0\},\cdot,\leq)$

c) $(\mathbb{R},+,\geq)$

d) Angeordnete Gruppen, die Axiom (III) erfüllen

e) Ein lexikographisches Produkt $(G,*,\preceq)$ mit $G = \prod_{\lambda\in\Lambda} G_\lambda$, $* = (*_\lambda)$ und der lexikographischen Ordnung $\preceq$ erfüllt genau dann (I)-(IV), wenn jedes $(G_\lambda,*_\lambda,\leq_\lambda)$ eine angeordnete Gruppe ist, die Axiom (III) erfüllt.(vgl. [3]).

Für das Folgende sei angenommen, daß $(S,*,\leq)$ stets die Axiome (I)-(IV) erfülle.

Axiom (III) garantiert, daß die innere Verknüpfung und die Ordnungsrelation kompatibel sind. Dies ist wichtig für Algorithmen, die wechselweise Verknüpfungen und Vergleiche bezüglich "$\leq$" durchführen. Aufgrund von Axiom (III) gilt:

Satz 1: Sei $B \subseteq S$ und B besitze ein kleinstes Element b_o. Dann gilt

$$\min_{b\in B} (a*b) = a * \min_{b\in B} b$$

Beweis: Da für alle $b \in B$ stets $b_o \leq b$ gilt, folgt aus (III)

$$a * b_o \leq a * b.$$

Axiom (IV) ermöglicht eine abgeschwächte Form von Subtraktionen: Häufig werden in den Algorithmen Reduktionsschritte durchgeführt, d.h. ist $a \leq b$ und $a * c = b$, so kann das größere Element b durch c ersetzt werden, während die Zielfunktion mit dem Element a transformiert wird.

Aufgrund von Axiom (IV) gibt es zu jedem $a \in S$ ein a_o mit $a * a_o = a$.

Definition 1. Ein Element $a_o \in S$ heißt von a dominiert, wenn $a * a_o = a$ gilt. dom a ist die Menge der von a dominierten Elemente.

Ist $(S,*)$ eine Gruppe, so enthält dom a nur das neutrale Element der Gruppe. Für die Verknüpfung "max" ist jedoch dom a $= \{a_o \mid a_o \leq a\}$. Die Mengen dom a besitzen folgende leicht nachweisbaren Eigenschaften. Es sei $a, b \in S$. Dann gilt

(2.1) dom a $\subseteq$ dom $(a * b)$

(2.2) $a \leq b \Rightarrow$ dom a $\subseteq$ dom b

(2.3) Es sei $\bar{S}$ eine nach unten beschränkte Teilmenge von S. Dann gilt

$$\bigcap_{a\in\bar{S}} \text{dom } a \neq \emptyset$$

Die Menge dom a ermöglicht es, positive Elemente relativ zu a zu erklären.

Definition 2. Sei $a, b \in S$. a heißt b-positiv [b-negativ], wenn ein $b_o \in$ dom b existiert mit $b_o \leq a$ [$b_o \geq a$]. pos b [neg b] ist die Menge der bezüglich b positiven [negativen] Elemente.

Es läßt sich nun unschwer zeigen:

(2.4) $a \in \text{pos } b \Leftrightarrow b \leq b * a$

$a \in \text{neg } b \Leftrightarrow b \geq b * a$

(2.5) $\text{dom } b = \text{neg } b \cap \text{pos } b$

(2.6) Die Mengen $(\text{neg } b \setminus \text{dom } b)$, $\text{dom } b$ und $(\text{pos } b \setminus \text{dom } b)$ bilden eine Partition von S.

Definieren wir für Teilmengen $A, B \subseteq S$ eine Ordnung $A \leq B$ durch $\bigwedge_{a \in A} \bigwedge_{b \in B} a \leq b$, so gilt:

(2.7) $(\text{neg } b \setminus \text{dom } b) < \text{dom } b < (\text{pos } b \setminus \text{dom } b)$

(2.8) Für alle $b \in S$ gilt:

$\text{pos } a \subseteq \text{pos } (a*b)$, $\text{neg } a \subseteq \text{neg } (a*b)$

aber:

$\text{pos } (a*b) \setminus \text{dom } (a*c) \subseteq \text{pos } a \setminus \text{dom } a$

$\text{neg } (a*b) \setminus \text{dom } (a*b) \subseteq \text{neg } a \setminus \text{dom } a$

Ferner gilt:

<u>Satz 2</u>. Sei $a \in S$. Dann erfüllen die Systeme $(\text{dom } a, *, \leq)$ und $(\text{pos } a, *, \leq)$ die Axiome (I)-(IV).

Ein Beweis dieses Satzes mit Hilfe von Fallunterscheidungen findet sich ebenfalls in [3]. Man beachte, daß eine analoge Aussage für das System $(\text{neg } a, *, \leq)$ nicht gilt infolge der Unsymmetrie bei Axiom (III) und (IV).

Bei Problemen, die mit Hilfe reeller Variablen formuliert werden, tritt nun noch ein Variablenbereich Ω hinzu. Im Falle eines Systems $(S,*,\leq;\Omega)$ wollen wir stets voraussetzen, daß $(S,*)$ ein neutrales Element besitzt. Es gelte nun

(V) $0 \in \Omega \subseteq \mathbb{R}^+$ und $(\Omega,+)$, $(\Omega,\cdot)$ abgeschlossen

Im allgemeinen muß Ω noch weitere Bedingungen erfüllen, z.B. Axiom (III) bezüglich der Addition. Da diese aber von der Struktur des Problems bzw. der Restriktionen und nicht von der Zielfunktion abhängen, soll hier nicht darauf eingegangen werden. Die äußere Verknüpfung $\Box$ zwischen Ω und S muß jedoch folgende Verträglichkeitsaxiome erfüllen:

(VI) Für alle $\omega_1, \omega_2 \in \Omega$ und $a \in S$ gilt $\omega_1 \Box (\omega_2 \Box a) = (\omega_1 \cdot \omega_2) \Box a$

(VII) Für alle $\omega_1, \omega_2 \in \Omega$ und $a, b \in S$ gilt:

$$\omega_1 \,\square\, (a*b) = (\omega_1 \square a) * (\omega_1 \square b)$$

$$(\omega_1+\omega_2) \,\square\, a = (\omega_1 \square a) * (\omega_2 \square b)$$

(VIII) Für alle $\omega_1, \omega_2 \in \Omega$ und $a, b \in S$ gilt:

$$\omega_1 \le \omega_2 \Rightarrow \omega_1 \square a \le \omega_2 \square a$$

$$a \le b \Rightarrow \omega_1 \square a \le \omega_1 \square b$$

(IX) Für alle $a \in S$ gilt

$$0 \,\square\, a = e$$

Beispiele für Systeme $(S,*,\le;\Omega)$ sind etwa

a) $(S,+,\le;\Omega)$ mit $\omega \,\square\, a = \omega \cdot a$, $S \subseteq \mathbb{R}^n$, $\Omega \subseteq \mathbb{R}^+_m$

b) $(S,\cdot,\le;\Omega)$ mit $\omega \,\square\, a = a^\omega$, $S \subseteq \mathbb{R}$, $\Omega \subseteq \mathbb{R}^+$

c) $(S,\max,\le;\Omega)$ mit $S \subseteq \mathbb{R}$, $\Omega \subseteq \mathbb{R}^+$ und

$$\omega \,\square\, a = \begin{cases} a, & \text{falls } \omega > 0 \\ e, & \text{falls } \omega = 0 \end{cases}$$

In $(S,*,\le;\Omega)$ läßt sich ein Paar von zueinander dualen Optimierungsaufgaben formulieren. Es sei $c \in S^n$, $A \in M_{mn}(\Omega)$, $b \in \Omega^m$. Die Optimierungsaufgabe

$$(2.9) \qquad \min_{x\in\Omega^n} \left\{ \mathop{\ast}_{j\in N} x_j \square c_j \mid Ax = b \right\}$$

sei das primale Problem. Als dazu duale Aufgabe setzen wir

$$(2.10) \qquad \max_{y\in S^m} \left\{ \mathop{\ast}_{1\le i\le m} b_i \square y_i \mid \mathop{\ast}_{1\le i\le m} (a_{ji} \square y_i) \le c_j,\ 1\le j\le n \right\}$$

Während beim primalen Problem nichtnegative reelle Zahlen x_j gesucht werden, sind die Variablen des dualen Problems Elemente der Halbgruppe S. Es gilt

Satz 3 (Schwacher Dualitätssatz)

Ist $x \in \Omega^n$ eine Lösung von (2.9) und $y \in S^m$ eine Lösung von (2.10), dann gilt

$$\mathop{\ast}_{1\le j\le n} x_j \square c_j \ge \mathop{\ast}_{1\le i\le m} b_i \square y_i$$

Beweis:
$$\mathop{\ast}_{1\le i\le m} b_i \square y_i = \mathop{\ast}_{1\le i\le m} \left(\sum_{1\le j\le n} a_{ij} x_j \right) \square y_i =$$

$$= \mathop{\ast}_{1\le i\le m} \mathop{\ast}_{1\le j\le n} (a_{ij} x_j) \square y_i = \mathop{\ast}_{1\le j\le n} x_j \square \left(\mathop{\ast}_{1\le i\le m} a_{ij} \square y_i \right) \le$$

$$\le \mathop{\ast}_{1\le j\le n} x_j \square c_j$$

Q.E.D.

Unter welchen Voraussetzungen sich ein dem starken Dualitätssatz analoges Resultat beweisen läßt, ist bis jetzt nicht bekannt.

3. Verallgemeinerte Zuordnungprobleme

Verallgemeinerte Zuordnungprobleme wurden erstmals von Burkard, Hahn und Zimmermann [3] untersucht. Dort wurde auch ein Lösungsverfahren für sie angegeben.

Gegeben sei ein System $(S,*,\le)$, das die Axiome (I)-(IV) erfüllt. Ferner sei $N = \{1,2,\dots,n\}$. Ein verallgemeinertes lineares Zuordnungsproblem läßt sich folgenderweise formulieren:

> Seien $c_{ij} \in S$ $(i,j \in N)$. Gesucht wird eine Permutation φ von N, die $\mathop{\ast}_{i\in N} c_{i\varphi(i)}$ minimiert.

Verallgemeinerte Zuordnungsprobleme können durch sukzessive Transformationen gelöst werden.

Definition 3: Eine Transformation von $C = (c_{ij})$ auf $\bar{C} = (\overline{c_{ij}})$ heißt zulässig, wenn es eine Konstante $a \in S$ gibt, so daß für jede Permutation φ von N gilt:

$$\mathop{\ast}_{i\in N} c_{i\varphi(i)} = a * \mathop{\ast}_{i\in N} \overline{c_{i\varphi(i)}}$$

a heißt der Index der Transformation.

Für das System $(R,+,\leq)$ sind Matrizentransformationen, wie sie etwa bei der Ungarischen Methode durchgeführt werden, zulässige Transformationen. Wir zeigen, daß hier ein ähnliches Resultat gilt

Satz 4 Es sei $C \in M_{nn}(S)$ und $I \subseteq N$, $J \subset N$ mit $|I| > |J|$

Ferner sei

$$\alpha = \min \left\{ c_{ij} \mid i \in I,\ j \notin J \right\}$$

Dann ist die Transformation $C \rightsquigarrow \bar{C}$, wobei die Koeffizienten $\overline{c_{ij}}$ definiert sind durch

$$\alpha * \overline{c_{ij}} = c_{ij} \qquad (i \in I,\ j \notin J)$$

(3.1)
$$\overline{c_{ij}} = c_{ij} \qquad (i \in I,\ j \in J \text{ und } i \notin I,\ j \notin J)$$

$$\overline{c_{ij}} = \alpha * c_{ij} \qquad (i \notin I,\ j \in J)$$

eine zulässige Transformation mit dem Index $\alpha^{|I|-|J|}$

Bemerkung: Ist $I = \{i\}$, $J = \emptyset$, so handelt es sich um eine Reduktion der Zeile i; ist $I = N$, $J = N \setminus \{j\}$, so ist (3.1) eine Reduktion der Spalte j.

Beweis: Es sei φ eine beliebige Permutation von N. Wir bezeichnen mit $n_o(\varphi)$ die Anzahl der Paare $(i,\varphi(i))$ mit $i \in I$, $\varphi(i) \notin J$, mit $n_1(\varphi)$ die Anzahl der Paare $(i,\varphi(i))$ mit $i \in I$, $\varphi(i) \in J$ und $i \notin I$, $\varphi(i) \notin J$ sowie mit $n_2(\varphi)$ die Anzahl jener $(i,\varphi(i))$ mit $i \notin I$, $\varphi(i) \in J$. Dann gilt

(3.2)
$$n_o(\varphi) = m + n_2(\varphi) \qquad \text{mit } m = |I|-|J|$$

Nun ist

(3.3)
$$\underset{i \in N}{\ast} c_{i\varphi(i)} = \underset{i \in I}{\ast} c_{i\varphi(i)} * \underset{i \notin I}{\ast} c_{i\varphi(i)} =$$

$$= \alpha^{n_o} * \underset{i \in I}{\ast} \overline{c_{i\varphi(i)}} * \underset{i \notin I}{\ast} c_{i\varphi(i)}$$

Infolge (3.2) gilt

$$\alpha^{n_o} = \alpha^{m} * \alpha^{n_2}$$

und aus (3.1) folgt

$$\alpha^{n_2} * \underset{i \notin I}{\ast} c_{i\varphi(i)} = \underset{i \notin I}{\ast} \overline{c_{i\varphi(i)}}$$

daher folgt aus (3.3) die Behauptung des Satzes. Q.E.D.

Es ist möglich, ein Lösungsverfahren für verallgemeinerte lineare Zuordnungsprobleme anzugeben, das nach höchstens $\frac{1}{2}(n^2+3n-2)$ Transformationen der Form (3.1) die Optimallösung ergibt [3]. Jede Transformation der Gestalt (3.1) erhöht den Wert der Zielfunktion um α^m mit $m = |I|-|J|$. Dadurch wird bei Bottleneckproblemen bei jeder derartigen Transformation die Anzahl der vom Zielfunktionswert z dominierten Elemente größer, während etwa bei Summenproblemen nur die Nullelemente vom Zielfunktionswert dominiert werden. Da das Verfahren dann abbricht, wenn n voneinander unabhängige Elemente gefunden sind, die vom Zielfunktionswert dominiert werden, erhält man bei Bottleneckproblemen wesentlich schneller die Optimallösung als bei Summenproblemen.

Verallgemeinerte quadratische Zuordnungsprobleme sind Probleme der Gestalt

$$(3.4) \qquad \min_{\varphi} \underset{i \in N}{\ast} \underset{p \in N}{\ast} d_{i\varphi(i)p\varphi(p)}$$

Allgemeine quadratische Zuordnungsprobleme lassen sich durch eine verallgemeinerte Störungsmethode [1] lösen.

4. Flußprobleme mit minimalen Kosten

Gegeben sei ein gerichteter Graph $G = (V,E)$, in dem zwei voneinander verschiedene Knoten ausgezeichnet werden: die Quelle q und die Senke s. Die Abbildung $c : E \to \mathbb{R}^+$ vermittle eine Bewertung der Knoten des Graphen. Wir bezeichnen mit $V(x) = \{y \mid (y\ x) \in E\}$ die Vorgängerknoten des Knotens $x \in V$ und mit $N(x) = \{y \mid (x,y) \in E\}$ die Nachfolgerknoten von x. Ein Fluß mit dem Wert v von q nach s ist eine Abbildung $f : E \to \mathbb{R}^+$ mit folgenden Eigenschaften

$$\sum_{y \in N(x)} f(x,y) - \sum_{y \in V(x)} f(y,x) = \begin{cases} v & x=q \\ 0 & x \neq q,\ x \neq s \\ -v & x=s \end{cases}$$

$$\bigwedge_{(x,y) \in E} f(x,y) \leq c(x,y)$$

Zur Bestimmung eines maximalen Flusses von q nach s steht zum Beispiel der Markierungsalgorithmus von Ford und Fulkerson [5] zur Verfügung. Es sei nun eine weitere Abbildung $a : E \to S$ gegeben. Wir nennen $a(x,y)$ die Kosten für eine Flußeinheit in der Kante (x,y). Ein Flußproblem mit minimalen verallgemeinerten Kosten läßt sich nun so formulieren:

Gegeben sei ein gerichteter Graph $G = (V,E)$ mit Quelle q, Senke s und Bewertung c. Ferner sei $(S,*,\leq;\Omega)$ eine Halbgruppe, die die Axiome (I)-(IX) erfüllt und $a : E \to S$ sei eine Abbildung, die jeder Kante "Kosten" $a(x,y)$ zuordnet. Gesucht ist ein maximaler Fluß von q nach s, für den

$$\mathop{\ast}_{(x,y) \in E} f(x,y) \,\square\, a(x,y)$$

minimal wird.

Ist $(S,*,\leq;\Omega)$ das System $(\mathbb{R},+,\leq;\mathbb{R}^+)$ und $\square$ die Multiplikation, so erhält man ein maximales Flußproblem mit minimalen Kosten im herkömmlichen Sinne. Ein Algorithmus zur Lösung dieses Problems findet sich ebenfalls in [5].

Die Idee des nachfolgenden Verfahrens zur Lösung des maximalen Flußproblems mit minimalen allgemeinen Kosten besteht im folgenden:

Es seien $f_1, f_2, \dots, f_k$ Teilflüsse, deren Summe einen maximalen Fluß ergibt. Nach Axiom (VII) gilt

$$\mathop{\ast}_{(x,y) \in E} \left(\sum_{1 \leq i \leq k} f_i(x,y) \right) \square\, a(x,y) = \mathop{\ast}_{1 \leq i \leq k} \left(\mathop{\ast}_{(x,y) \in E} f_i(x,y) \,\square\, a(x,y) \right)$$

Zur Bestimmung eines maximalen Flusses mit minimalen Kosten werden zunächst die Kapazitätsrestriktionen außer acht gelassen und es wird ein Fluß von q nach s mit dem Wert 1 konstruiert, der minimale Kosten z_1 hat. Jene Kanten, in denen der Fluß mit den Kosten z_1 fließen kann, definieren einen Teilgraphen G_1 von G. In diesem Teilgraphen bestimmt man nun unter Berücksichtigung der gegebenen Kapazitäten einen maximalen Fluß f_1. Dieser besitzt die Kosten $v_1 \square z_1$. Der maximale Fluß f_1 liefert einen minimalen Schnitt $(X, \bar{X})$ mit

$$X \cap \bar{X} = \emptyset, \quad q \in X, \quad s \in \bar{X}, \quad X \cup \bar{X} = V$$

Man streicht in G die Kanten (x,y) mit $(x,y) \in G_1$, $x \in X$, $y \in \bar{X}$ sowie $x \in \bar{X}$, $y \in X$ und sucht im restlichen Teilgraphen erneut - ohne Berücksichtigung der Kapazitäten - einen Fluß der Größe 1 mit minimalen Kosten z_2. Im zugehörigen Teilgraphen G_2 bestimmt man nun einen maximalen Fluß f_2 mit den Kosten $v_2 \square z_2$. Aufgrund der Verträglichkeitsaxiome (VIII) gilt:

$f_1 + f_2$ ist unter den Flüssen mit dem Wert $v_1 + v_2$ minimal.

Beweis: Würde es einen Fluß f der Größe $v_1 + v_2$ geben, der kleinere Kosten ergäbe, so wäre

$$(4.1) \qquad (v_1 \square z_1) * (v_2 \square z_2) > (v_1 + v_2) \square z$$

Wegen der Wahl von z_1 wäre $z > z_1$ und wegen der Wahl von z_2 wäre auch $z > z_2$. Daraus folgt nach Axiom (VIII)

$$(v_1 \square z) * (v_2 \square z) \geq (v_1 \square z_1) * (v_2 \square z_2)$$

im Widerspruch zu (4.1). \hfill Q.E.D.

Somit läßt sich der maximale Fluß von q nach s mit minimalen allgemeinen Kosten in der oben beschriebenen Weise aufbauen.
Einheitsflüsse mit minimalen Kosten lassen sich einfach bestimmen, wenn man den Knoten von G sogenannte "Knotenzahlen" $\pi(x) \in S$ zuordnet. Anfangs wählt man für alle $x \in V$:

$$\pi(x) \in \bigcap_{(x,y) \in E} dom\ a(x,y)$$

und erklärt alle Kanten für zulässig, für die

$$\pi(x) * a(x,y) = \pi(y)$$

gilt. Im Teilgraphen der zulässigen Kanten bestimmt man nun einen maximalen Fluß und einen dadurch bestimmte minimalen Schnitt $(X,\bar{X})$. Da jeder Fluß in G über eine Kante des Schnittes läuft, lassen sich die Flüsse durch die Kanten des Schnittes klassifizieren. Gilt für eine Kante (x,y) mit $x \in X$, $y \in \bar{X}$

$$\pi(x) * a(x,y) \leq \pi(y) \tag{4.2}$$

oder für eine Kante (x,y) mit $x \in \bar{X}$, $y \in X$

$$\pi(x) \leq a(x,y) * \pi(y) \tag{4.3}$$

so würde ein Fluß durch diese Kanten nicht größere Einheitskosten verursachen als ein Fluß, der auf den Schnitt $(X,\bar{X})$ führte. Daher müssen nach obigen Überlegungen die Flüsse in diesen Kanten bereits fixiert sein, nämlich Kanten, für die (4.2) gilt, sind saturiert, d.h. $f(x,y) = c(x,y)$, und Kanten, für die (4.3) gilt, sind flußlos, d.h. $f(x,y) = 0$.

Andrerseits verursacht ein Einheitsfluß durch eine Kante (x,y) mit $x \in X$, $y \in \bar{X}$ und

$$\pi(x) * a(x,y) > \pi(y)$$

Mehrkosten in der Größe $\bar{a}(x,y)$, wobei $\bar{a}(x,y)$ bestimmt ist durch

$$\overline{a(x,y)} * \pi(y) = \pi(x) * a(x,y)$$

Analog verursacht ein Einheitsfluß durch eine Kante (x,y) mit $x \in \bar{X}$, $y \in X$ und

$$\pi(x) > a(x,y) * \pi(y)$$

Mehrkosten $\bar{a}(x,y)$ mit

$$\bar{a}(x,y) * a(x,y) * \pi(y) = \pi(x)$$

Man beachte, daß die Mehrkosten $\bar{a}(x,y)$ infolge (4.2), (4.3) und Axiom (IV) wohl definiert sind.

Jetzt wird das Minimum δ der Größen $\bar{a}(x,y)$ bestimmt und die Knotenzahlen $\pi(y)$, $y \in \bar{X}$ werden durch $\pi(y) * \delta$ ersetzt. Dadurch wird mindestens

eine Kante des Schnittes zulässig. Der Fluß mit der Größe 1 durch diese Kante bringt unter allen noch nicht untersuchten Flüssen den kleinsten Kostenzuwachs.
Somit ergibt sich folgender Algorithmus zur Bestimmung maximaler Flüsse mit minimalen Kosten. Es soll dabei noch vorausgesetzt werden:

$$(4.4) \qquad z \in \bigcap_{(x,y)\in E} \operatorname{dom} a(x,y) \Rightarrow \bigwedge_{(x,y)\in E} z \leq a(x,y)$$

(Dies ist dann erfüllt, wenn $a(x,y) \in \operatorname{pos} z$ gilt)

<u>Algorithmus</u>

Anfangswerte:

$$\bigwedge_{(x,y)\in E} f(x,y) = 0$$

$$\bigwedge_{x\in V} \pi(x) \in \bigcap_{(x,y)\in E} \operatorname{dom} a(x,y)$$

(dann gilt nach (4.4) : $a(x,y) \geq \pi(x)$)

1. Zulässige Kanten sind solche, für die $\pi(x) * a(x,y) = \pi(y)$ gilt.
2. Man bestimme einen maximalen Fluß im Teilgraphen der zulässigen Kanten und einen zugehörigen minimalen Schnitt $(X,\bar{X})$.
3. Man bestimme

   $$L_1 := \left\{(x,y) \mid x \in X,\ y \notin X,\ \pi(x) * a(x,y) > \pi(y)\right\}$$

   $$L_2 := \left\{(x,y) \mid x \notin X,\ y \in X,\ \pi(x) * a(x,y) < \pi(y)\right\}$$

   Ist $L_1 = L_2 = \emptyset$, so terminiere man. Der augenblickliche Fluß ist maximal und hat minimale Kosten.
   Andernfalls gehe man zu 4.
4. Man bestimme $\bar{a}(x,y)$ durch

   $$\bar{a}(x,y) * \pi(y) = a(x,y) * \pi(x) \quad \text{für } (x,y) \in L_1$$

   $$\bar{a}(x,y) * a(x,y) * \pi(x) = \pi(y) \quad \text{für } (x,y) \in L_2$$

   sodann bestimme man

   $$\delta := \min\left\{\bar{a}(x,y) \mid (x,y) \in L_1 \cup L_2\right\}$$

5. Man definiere neue Knotenzahlen

$$\pi(x) := \begin{cases} \pi(x) & x \in \underline{X} \\ \pi(x) * \delta & x \in \bar{X} \end{cases}$$

und gehe zu 1.

Zur Illustration des angegebenen Algorithmus sei ein Zeittransportproblem als Beispiel gelöst

Beispiel

Gesucht werden Zahlen $x_{ij} \in \mathbb{R}^+$, so daß

$$\max_{i,j} x_{ij} \,\square\, a_{ij}$$

minimal wird unter den Restriktionen

$$\sum_{1 \leq j \leq 4} x_{ij} = a_i \text{ mit } a_i = 4 \quad (i=1,2,3)$$

$$\sum_{1 \leq i \leq 3} x_{ij} = b_j \text{ mit } b_1 = 2,\ b_2 = b_3 = 3,\ b_4 = 4$$

wobei $x_{ij} \,\square\, a_{ij} = \begin{cases} a_{ij}, & \text{für } x_{ij} > 0 \\ 0, & \text{sonst} \end{cases}$ sei und die

Zeiten folgende Werte annehmen:

$$(a_{ij}) = \begin{pmatrix} 1 & 2 & 3 & 2 \\ 4 & 3 & 4 & 2 \\ 1 & 4 & 3 & 2 \end{pmatrix}$$

Diesem Transportproblem kann in üblicher Weise ein Graph zugeordnet werden. Dabei haben die von q ausgehenden und die in s einmündenden Kanten die Kosten 0 und die Kapazitäten a_i bzw. b_j. Alle übrigen Kanten haben unbeschränkte Kapazität.
Ein positiver Fluß mit kleinsten Kosten hat die Größe 2 und geht etwa von q über i_1, j_1 nach s. Der erste Schnitt ist $X = \{q, i_1, i_2, i_3, j_1\}$, $\bar{X} = \{j_2, j_3, j_4, s\}$.

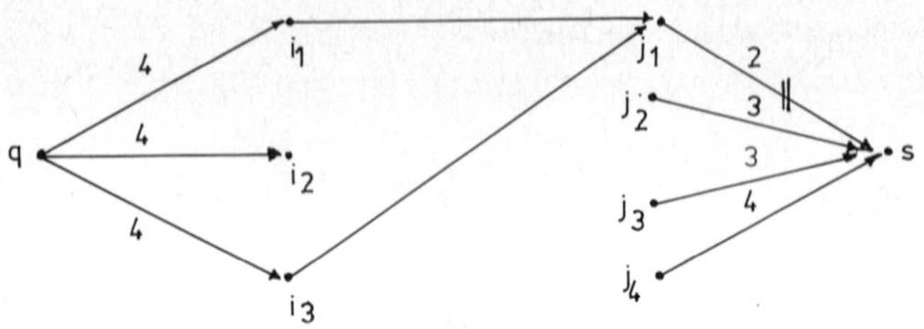

Abb. 1: Erster zulässiger Teilgraph mit Kapazitäten. Die Kante (j_1,s) wird saturiert und ergibt daher den ersten minimalen Schnitt $(X,\bar{X})$.

Man bestimmt

$L_1 = \left\{(i,j) \mid i \in \{i_1,i_2,i_3\},\ j \in \{j_2,j_3,j_4\}\right\}$

$L_2 = \emptyset$

Da sich die reduzierten Kosten $\bar{a}(x,y)$ aus

$$\max(1,\bar{a}_{ij}) = \max(1,a_{ij}) \text{ mit } a_{ij} > 1$$

berechnen, ist $\overline{a_{ij}} = a_{ij}$ und daher

$$\delta = \max_{(x,y)\in L_1} a_{ij} = 2$$

Damit erhält man über die neuen Knotenzahlen folgenden neuen Teilgraphen

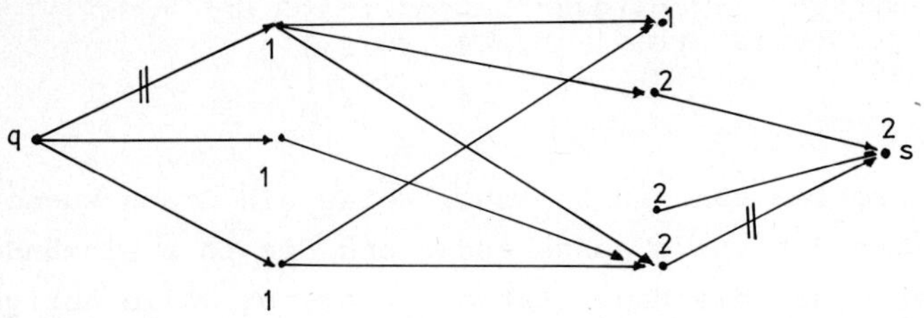

Abb. 2: Zweiter zulässiger Teilgraph mit neuen Knotenzahlen. Der neue minimale Schnitt wird durch die Kanten (q,i_1) und (j_4,s) erzeugt.

Im nächsten Schritt erhält man einen maximalen Fluß der Größe 12, dieser ist somit optimal. Damit lautet die Optimallösung

$$x_{11} = 2,\ x_{12} = 2,\ x_{22} = 1,\ x_{24} = 3,\ x_{33} = 3,\ x_{34} = 1$$

mit dem Wert 3 für die Zielfunktion.

Literatur

[1] BURKARD, R.E.: Quadratische Bottleneckprobleme. Op. Res. Verfahren 18 (1974), 26-41

[2] BURKARD, R.E.: Numerische Erfahrungen mit Summen- und Bottleneckzuordnungsproblemen. Erscheint: Numerische Methoden bei kombinatorischen und graphentheoretischen Problemen, hrsg. von L. Collatz und H. Werner, Birkhäuser Verlag, Basel 1975

[3] BURKARD, R.E., W. HAHN und U. ZIMMERMANN: An Algebraic Approach to Assignment Problems. Report 1974-1, Mathematisches Institut der Universität Köln, Nov. 1974

[4] CARRE, B.A.: An Algebra for Network Routing Problems. J. Inst. Maths. Applics 7 (1971), 273-294

[5] FORD, L.R. und D.R. FULKERSON: Flows in Networks. Princeton Univ. Press, Princeton N.J., 6. Auflage 1971

[6] GONDRAN, M.: Algèbre des Chemins et Algorithmes. Erscheint in Proceedings Summer School in Combinatorial Programming, (Versailles 1974)

[7] KARP, R.M.: Reducibility Among Combinatorial Problems. In: Complexity of Computer Computations, hrsg. von R.E. Miller und J.W. Thatcher, Plenum Press, New York 1972, 85-104

[8] ROY, B.: Chemins et Circuits: Enumération et Optimisation. Erscheint in Proceedings Summer School in Combinatorial Programming, (Versailles 1974)

On Minkowski Matrices and the Linear Complementarity Problem

Richard W. Cottle[1)]
Stanford University
Stanford, California 94305/USA

1. Introduction. The genesis of this paper is an article by DE DONATO and MAIER [4] in which there appears a special case of a model that combines elements of the linear programming problem and the (parametric) linear complementarity problem. For the present purposes, the model can be stated as follows:

(A) maximize α

$$\text{subject to} \quad q + \alpha p + Mz \geq 0 \tag{1}$$

$$0 \leq z \leq a \tag{2}$$

$$z^T[q + \alpha p + Mz] = 0 \tag{3}$$

In this problem, it is assumed that M is a given $n \times n$ matrix having positive principal minors. The vectors a, q, p in R^n are assumed to satisfy the relations

$$a > 0, \qquad q \geq 0, \qquad p \not\geq 0 \tag{4}$$

(The reason for specifying $p \not\geq 0$ is that otherwise the maximum does not exist.) The application DE DONATO and MAIER have in mind typically involves symmetric positive definite matrices, though it would be advantageous to be able to handle the positive semi-definite case as well.

DE DONATO and MAIER identify their problem (A) as a special case of a model studied earlier by KIRCHGÄSSNER [7] and HOSCHKA [5] in which one wishes to solve a linear programming problem containing an "alternative condition" among a subset of the variables. It should also be mentioned that a "bounded-variable linear complementarity problem" of a different sort arises also in a paper by AVI-ITZHAK [1].

---

1) Research partially supported by NSF Grant MPS 71-03341 A03 (formerly GP 31393X) and the Office of Naval Research under Contract N 00014-67-A-0112-0011.

Under the hypothesis imposed above, problem (A) is not really difficult to solve. Indeed, one can solve the parametric linear complementarity problem

(B)
$$q + \alpha p + Mz \geq 0 \tag{5}$$

$$\alpha \geq 0, \qquad z \geq 0 \tag{6}$$

$$z^T[q + \alpha p + Mz] = 0 \tag{7}$$

by the method proposed in [2]. In doing so, one suppresses the "monotonicity checks" and concentrates on the development of the piecewise linear solution mapping $\bar{z}(\alpha)$. Indeed, problem (B) has a unique solution for each value of α. Thus there is a well-defined mapping $\bar{z} : R^1_+ \to R^n$ where $\bar{z}(\alpha)$ denotes the solution of (B) for $\alpha = \bar{\alpha}$. The components $\bar{z}_i$ of this mapping are piecewise linear functions of α having only finitely many breakpoints. Solving (A) amounts to finding the largest value of α for which (B) has a solution $\bar{z}$ satisfying the restriction $\bar{z} \leq a$. This can be done by keeping track of the largest such α to date. To accomplish this it is not really necessary to record the complete description of each piecewise linear function $\bar{z}_i$.

However, another seemingly more direct method for solving (A) is proposed by DE DONATO and MAIER [4, p. 315]. Let

$$w := q + \alpha p + Mz \tag{8}$$

Then their approach can be summarized as follows: Maximize α subject to conditions (1) and (2) by the simplex method for linear programming, and enforce condition (3) by imposing the restriction that not both w_i and z_i may simultaneously be basic (at a positive level). This would mean that the linear programming bases are complementary. We shall refer to the DE DONATO-MAIER proposal as the <u>restricted basis simplex method</u>.

The present work stems from an investigation into the validity of the restricted basis simplex method for (A). Since DE DONATO and MAIER gave no mathematical justification for their proposal, one may certainly wonder whether it works under the given hypotheses. Validity of the restricted basis simplex method would imply that examination of the solution mapping is unnecessary, and in the case where there are many basis changes in the parametric linear complementarity problem after the solution of (A) is found, this could represent a significant computational advantage.

Unfortunately, as will be shown in this paper, the restricted basis simplex method does not always perform in the manner intended by DE DONATO and MAIER. A numerical example will substantiate this claim. It then becomes interesting to characterize the circumstances under which the restricted basis simplex method actually does achieve its goal. The purpose of this paper is to provide a characterization of this nature.

2. Numerical examples. In this section we shall work out two numerical examples that illustrate what can happen when one attempts to solve problem (A) by the restricted basis simplex method.

Example 1. Consider the following data for problem (A):

$$M = \begin{pmatrix} 2 & 1 & 1 \\ 1 & 2 & 1 \\ 1 & 1 & 2 \end{pmatrix}, \quad a = \begin{pmatrix} 1 \\ 1 \\ 1 \end{pmatrix}, \quad q = \begin{pmatrix} 2/5 \\ 31/5 \\ 31/5 \end{pmatrix}, \quad p = \begin{pmatrix} -2/5 \\ -1 \\ -1 \end{pmatrix}$$

Although it is not really necessary to do so, we shall handle the upper bounds (see eq. (2)) explicitly by introducing the nonnegative slack variables u_i for which

$$z_i + u_i = a_i \qquad (i = 1, \ldots, n) \quad .$$

Then the initial linear programming "simplex tableau" can be written as follows:

	w_1	w_2	w_3	u_1	u_2	u_3	z_1	z_2	z_3	α	RHS
Basic variables	0	0	0	0	0	0	0	0	0	1	0
w_1	1	0	0	0	0	0	-2	-1	-1	2/5	2/5
w_2	0	1	0	0	0	0	-1	-2	-1	1	31/5
w_3	0	0	1	0	0	0	-1	-1	-2	1	31/5
u_1	0	0	0	1	0	0	0	0	0	0	1
u_2	0	0	0	0	1	0	0	0	0	0	1
u_2	0	0	0	0	0	1	0	0	0	0	1

In the first change of basis, α replaces w_1. After the second change of basis, one arrives at the tableau

Basic variables	w_1	w_2	w_3	u_1	u_2	u_3	z_1	z_2	z_3	α	RHS
	-5/2			-5				5/2	5/2		-6
α	5/2			5				-5/2	-5/2	1	6
w_2	-5/2	1		-4				1/2	3/2		6/5
w_3	-5/2		1	-4				3/2	1/2		6/5
z_1				1			1				1
u_2					1			1			1
u_3						1			1		1

in which only the nonzero entries are shown. Here the only choice is to make either z_2 or z_3 basic, and each of these leads to a basic feasible solution of the linear program which violates the orthogonality (or complementarity) condition $z^T w = 0$. According to the proposed method, one would be forced to terminate the procedure. However, by continuing with the simplex method and ignoring the noncomplementary nature of some intermediate bases, one obtains after three more pivots

Basic variables	w_1	w_2	w_3	u_1	u_2	u_3	z_1	z_2	z_3	α	RHS
	5/8	-15/8	5/8		-5/2						-10
α	-5/8	15/8	-5/8							1	10
z_3		1	-1						1		1
z_2								1			1
z_1	-5/8	- 1/8	3/8				1				4/5
u_1	5/8	1/8	-3/8	1	1/2						1/5
u_3		- 1	1		-1	1					0

This tableau reveals a nonoptimal solution to the "pure" linear programming problem. The solution is complementary and has a larger value of α than in the preceding tableau. As it turns out, this is the largest value of α for which the linear program has a complementary basis. After two more basis changes, one obtains the optimal solution of the linear program.

While the example above shows that the restricted basis simplex method does not always work, it does not reveal very clearly what went wrong. At present, we have only a partial characterization of the circumstances under which the method is generally valid (i.e. works for

<u>arbitrary</u> $q \geq 0$, $p \not\geq 0$). To explain these, we introduce the notation $(r,M)_a$ for the problem of finding a solution to the system

$$\text{(C)} \qquad r + Mz \geq 0 \tag{9}$$

$$0 \leq z \leq a \tag{10}$$

$$z^T[r + Mz] = 0 \tag{11}$$

(When $a_i = \infty$ for all i we write (r,M).) We now define $\mathscr{R}(a,M)$ to be the set of all $r \in R^n$ such that $(r,M)_a$ has a solution. (Since M is assumed to have positive principal minors, a solution to $(r,M)_a$ must be unique.)

<u>Proposition</u>. The restricted basis simplex method will solve $(r,M)_a$ if $\mathscr{R}(a,M)$ is convex.

<u>Proof</u>. The method works as long as it never becomes necessary to choose a noncomplementary basis between complementary bases. Suppose the linear program has complementary bases B^0, B^1 with corresponding objective function values α_0, α_1 where $\alpha_0 < \alpha_1$. If $\mathscr{R}(a,M)$ is convex, then for each value of $\alpha \in (\alpha_0, \alpha_1)$, the problem $(a + \alpha p, M)_a$ has a solution. In particular if the successor to B^0 is $\tilde{B}$ in the linear programming simplex method sense and $\tilde{\alpha} > \alpha_0$, then there must be a complementary alternative to $\tilde{B}$ by which the value $\tilde{\alpha}$ can be attained.

Although the convexity of $\mathscr{R}(a,M)$ is a sufficient condition for the restricted basis simplex method to work, it is not exactly necessary. This can be seen in the following example.

<u>Example 2</u>. Let

$$M = \begin{pmatrix} 2 & 1 \\ 1 & 2 \end{pmatrix}, \quad a = \begin{pmatrix} 1 \\ 1 \end{pmatrix}, \quad q = \begin{pmatrix} 1 \\ 15 \end{pmatrix}, \quad p = \begin{pmatrix} -1 \\ -5 \end{pmatrix}$$

As shown in Figure 1, the region $\mathscr{R}(a,M)$ is nonconvex. However, the restricted basis simplex method encounters only complementary bases until the maximal value $\alpha = 32/9$ (which is optimal for problem (A)) is reached. Hence the convexity of $\mathscr{R}(a,M)$ is not a necessary condition for the restricted basis method to work.

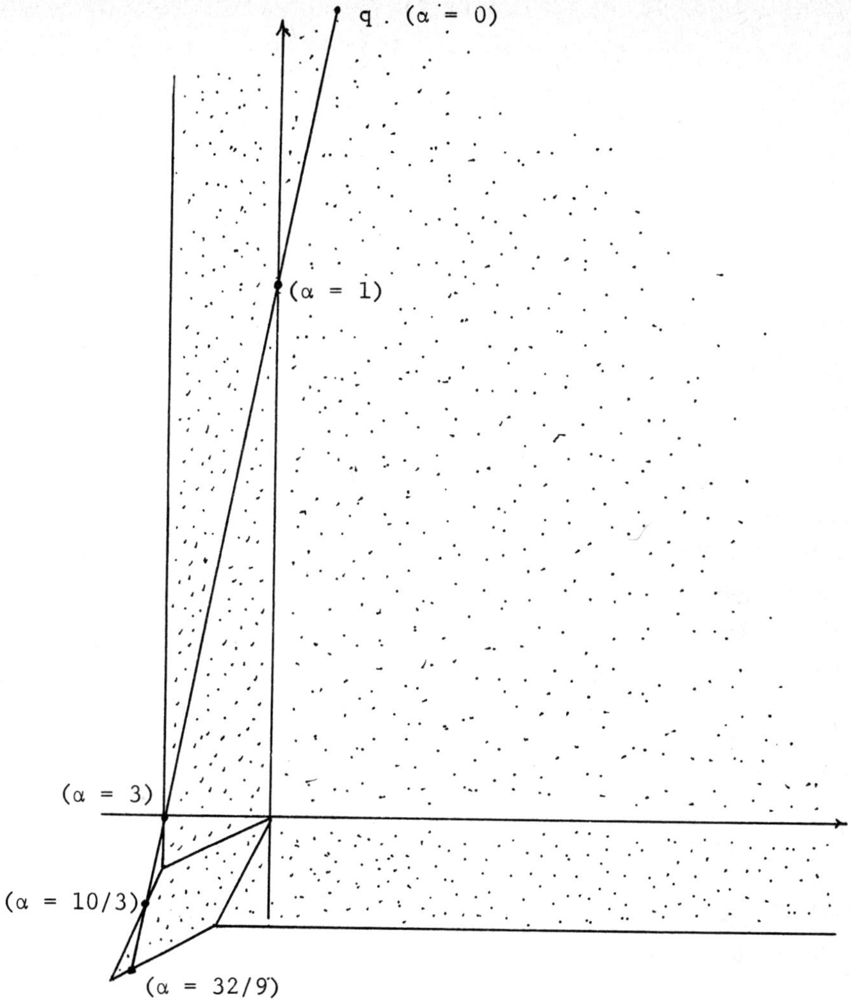

Figure 1

In speaking of the validity of the method one assumes that $q \geq 0$ and $p \ngeq 0$ are completely arbitrary. The argument that the restricted basis simplex method works when $\mathcal{R}(a,M)$ is convex can be applied to another type of situation. We may express (A) as

Find $\bar{\alpha} = \sup\{\alpha | q + \alpha p \in \mathcal{R}(a,M)\}$
and find the solution of $(q + \bar{\alpha}p, M)$

If the line segment $\{q + \alpha p | \alpha \in [0,\bar{\alpha}]\}$ lies within $\mathcal{R}(a,M)$ the method presents no difficulties.

These observations prompt several questions such as

1. When is $\mathcal{R}(a,M)$ convex?

2. What is the set of points $q \geq 0$ such that for all $p \not\geq 0$,

$$\{q + \alpha p \mid \alpha \in [0,\bar{\alpha}]\} \subset \mathscr{R}(a,M) \qquad ?$$

3. What is the set of points $p \not\geq 0$ such that for every $q \geq 0$,

$$\{q + \alpha p \mid \alpha \in [0,\bar{\alpha}]\} \subset \mathscr{R}(a,M) \qquad ?$$

In the next section, we shall answer question 1. The others are answered in a forthcoming paper by I. KANEKO [6].

3. <u>On the convexity of</u> $\mathscr{R}(a,M)$. We now wish to address the issue of when the set $\mathscr{R}(a,M)$ is convex. As noted in the preceding section, this set being convex will ensure the validity of the restricted basis simplex method.

<u>Theorem</u>. Let M be an $n \times n$ matrix with positive principal minors, and let $a \in R^n$ be a positive vector. Then $\mathscr{R}(a,M)$ is convex if and only if M has nonpositive off-diagonal elements (and hence is a Minkowski matrix).

<u>Proof</u>. Suppose M is a Minkowski matrix. Let r and s be arbitrary elements of $\mathscr{R}(a,M)$. Let x and y be the corresponding solutions of $(r,M)_a$ and $(s,M)_a$, respectively. Let $\lambda \in (0,1)$ and define

$$t = (1 - \lambda)r + \lambda s$$

$$\bar{z} = (1 - \lambda)x + \lambda y$$

As a convex combination of solutions of (10), $\bar{z}$ satisfies these conditions also. Moreover, $\bar{z}$ satisfies (9) since

$$(1 - \lambda)r + \lambda s + M[(1 - \lambda)x + \lambda y]$$
$$= (1 - \lambda)\,[r + Mx] + \lambda[s + My] \geq 0$$

Thus $\bar{z}$ is a "feasible" solution of $(t,M)_a$. It is, therefore, a fortiori feasible for (t,M). But because M is a Minkowski matrix the solution $\hat{z}$ of (t,M) is the <u>vector minimum</u> of all feasible solutions of (t,M). See [3, p. 246]. Hence in particular, $\hat{z} \leq \bar{z} \leq a$. This shows that $(t,M)_a$ is solvable or equivalently that $t \in \mathscr{R}(a,M)$. Hence $\mathscr{R}(a,M)$ is convex.

Conversely, suppose $\mathscr{R}(a,M)$ is convex and M is not a Minkowski matrix. Then M must have at least one positive off-diagonal element which we may without loss of generality assume to be m_{21}. Now define

$$r = -M_1a_1 - M_2a_2 + I_3 + \cdots + I_n$$

Let $$x = (a_1, a_2, 1, \ldots, 1)^T$$

and $$B = [-M_1, -M_2, I_3, \ldots, I_n]$$

Note that $r \in \mathscr{R}(a,M)$; in fact, $r \in \text{int pos } B$. Letting $C = B^{-1}$, we have $x = Cr$. Next define

$$s = I_1 - M_2a_2 + I_3 + \cdots + I_n = r + I_1 + M_1a_1$$

The vector s belongs to $\mathscr{R}(a,M)$. Now let λ be an arbitrary element of $(0,1)$ and define

$$t = (1 - \lambda)r + \lambda s \quad .$$

Since $\mathscr{R}(a,M)$ is assumed to be convex, $t \in \mathscr{R}(a,M)$. For λ sufficiently small, $t \in \text{int pos } B$, therefore

$$t = By, \qquad y > 0$$

has a solution. We therefore have

$$\begin{aligned} y = B^{-1}t &= C\,[\,(1 - \lambda)r + \lambda s] \\ &= C\,[\,(1 - \lambda)r + \lambda(r + I_1 + M_1a_1)] \\ &= C[r + \lambda I_1 + \lambda M_1a_1] \\ &= x + \lambda C_1 + \lambda I_1a_1 \quad . \end{aligned}$$

From the hypotheses and the definition of C, it follows that $c_{21} > 0$. The second component of C_1 is positive whereas that of I_1a_1 is zero. Hence $y_2 > a_2$ and this violates the convexity of $\mathscr{R}(a,M)$.

This theorem gives yet another characterization of Minkowski matrices in terms of the linear complementarity problem. It would be nice to be able to establish the result without the underlying assumption that M has positive principal minors; but this cannot be done, for if

$$M = \begin{pmatrix} 1 & -1 \\ -1 & 1 \end{pmatrix} \quad \text{and} \quad a = \begin{pmatrix} 1 \\ 1 \end{pmatrix}$$

the set $\mathscr{R}(a,M)$ is convex.

4. Acknowledgment. Part of the research reported in this paper was performed while the author was a Guest Professor at the Institut für Operations Research of the Eigenossische Technische Hochschule, Zürich. He wishes to express his gratitude for their hospitality.

REFERENCES

[1] B. AVI-ITZHAK, "Heavy traffic characteristics of a circular data network," Bell System Technical Journal 50 (1971), 2521-2549.

[2] R. W. COTTLE, "Monotone solutions of the parametric linear complementarity problem," Mathematical Programming 3 (1972), 210-224.

[3] R. W. COTTLE and A. F. VEINOTT, Jr., "Polyhedral sets having a least element," Mathematical Programming 3 (1972), 238-249.

[4] O. DE DONATO and G. MAIER, "Mathematical programming methods for the inelastic analysis of reinforced concrete frames allowing for limited rotation capacity," International Journal for Numerical Methods in Engineering 4 (1972), 307-329.

[5] P. HOSCHKA, "Einige Modelle für die Planung des Produktionsprogramms bei konkurrierender Produktion," Bericht Nr. 272, Deutsche Versuchsanstalt für Luft- und Raumfahrt E.V., Porz-Wahn (Rhld.), September 1963.

[6] I. KANEKO, "A maximization problem related to parametric linear complementarity," Technical Report 75-1, Department of Operations Research, Stanford University, January 1975.

[7] K. KIRCHGÄSSNER, "Ein Verfahren zur Maximierung linearer Funktionen in nichtkonvexen Bereichen," Zeitschrift für angewandte Mathematik und Mechanik 42 (1962), T22-T24.

Exact Penalty - Functions in Infinite Optimization

Hans CZAP

am Lehrstuhl für mathematische Verfahrensforschung und Datenverarbeitung Universität Göttingen.

Summary: For a class of penalty functions including those considered by Zangwill (7), Pietrzykowski (8) and Evans, Gould and Tolle (2) we show the essentialnecessary and sufficient properties for local exactness in an infinite dimensional optimization problem.

In particular we show that the local exactness property will follow by an inverse function theorem generalizing thus the results of Evans, Gould and Tolle. Next we generalize the results of Howe(5) and show in addition that the Kuhn-Tucker necessary conditions for a solution of the programming problem must be necessarily satisfied in order to have local exactness of our penalty function.

1. Introduction

We consider the mathematical programming problem (P)

$$\text{(P)} \qquad \begin{array}{l} \text{minimize} \quad f(x) \\ \quad x \in C:\ F(x) \in K \end{array}$$

where K is some closed convex cone centered at zero in a normed space Y, C some arbitrary set in a topological space X, $f: C \to Y$ an arbitrary function.

Definitions:

(1.1) Let U be a zero-neighborhood in Y, $||.||$ the norm of Y, $\rho: U \to R^+$ a nonnegative realvalued function with the properties: there is $m_1>0$, $m_2>0$: $m_1||u|| \leq \rho(u) \leq m_2||u||$ for all u in U .

In order to measure the distance of any point w in a neighborhood U+K of K to K we define

$$(1.2) \qquad q(w) := \inf_{y \in K} \rho(y-w) \quad (w-U), \quad w \in U+K := \{u + y \mid u \in U,\ y \in K\}.$$

(1.3) Let E be the set of all such pairs (q,U).

(1.4) <u>For</u> $(q,U) \varepsilon E$ <u>we define our penalty function</u> $G(x,M)$ <u>by</u>

$$G(x,M) := f(x) + Mq(F(x)) \ , \quad x \varepsilon C, \ F(x) \varepsilon U+K, \ M \geqq o.$$

The most important case is to take in (1.1) for ρ the norm in Y. Then U may be choosen to equal Y and we have $U+K = Y$, so that $G(x,M)$ is defined for all $x \varepsilon C$ and all $M \geqq 0$.

One easily checks that the penalty functions of Zangwill(7), response Pietrzykowski(8) and Evans, Gould and Tolle(2) are included in our formulation (1.4).

Penalty functions are considered in order to find an optimal solution of our programming problem (P).

Clearly, if x_o is locally optimal for (P), then

(1.5) $$f(x_o) = \min_{\substack{x \varepsilon C \cap V \\ F(x) \varepsilon U+K}} \ \sup_{M \geqq 0} G(x,M) \ , \quad \text{for some neighborhood } V \text{ of } x_o.$$

Our aim is to establish conditions guaranteeing

(1.6) $$f(x_o) = \min_{x \varepsilon C \cap V:\ F(x) \varepsilon U+K} G(x,\bar{M}) \quad \text{for some } \bar{M} \geqq 0 \ .$$

(1.7) <u>Definition:</u>

<u>For</u> $(q,U) \varepsilon E$ <u>we call</u> $G(x,M)$ <u>to be locally exact in</u> x_o <u>if and only if there is a vicinity</u> V <u>of</u> x_o <u>and a</u> $\bar{M} \geqq 0$, <u>such that</u>

$$G(x_o,\bar{M}) = G(x_o,M) = \min_{x \varepsilon C \cap V:\ F(x) \varepsilon U+K} G(x,M) \quad \underline{\text{for all}} \ M \geqq \bar{M}.$$

The relation (1.6) is a simple consequence of this definition.

(1.8) <u>Lemma:</u>

(i) <u>If</u> $G(x,M)$ <u>is locally exact in</u> x_o, <u>then</u> x_o <u>is a local optimal solution of our programming problem</u> (P).

(ii) $G(x,M)$ <u>is locally exact in</u> x_o <u>if and only if</u> x_o <u>is feasible</u>, i.e. $F(x_o) \varepsilon K$, <u>and there is a vicinity</u> V <u>of</u> x_o <u>and a</u> $\bar{M} \geqq 0$, <u>such that</u>

(1.9) $$f(x_o) \leqq f(x) + \bar{M}q(F(x)) \quad \underline{\text{for all}} \quad x \varepsilon C \cap V:\ F(x) \varepsilon U+K \ .$$

Proof:

Let $G(x,M)$ be locally exact in x_o. Then for $M > \bar{M}$ one has

$$G(x_o,\bar{M}) = f(x_o) + \bar{M}q(F(x_o)) \leq f(x_o) + Mq(F(x_o)) = G(x_o,M) = G(x_o,\bar{M}) ,$$

that is $q(F(x_o)) = 0$ or $F(x_o) \in K$, x_o feasible.

Since $f(x_o) = G(x_o,\bar{M}) \leq G(x,M)$ for all $M \geq \bar{M}$ and all $x \in C \cap V$: $F(x) \in U+K$,

we have relation (1.9) and it follows

$$f(x_o) \leq \sup_{M \geq 0} G(x,M) \quad \text{for all } x \in C \cap V\text{: } F(x) \in U+K$$

and hence

$$f(x_o) \leq \min_{\substack{x \in C \cap V \\ F(x) \in U+K}} \sup_{M \geq 0} G(x,M) = \min_{\substack{x \in C \cap V \\ F(x) \in U+K}} f(x) \leq \min_{x \in C \cap V:\ F(x) \in K} f(x) .$$

Thus x_o is a local optimal solution to (P).

In order to show for statement (ii) the opposite direction assume that (1.9) holds for a feasble x_o. Then

$$G(x_o,M) = f(x_o) \leq f(x) + \bar{M}q(F(x)) \leq f(x) + Mq(F(x)) = G(x,M)$$

$$\text{for all } M \geq \bar{M} \text{ and all } x \in C \cap V \cap F^{-1}(U+K),$$

therefore (1.7) holds.

2. Global Results

In the <u>topological dual space</u> Y^* <u>of</u> Y define the <u>polare cone of</u> K,

$$(2.1) \qquad K^- := \{1 \in Y^* \mid 1(y) \leq 0 \text{ for all } y \in K\}$$

and consider the <u>Lagrangian function</u> associated with problem (P),

$$(2.2) \qquad \phi(x,1) := f(x) + 1(F(x)) \quad x \in C,\ 1 \in K^-.$$

In (1) the author has linked the existence of a saddlepoint of the Lagrangian function to global exactness of the penalty function $G(x,M)$. The proofs of the following theorems will be found there.

(2.3) Theorem

If $\phi(x,l)$ has a saddlepoint $(x_0,\bar{l}) \in C \times K^-$, i.e.

$$\phi(x_0,l) \leq \phi(x_0,\bar{l}) \leq \phi(x,\bar{l}) \quad \text{for all } x \in C, \text{and all } l \in K^-,$$

then for given $(q,U) \in E$, $G(x,M)$ is globally exact.

the opposite direction of theorem (2.3) holds only if (P) is a convex optimization problem.

(2.4) Theorem

Let X be a linear space, C some convex subset of X, $f: C \to R$ convex and $F: C \to Y$ K-concave, that is

$$F(\Sigma\lambda_i x_i) - \Sigma\lambda_i F(x_i) \in K$$

for any finite sum Σ with $\Sigma\lambda_i = 1$, $\lambda_i \geq 0$ and $x_i \in C$.
Then one has for $(q,U) \in E$:
$G(x,M)$ is globally exact if and only if $\phi(x,l)$ has a saddlepoint in $C \times K^-$.

3. Conditions for Local Exactness

Examinating the relation (1.9)

$$f(x_0) \leq f(x) + \bar{M}q(F(x)) \qquad x \in C \cap V \cap F^{-1}(U+K)$$

one has to pay special attention to the level set

$$(3.0.1) \qquad L_0 := \{ x \in C \mid f(x) < f(x_0) \} .$$

Clearly, if x_0 is feasible, then $G(x,M)$ is in x_0 locally exact if and only if for sufficiently small V there is a $\bar{M} > 0$, such that

$$(3.0.2) \qquad \frac{q(F(x))}{f(x_0)-f(x)} \geq \frac{1}{\bar{M}} \qquad \text{for all } x \in V \cap L_0 .$$

The principial idea of the following section (3.1) is, that one may examine the limiting behavior of

$$\frac{F(x) - F(x')}{f(x') - f(x)} \quad \text{instead of} \quad \frac{q(F(x))}{f(x_0) - f(x)}$$

where x' is feasible and x and x' are in a neighborhood of x_0. This idea was earlier used by Evans, Gould and Tolle in (2).

3.1 Sufficiency for local Exactness by an Inverse Function Theorem

(3.1.1) Theorem (for a similiar result see Pietrzykowski(8))

In addition to the notations and definitions of section (1) assume X and Y to be Banach spaces and

$F: W \to Y$ to be a continuously differentiable function on some open set W of X containing C.

Let x_o be locally optimal for (P) and assume that Y is the direct product of two Banach spaces Y_1 and Y_2,

$Y = Y_1 \times Y_2$, such that for $F(x_o) = (F_1(x_o), F_2(x_o))$ it holds, that

$F_1(x_o) = 0_{Y_1}$ and

$F_2(x_o)$ is an element of the relative interior of $\pi_2(K)$ with respect to Y_2, where π_2 is the projection of Y onto Y_2.

Assume $DF_1(x_o): X \to Y_1$ to be onto Y_1 and let finally

$f: C \to R$ be Lipschitz continuous in a neighborhood V of x_o, that is

$$|f(x) - f(x')| \leq \alpha ||x - x'|| \quad \text{for all } x, x' \in V \text{ and some } \alpha > 0.$$

Then $G(x,M)$ is locally exact in x_o.

Proof:

(3.1.2) By the inverse function theorem (section 4) there is in Y_1 a ball $B(\delta,0)$ around 0_{Y_1} with radius δ such that the equation $F_1(x') = y'$ has a solution x' for any $y' \in B(\delta,0)$ and for x, s.th. $F_1(x) \in B(\delta,0)$ one has

$$||x - x'|| \leq \beta ||F_1(x) - F_1(x')|| \quad \text{for some constant } \beta > 0.$$

By continuity of F there is a ball $B(\varepsilon, x_o)$ with

$$\text{(3.1.3)} \qquad F_1(B(\varepsilon,x_o)) \subseteq B(\tfrac{\delta}{4},0) \quad \text{and}$$
$$F_2(B(\varepsilon,x_o)) \subseteq \pi_2(K) .$$

Since F_1 is continuously differentiable there is some constant $\gamma > 0$, and a $\varepsilon_1 > 0$ with

$$\text{(3.1.4)} \qquad ||F_1(x)|| \leq \gamma ||x - x_o|| \quad \text{whenever } ||x - x_o|| \leq \varepsilon_1 .$$

Let $\bar{\varepsilon} := \min\left(\frac{\varepsilon}{2}, \varepsilon_1, \frac{\varepsilon}{4\beta\gamma}\right)$ where β was given by (3.1.2) and choose as neighborhood V of x_o $B(\bar{\varepsilon}, x_o)$.

For $x \in V \cap L_o$, where L_o was given by (3.0.1), we have to show the existence of a $\bar{M} > 0$ which is independent of x such that (3.0.2) holds.

By the definition of q(.), (1.1) and (1.2), we conclude

$$(3.1.5)\quad q(F(x)) \geq m_1 \inf_{y\in K} ||F(x) - y|| \geq m_1 \inf_{y_1\in \pi_1(K)} ||F_1(x) - y_1||$$

$$\geq \frac{m_1}{2} ||F_1(x) - y'|| \quad \text{for some } y'\in\pi_1(K).$$

Since $0_{Y_1} \in \pi_1(K)$, it follows

$$(3.1.6)\quad ||F_1(x) - y'|| \leq 2 \inf ||F_1(x) - y_1|| \leq 2||F_1(x)|| \leq \frac{\delta}{2},$$

and therefore

$$||y'|| \leq ||y' - F_1(x)|| + ||F_1(x)|| \leq \frac{\delta}{2} + \frac{\delta}{4} < \delta.$$

By (3.1.2) there is a x' with $F_1(x') = y'$ and

$$||x - x'|| \leq \beta||F_1(x) - F_1(x')|| .$$

By (3.1.6),(3.1.4) and the definition of $\bar{\varepsilon}$ it follows

$$||x - x'|| \leq 2\beta||F_1(x)|| \leq 2\beta\gamma||x - x_o|| \leq 2\beta\gamma\bar{\varepsilon} \leq \frac{\varepsilon}{2} .$$

Therefore

$$||x' - x_o|| \leq ||x' - x|| + ||x - x_o|| \leq \frac{\varepsilon}{2} + \frac{\varepsilon}{2} = \varepsilon$$

and hence by (3.1.3) $\quad F_2(x') \in \pi_2(K)$.

Since by construction $y' = F_1(x') \in \pi_1(K)$ it now follows that x' is a feasible point.

By optimality of x_o, $f(x_o) \leq f(x')$. Using now the Lipschitzcontinuity of f, (3.1.5) and again (3.1.2) we may conclude

$$\frac{q(F(x))}{f(x_o)-f(x)} \geq \frac{q(F(x))}{f(x')-f(x)} \geq \frac{m_1}{2}\,\frac{||F_1(x)-y'||}{f(x')-f(x)} \geq \frac{m_1}{2\beta}\,\frac{||x - x'||}{f(x')-f(x)} \geq \frac{m_1}{2\alpha\beta} .$$

Thus $\bar{M}$ may be taken to be $\dfrac{2\alpha\beta}{m_1}$.

3.2 Cones of Tangents

In order to get a better insight into the geometrical meaning of local exactness and to derive some further conditions we consider the concept of cones of tangents.

(3.2.1) Definition

Let Z be a normed space, $A \subseteq Z$ and $a \in \bar{A}$ ($\bar{A}$ = closed hull of A). The cone of weak tangents $T_w(A,a)$ to A of a is the set of all $z \in Z$ with the property that there is a sequence (a_n) in A converging strongly to a and a sequence of nonnegative numbers (α_n) such that

$$\alpha_n(a_n - a) \xrightarrow{w} z$$

(where " $\xrightarrow{w}$ " means weak convergency).

The subset $T(A,a)$ of $T_w(A,a)$ consisting of all those z with $\alpha_n(a_n - a) \longrightarrow z$ (" $\longrightarrow$ " means strong convergency) will be called the cone of tangents to A of a.

(3.2.2) Lemma

For the closed convex cone K contained in the normed linear space Y one has with $F(x_o) \in K$

(a) $\quad K - F(x_o) \subseteq T(K,F(x_o))$

(b) $\quad K \subseteq T(K,F(x_o))$

(c) $\quad - F(x_o) \in T(K,F(x_o))$.

Proof:

Let $y \in K$. Then by convexity $y_n := \frac{1}{n}y + (1-\frac{1}{n})F(x_o) \in K$ $(n=1,2,\ldots)$. With $n(y_n - F(x_o)) = y - F(x_o)$ statement (a) follows. (b) follows by $K = K + F(x_o) - F(x_o) \subseteq K - F(x_o) \subseteq T(K,F(x_o))$ and (c) is a simple consequence of (a) and $0_Y \in K$.

(3.2.3) Lemma:

Asumptions as in (3.2.2)

$$T^-(K,F(x_o)) = \{ 1 \in K^- \mid 1(F(x_o)) = 0 \} .$$

Proof:

Since $K \subseteq T(K,F(x_o))$ we have $T^-(K,F(x_o)) \subseteq K^-$, so that for $1 \in T^-(K,F(x_o))$ we have to show the "complementary slackness condition" $1(F(x_o)) = 0$.

By (3.2.2) $\pm F(x_o) \in T(K,F(x_o))$ and therefore $\pm 1(F(x_o)) \leq 0$, that is $1(F(x_o)) = 0$.

To show the opposite direction assume $1 \in K^-$ and $1(F(x_o)) = 0$.

Take $\eta \in T(K,F(x_o))$ arbitrarily. There are sequences (y_n) in K and $\lambda_n \geq 0$:

$$y_n \longrightarrow F(x_o) \quad \text{and} \quad \lambda_n(y_n - F(x_o)) \longrightarrow \eta \quad \text{for} \quad n \longrightarrow \infty .$$

For every n $1(\lambda_n(y_n - F(x_o)) = \lambda_n 1(y_n) \leq 0$ holds and by continuity of 1

it follows $l(\eta) \leq 0$. Therefore $l \in T^-(K,F(x_o))$.

The following theorem establishes the main result of this section(3.2).

(3.2.4) Theorem

Let x_o be locally optimal for (P), assume f to be continuous in x_o and take X and Y as normed linear spaces. Then one has:

G(x,M) is locally exact in x_o if and only if there is an $\alpha > 0$ such that for every sequence (x_n) in L_o converging to x_o,

$$\inf_{\eta \in T(K,F(x_o))} \left\| \eta - \frac{F(x_n) - F(x_o)}{f(x_o) - f(x_n)} \right\| \geq \alpha \quad \text{holds for allmost all } n.$$

Proof:

Part I) Suppose there exists an $\alpha > 0$ with the wished properties. Then for any sequence (x_n) in L_o, $x_n \longrightarrow x_o$ one has by (1.1) and (1.2) for allmost all n,

$$\frac{q(F(x_n))}{f(x_o) - f(x_n)} \geq \frac{m_1}{f(x_o) - f(x_n)} \inf_{y \in K} \| y - F(x_o) - (F(x_n) - F(x_o)) \|$$

$$= \frac{m_1}{f(x_o) - f(x_n)} \inf_{z \in K-F(x_o)} \| z - (F(x_n) - F(x_o)) \|$$

$$\geq \frac{m_1}{f(x_o) - f(x_n)} \inf_{\eta \in T(K,F(x_o))} \| \eta - (F(x_n) - F(x_o)) \| \quad \text{(by lemma (3.2.2))}$$

$$= m_1 \inf_{\eta \in T(K,F(x_o))} \left\| \eta - \frac{F(x_n) - F(x_o)}{f(x_o) - f(x_n)} \right\| \quad \text{(since } T(K,F(x_o)) \text{ is a cone)}$$

$$\geq \alpha \quad \text{by hypothesis.}$$

Since α is independent of the sequence x_n the locally exactness of G(x,M) follows.

Part II) To show the opposite direction define the set

$$S(K,F(x_o)) := \{\eta \in T(K,F(x_o)) \mid \exists\, \bar{\varepsilon} = \bar{\varepsilon}(\eta) > 0 \text{ s.th. } \bar{\varepsilon}\eta + F(x_o) \in K \}.$$

Observe that by convexity of K, for $0 \leq \varepsilon \leq \bar{\varepsilon}$,

$$\varepsilon\eta + F(x_o) \in K \qquad \text{if} \quad \bar{\varepsilon}\eta + F(x_o) \in K .$$

Let (x_n) be any sequence in L_o converging to x_o and choose $\eta \in S(K,F(x_o))$ arbitrarily. By continuity of f there exists a $n_o = n_o(\eta)$, s.th.

$$\vartheta_n := (f(x_o) - f(x_n))\eta + F(x_o) \in K \quad \text{for all} \quad n \geq n_o$$

By (1.1) and (1.2) we now have

$$\frac{q(F(x_n))}{f(x_o) - f(x_n)} \leq \frac{m_2}{f(x_o) - f(x_n)} \inf_{y \in K} ||y - F(x_n)||$$

$$\leq m_2 \left|\left| \frac{\vartheta_n - F(x_n)}{f(x_o) - f(x_n)} \right|\right| = m_2 \left|\left| \eta - \frac{F(x_n) - F(x_o)}{f(x_o) - f(x_n)} \right|\right| .$$

Assume G(x,M) to be locally exact. Then there is a $\bar{M} > 0$ such that for $\alpha := \frac{1}{\bar{M}}$ and allmost all n

$$\alpha \leq \frac{q(F(x_n))}{f(x_o) - f(x_n)} \qquad \text{holds.}$$

Combining these inequalities and since $\eta \in S(K,F(x_o))$ was arbitrarily choosen we have

$$\alpha \leq m_2 \inf_{\eta \in S(K,F(x_o))} \left|\left| \eta - \frac{F(x_n) - F(x_o)}{f(x_o) - f(x_n)} \right|\right|$$

The conclusion follows if we have shown

$$(+) \qquad \overline{S(K,F(x_o))} = T(K,F(x_o)) .$$

Assume (+) to be wrong. Then there is a $y \in T(K,F(x_o))$ and an open convex neighborhood U of zero in Y such that

$$(y + U) \cap S(K,F(x_o)) = \emptyset .$$

By definition of $S(K,F(x_o))$ we have $(\varepsilon(y+U) + F(x_o)) \cap K = \emptyset$ for all $\varepsilon>0$ or äquivalently

$$(y + U + \tfrac{1}{\varepsilon}F(x_o)) \cap K = \emptyset \quad \text{for all } \varepsilon > 0 .$$

Therefore the open convex set $\{ y+u+ \frac{1}{\varepsilon}F(x_o) \mid u \in U, \varepsilon > 0 \}$ does not meet the closed convex cone K and may therefore be strictly seperated by a closed hyperplane, that is

$$\exists\, l \in K^-: \quad l(y + u + \tfrac{1}{\varepsilon}F(x_o)) > 0 \quad \text{for all} \quad u \in U \quad \text{and} \quad \varepsilon > 0.$$

Since $F(x_o) \in K$, it follows $l(F(x_o)) \leq 0$.

$l(F(x_o)) < 0$ yields a contradiction because of

$$-\frac{1}{\varepsilon}l(F(x_o)) < l(y+u) \quad \text{for all } \varepsilon > 0 .$$

If $l(F(x_o)) = 0$, then by lemma (3.2.3) $l \in T^-(K,F(x_o))$ follows and therefore $l(y) \leq 0$ in contradiction to $l(y+u) > 0$ for all $u \in U$. This finishes the proof of theorem (3.2.4).

For those sequences (x_n) in L_o yielding a limit of

$$\frac{F(x_n) - F(x_o)}{f(x_o) - f(x_n)}$$

we may note a consequence of theorem (3.2.4).

(3.2.5) Theorem

Take X and Y as Banach spaces, let $DF(x_o)$ resp. $Df(x_o)$ be the (continuous) Fréchet derivative of F resp. f in x_o and assume x_o to be optimal for (P).

(i) Let (a) $\xi \in T(L_o,x_o)$
or (b) $\xi \in T_w(L_o,x_o)$ and $DF(x_o)$ completely continuous.
Assume $Df(x_o)(\xi) \neq 0$.

If $G(x,M)$ is locally exact in x_o, then it follows

$$DF(x_o)(\xi) \notin T(K,F(x_o)) .$$

(ii) Let X be reflexive.

If for all $\xi \in T_w(L_o,x_o)$, $\xi \neq 0$, $DF(x_o)(\xi) \notin T(K,F(x_o))$ holds, then locally exactness of $G(x,M)$ in x_o follows.

Proof:
(i) To $\xi \in T(L_o,x_o)$ exists some sequence (x_n) in L_o, $x_n \to x_o$ and $\lambda_n \geq 0$, s.th. $\xi_n := \lambda_n(x_n - x_o) \longrightarrow \xi$. Then $DF(x_o)(\xi_n) \longrightarrow DF(x_o)(\xi)$ by continuity.
In case (b), ξ_n converges weakly to ξ . Since $DF(x_o)$ is by hypothesis a completely continuous linear operator, the image $DF(x_o)(\xi_n)$ of the weakly convergent sequence (ξ_n) is strong convergent with limit $DF(x_o)(\xi)$ (see for example Smirnov(3), §106).
In any case (a) or (b),

$$\lambda_n(F(x_n) - F(x_o)) = DF(x_o)(\lambda_n(x_n - x_o)) + \frac{\Theta(x_n - x_o)}{||x_n - x_o||}||\lambda_n(x_n - x_o)||$$

$$\longrightarrow DF(x_o)(\xi) \quad \text{for } n \to \infty \text{ since } \frac{\Theta(x_n - x_o)}{||x_n - x_o||} \to 0_Y \quad \text{for } n \to \infty.$$

Similiarily

$$(+) \qquad 0 \leq \lambda_n(f(x_o) - f(x_n)) \longrightarrow -Df(x_o)(\xi) \quad \text{for } n \to \infty \quad .$$

By hypothesis $Df(x_o)(\xi) \neq 0$, that is $Df(x_o)(\xi) < 0$.
Suppose now that $G(x,M)$ is locally exact in x_o. Then by theorem(3.2.4)

$$\left\| \eta - \frac{F(x_n) - F(x_o)}{f(x_o) - f(x_n)} \right\| \geq \alpha > 0 \quad \text{for all } \eta \in T(K,F(x_o)) \text{ and allmost all } n.$$

In the limit we get

$$\left\| \eta - \frac{DF(x_o)(\xi)}{-Df(x_o)(\xi)} \right\| \geq \alpha \quad \text{for all } \eta \in T(K,F(x_o))$$

and since $-Df(x_o)(\xi) > 0$, $DF(x_o)(\xi) \notin T(K,F(x_o))$ follows.

(ii) Assume $G(x,M)$ is not locally exact in x_o. By theorem (3.2.4) there exists a sequence (x_n) in L_o, $x_n \to x_o$, s.th.

$$\lim_{n\to\infty} \inf_{\eta \in T(K,F(x_o))} \left\| \eta - \frac{F(x_n) - F(x_o)}{f(x_o) - f(x_n)} \right\| = 0 .$$

Since $T(K,F(x_o))$ is a cone this holds if and only if

$$(+) \qquad \lim_{n\to\infty} \left(\frac{\|x_n - x_o\|}{f(x_o) - f(x_n)} \inf_{\eta \in T(K,F(x_o))} \left\| \eta - \frac{F(x_n) - F(x_o)}{\|x_n - x_o\|} \right\| \right) = 0 .$$

By reflexivity of X the bounded sequence $\xi_n := \dfrac{x_n - x_o}{\|x_n - x_o\|}$ has a weakly convergent subsequence, $\xi_{n_j} \xrightarrow{w} \xi \neq 0$ and clearly $\xi \in T_w(L_o,x_o)$ holds. Without loss of generality let's assume $n = n_j$.

Because of
$$\lim_{n\to\infty} \frac{f(x_o) - f(x_n)}{\|x_n - x_o\|} = -Df(x_o)(\xi) \geq 0 \quad \text{there is an } \varepsilon>0$$
and a $n_o = n_o(\varepsilon)$, s.th.

$$\frac{\|x_n - x_o\|}{f(x_o) - f(x_n)} \geq \varepsilon \quad \text{for all } n \geq n_o.$$

By (+)

$$\lim_{n\to\infty} \inf_{\eta \in T(K,F(x_o))} \left\| \eta - \frac{F(x_n) - F(x_o)}{\|x_n - x_o\|} \right\| = 0$$

must be valid.
Since for any $l \in T^-(K,F(x_o))$, $l \neq 0$

$$\left\| \eta - \frac{F(x_n) - F(x_o)}{\|x_n - x_o\|} \right\| \geq \frac{1}{\|l\|} \left| l(\eta) - l\left(\frac{F(x_n) - F(x_o)}{\|x_n - x_o\|} \right) \right|$$

and $l(\eta) \leq 0$ for all $\eta \in T(K,F(x_o))$, we must have

$$\limsup_{n \to \infty} \left(l\left(\frac{F(x_n) - F(x_o)}{\| x_n - x_o \|} \right) \right) \leq 0 \quad .$$

By continuity of the linear functional $l \circ DF(x_o)$ it follows

$$l \circ DF(x_o)(\xi) \leq 0$$

Since l was arbitrary this holds for all $l \in T^-(K,F(x_o))$. Hence

$$DF(x_o)(\xi) \in T^{--}(K,F(x_o)) = T(K,F(x_o))$$

by convexity, closedness and the cone property of $T(K,F(x_o))$ (see for example Varaiya(4)).
Thus the proof of (3.2.5) is complete.

(3.2.6) Remark

Part (ii) of theorem (3.2.5) is a generalization of a result given by Howe(5). He uses a condition (G) which may be generalized to infinite dimensional spaces.

(G) $\xi \in T_w(C,x_o)$, $\xi \neq 0$ and $DF(x_o)(\xi) \in T(K,F(x_o))$ imply $Df(x_o)(\xi) > 0$.

Clearly if f and F have a continuous Fréchet derivative in x_o and X is reflexive, if condition (G) holds, the asumptions of theorem (3.2.5 ii) are met, since
$T_w(L_o,x_o) \subseteq T_w(C,x_o)$ and $Df(x_o)(\xi) \leq 0$, for all $\xi \in T_w(L_o,x_o)$.
Thus $G(x,M)$ must be locally exact.

3.3 Necessarity of a Minimum Principle

Since all sufficient conditions for local exactness of $G(x,M)$ given so far imply as well the Kuhn-Tucker necessary optimality conditions the question arises whether or not the Kuhn-Tucker necessary conditions for optimality imply the local exactness property. In (5) Howe has given an example where this implication is not valid.
In this section we show the opposite:
If $G(x,M)$ is locally exact in x_o, then the Kuhn-Tucker necessary conditions hold in x_o - to be precise a strong minimumprinciple holds in x_o.

(3.3.1) Theorem:

Take X and Y as normed linear spaces. Let f and F be (continuous) Fréchet differentiable in x_o, $G(x,M)$ locally exact there and assume there is some neighborhood U of x_o, such that $C \cap U$ is convex, where C is the constraint set.

Then there is a $\bar{l} \in K^-$ with the properties

$$\bar{l}(F(x_o)) = 0 \quad \text{and}$$
$$0 \leq Df(x_o)(\xi) + \bar{l}(DF(x_o)(\xi) \quad \text{for all} \quad \xi \in T_w(C,x_o).$$

Proof:

By local convexity of C in x_o the closed cone $T(C,x_o)$ is convex and equals thus his weak closure $T_w(C,x_o)$.

For $\xi \in T(C,x_o)$, $||\xi|| = 1$ we have a sequence (x_n) in C, $x_n \to x_o$, such that $\dfrac{x_n - x_o}{||x_n - x_o||} \longrightarrow \xi$.

By local exactness (see (1.9)) there is a n_o with

$$0 \leq f(x_n) - f(x_o) + \bar{M}q(F(x_n)) \quad \text{for all} \quad n \geq n_o ,$$

and consequently as in the proof of theorem (3.2.4), part II,

$$0 \leq \frac{f(x_n) - f(x_o)}{||x_n - x_o||} + m_2\bar{M} \inf_{\eta \in T(K,F(x_o))} || \eta - \frac{F(x_n) - F(x_o)}{||x_n - x_o||} || , \quad n \geq n_o .$$

Going to the limit we get

$$(+) \qquad 0 \leq Df(x_o)(\xi) + m_2\bar{M} \inf_{\eta \in T(K,F(x_o))} || \eta - DF(x_o)(\xi) || .$$

This inequality holds as well for $\xi = 0$ and thus for all $\xi \in T(C,x_o)$.

Hence the convex optimization problem

$$(\bar{P}) \qquad \text{minimize} \quad Df(x_o)(\xi)$$
$$\xi \in T(C,x_o): \; DF(x_o)(\xi) \in T(K,F(x_o))$$

has as optimal solution $\xi_o = 0_X$ and the penalty function

$$\bar{G}(\xi,M) := Df(x_o)(\xi) + M \inf_{\eta \in T(K,F(x_o))} || \eta - DF(x_o)(\xi) ||$$

is by (+) globally exact.

By theorem (2.4) the existence of $\bar{l} \in T^-(K,F(x_o))$ follows together with the saddlepoint relation for $(\xi_o,\bar{l}) = (0_X,\bar{l})$,

$$0 \leq Df(x_o)(\xi) + \bar{l}DF(x_o)(\xi) \quad \text{for all } \xi \in T(C,x_o).$$

By lemma (3.2.3) $\bar{l} \in T^-(K,F(x_o))$ is äquivalent to $\bar{l} \in K^-$ and $\bar{l}(F(x_o))=0$. This finishes our proof.

4. Appendix: Inverse Function Theorem

The following generalized form of an inverse function theorem is a slight modification of a theorem given by Ljusternik - Sobolev(6), which will be shown by a minor change of the proof given by the authors above.

Theorem:

Let T be a continuous Fréchet differentiable function of an open set V of the Banach space X into the Banach space Y. In $x_o \in V$ let $DT(x_o)$ onto Y.

Then there is a neighborhood U of the point $y_o = T(x_o)$, such that the equation $T(x) = y$ has a solution for every $y \in U$ and there is a constant $\beta > 0$, such that for $y=T(x)$, $y'=T(x')$, y and y' in U, we have $\| x - x' \| \leqq \beta \| y - y' \|$.

Proof:

Let L_o be the nullspace of the map $DT(x_o)$. By continuity L_o is closed and therefore the quotient space X/L_o is a Banach space with the norm

$$(4.1) \qquad \| L_x \| = \inf_{x \in L_x} \|x\| , \qquad L_x \in X/L_o$$

The operator $A: X/L_o \longrightarrow Y$ defined by $A(L_x) = DT(x_o)(x)$ for some $x \in L_x$ is linear, continuous, one-to-one and onto; hence by the Banach inverse theorem, A has a continuous linear inverse A^{-1}.

(4.2) By continuity of $DT(x)$ in x_o to $\varepsilon := \dfrac{1}{4\|A^{-1}\|}$ there exists $r > 0$, such that for $\| x - x_o\| \leqq 2r$ the implication $\| DT(x) - DT(x_o) \| \leqq \varepsilon$ follows.

(4.3) Designate the ball around $y_o = T(x_o)$ with radius r by $B(r,y_o)$.

By continuity of T in x_o there is a neighborhood V of x_o with

$$T(V) \subseteq B(\tfrac{r\varepsilon}{2}, y_o) .$$

For $y \in B(\frac{r\varepsilon}{2}, y_o)$ and $x' \in V \cap B(r,x_o)$ we construct a sequence (L_n) of

(L_n) of elements of the quotient space $X/_{L_o}$ and a sequence (g_n) of elements in L_n iteratively:

$$L_o := 0_{X/_{L_o}} \quad , \; g_o = 0_X$$

(4.4) $$L_n - L_{n-1} = A^{-1}(\, y - T(x' + g_{n-1})\,)$$

and select $g_n \in L_n$, such that

$$\| g_n - g_{n-1} \| \leq 2 \| L_n - L_{n-1} \| \;, \quad n = 1,2,\ldots$$

which is possible by (4.1).
Rewriting (4.4) slightly, we have

$$L_n = A^{-1}(\, y - T(x' + g_{n-1}) + DT(x_o)(g_{n-1})\,)$$

and thus for $n \geq 2$

$$L_n - L_{n-1} = -A^{-1}(\, T(x' + g_{n-1}) - T(x' + g_{n-2}) - DT(x_o)(g_{n-1} - g_{n-2})\,).$$

Applying now the mean value theorem on the Fréchet differentiable function

$$\Gamma(x) = -A^{-1}(\, T(x) - DT(x_o)(x)\,),$$

that is $\| \Gamma(x + h) - \Gamma(x) \| \leq \|h\| \sup\limits_{0<\alpha<1} \| D\Gamma(x + \alpha h) \|$,

we get for $x = x' + g_{n-2}$ and $h = g_{n-1} - g_{n-2}$

(4.5) $$\| L_n - L_{n-1} \| \leq \|A^{-1}\| \; \|g_{n-1} - g_{n-2}\| \sup_{0<t<1} DT(x' + g_t^{(n)}) - DT(x_o),$$

(4.6) where $$g_t^{(n)} := t g_{n-1} + (1-t) g_{n-2} \,.$$

Because of

$$\begin{aligned} \|g_1\| &\leq 2\|L_1\| \leq 2\|A^{-1}\| \; \|y - T(x')\| \\ &\leq 2\|A^{-1}\|(\|y - y_o\| + \|T(x') - y_o\|\,) \\ &= 2\|A^{-1}\|(\,\tfrac{r\varepsilon}{2} + \tfrac{r\varepsilon}{2}\,) = 2r\varepsilon\|A^{-1}\| = \tfrac{r}{2} \end{aligned}$$

we obtain

$$\|g_t^{(2)}\| \leq \tfrac{r}{2} < r \quad \text{for all } t,\; 0 < t < 1\,.$$

Suppose we had shown for all $n \leq m$, $n \geq 2$, that $g_t^{(n)} \leq r$.
Then by (4.2) and because of

$$\|x' + g_t^{(n)} - x_o\| \leq \|x' - x_o\| + \|g_t^{(n)}\| \leq r + r = 2r$$

the relation

$$\sup_{0<t<1} \| DT(x' + g_t^{(n)}) - DT(x_o) \| \leq \varepsilon = \frac{1}{4\|A^{-1}\|}$$

follows. By (4.5) and the construction of the sequence (g_n) we obtain

$$(4.7) \quad \|g_n - g_{n-1}\| \leq 2\|L_n - L_{n-1}\| \leq \tfrac{1}{2}\|g_{n-1} - g_{n-2}\|$$

and thus

$$(4.8) \quad \|g_n\| = \|g_1 + (g_2 - g_1) + \dots + (g_n - g_{n-1})\| \leq \left(1 + \frac{1}{2} + \dots + \frac{1}{2^{n-1}}\right)\|g_1\| \leq 2\|g_1\| \leq r, \quad 2 \leq n \leq m.$$

In order to verify (4.7) and (4.8) for all natural numbers n we must show $\|g_t^{(m+1)}\| \leq r$. But this follows by

$$\|g_t^{(m+1)}\| = \|tg_m + (1-t)g_{m-1}\| \leq t\|g_m\| + (1-t)\|g_{m-1}\| \leq r.$$

By (4.7) the sequence (g_n) is a Cauchy sequence in the Banach space X and similiarily (L_n) in the Banach space X/L_o. Let their respective limits be g and L.
By (4.4) we conclude

$$0 = A^{-1}(y - T(x' + g))$$

that is $T(x' + g) = y$ and

$$\|g\| \leq 2\|g_1\| \leq 2\|A^{-1}\| \, \|y - T(x')\| .$$

With $x = x' + g$, $y' = T(x')$ and $\beta = 2\|A^{-1}\|$ our statement now follows.

Acknowledgement: I wish to thank Prof.Dr.J.Stoer for his advices on parts of this paper when preparing my thesis at the University Würzburg.

References:

(1) Czap H. Exact Penalty-Functions and Duality. Discussion paper Nr. 7402; Lehrstuhl für math. Verfahrensforschung(OR) und DV,Universität Göttingen.

(2) Evans J.P., Gould F.J.,Tolle J.W. Exact Penalty Functions in Nonlinear Programming. Math. Progr. 4 (1973), 72 - 97.

(3) Smirnov W.I. Lehrgang der Höheren Mathematik Teil V.Deutscher Verlag der Wissenschaften, Berlin 1967 (2.Aufl.).

(4) Varaiya P.P Nonlinear programming and optimal control. ERL Tech. Memo M - 129, Univ. of California, Berkley 1965.

(5) Howe S. New Conditions for Exactness of a simple Penalty Function. Siam J. Control 11 (1973) 378 - 381.

(6) Ljusternik L.A., Sobolev W.I. Elemente der Funktionalanalysis, Akademie Verlag Berlin, 1955.

(7) Zangwill W.I. Nonlinear Programming via Penalty-functions. Management Science Vol. 13 (1967) 448 - 460.

(8) Pietrzykowski T. An exact potential method for constrained maxima. Siam J. Num. Anal. 6 (1969) 299 - 304. Erratum: Vol.8 (1971) p.481.

Application of Dubovitskii-Milyutin Formalism to Optimal Settling Problem with Constraints

P. C. DAS

(Institut für Mathematik I   Freie Universität Berlin)

The problem of optimal control for system of differential equations with timelag has been considered by many authors. For a brief survey of these results one could refer to Banks [1]. In the earlyworks in this area, the target sets were assumed to lie in a finite demensional Euclidean space. But it is recognized already for many years that the natural state space for such systems is a suitable function space. Therefore it is natural to consider target sets in function spaces. There are already several results in this direction such as those of Banks and Kent [3], Jacobs and Kao [8], and Banks and Jacobs [9]. In the first of the above three articles an approach based on the work Neustadt [1] was adopted for a sufficiently general class of delay differential equations without phase constraints, while in the second one, the classical Langrange multiplier theory was used for a less general class of systems in a Sobolevspace without control or phase constraints. In the work of Banks and Jacobs a control problem was considered for a system of linear neutral differential equations where the objective functional was given through an integrand in a quadratic form. The necessary condition was derived by considering the attainable set and directly applying separation theorems of convex sets.

In the present article, we use the advantage of this choice of space for applying Dubovitskii-Milyutin formalism in the derivation of the necessary conditions for a settling problem. These results include as particular cases, the results of Jacobs and Kao [8] and Banks and Jacobs [2]. Further our problem contains both phase constraints and control constraints. We follow Hale [7] for notations in timelag systems.

Let $L_2([\alpha,\beta],R^n)$ denote the Hilbert space of square integrable functions mapping $[\alpha,\beta]$ to R^n and $W_2^1([\alpha,\beta],R^n)$ the set of all absolutely continuous functions $z:[\alpha,\beta] \to R^n$ such that $\dot{z}$ $L_2([\alpha,\beta],R^n)$. Here $\dot{z}$ denotes derivative with respect to t.

An inner product can be introduced in this space through

$$\langle z_1, z_2 \rangle = (z_1(\alpha), z_2(\beta)) + \int_\alpha^\beta (\dot{z}_1(s), \dot{z}_2(s))ds$$

The space $W_2^1([\alpha,\beta], R^n)$ with this inner product is a Hilbert space. Let $x: [a - \sigma, b] \to R^n$ and let t $[a,b]$. We shall denote by x_t a map from $[-\sigma, 0] \to R^n$ defined by $x_t(\theta) = x(t + \theta)$ $-\sigma \leq 0 \leq 0$.

In this article we consider the following problem: Minimize

$$J(x, u) = \int_a^b L(t, x(t), u(t))dt$$

subject to the following constraints:

$$\frac{dx}{dt} = A(t)\dot{x}(t - \sigma) + f(t, x(t), x(t - \sigma), u(t)) \qquad (1)$$

$$x_a = \varphi_1,\ x_b = \varphi_2,\ G(x(t), t) \leq 0 \text{ for all } t \in [a,b] \qquad (2)$$

$u \in U \subset L_2([a,b], R^n)$ where U is a closed convex set and $\overset{o}{U} \neq \emptyset$. Further solutions of (1) with constraints (2) are supposed to be in $W_2^1([a - \sigma, b], R^n)$. It should be remarked that a control variable with lag could be introduced in (1) with insubstantial changes in the deliberations. Further the constraints on the controls could be expressed through a range set of the control values. But in this case, the pattern of proof will be substantially altered and hence we do not deal with it here. For the sake of brevity we shall denote

$W_2^1([a - \sigma, b], R^n$ by X; $W_2^1([-\sigma, 0], R^n)$ by X and $L_2([a - \sigma, b], R^n)$ by L_2.

§ 1 General assumptions and results from Dubovitskii-Milyutin theory.

In provingthe main theorem on necessary conditions for optimality, we shall make the general assumptions $(\alpha)(\beta)$ and (γ). For the sake of easy comparison with the work of Jacobs and Kao [8] we shall follow many for of their natations.

$$\text{Denote } H(x) = \max_{t \in [a - \sigma, b]} G(x(t), t)$$

Let there exist a function $M: R x R^n x R^n \to R$, which is bounded on bounded sets such that

(α) $\| f(t,r_1,r_2,r_3) - f(t,r_1,r_2,r_3')\| \leq M(t,r_1 r_2) \quad \| r_3 - r_3'\|$

$\| D_i f(t,r_1,r_2,r_3) - D_i f(t,r_1,r_2,r_3')\| \leq M(t,r_1,r_2) \quad \| r_3 - r_3'\|$

for $i = 1,2, r_1, r_2 \in R^n$ and $r_3, r_3' \in R^m$.

$\| D_3 f(t,r_1,r_2,r_3)\| \leq M(t,r_1,r_2)$ for $r_1, r_2 \in R^n$ and $r_3 \in R^m$.

(β) A is a continuous matrixfunction in $[a,b]$.

(i) $L(\cdot,x(\cdot),u(\cdot))$ is integrable for every $x \in X$ and $u \in L_2$.

(ii) L is continuously partially differentiable with respect to x and u.

(iii) In every bounded set $\Omega \subset X x L_2$, there exist $M_\Omega \in L_2$ such that $|D_1 L(t,x(t),u(t)| \leq M_\Omega(t)$ and $|D_2 L(t,x(t),u(t)| \leq M_\Omega(t)$.

Here $D_i f$ (respectively $D_i L$) denotes the partial derivative with respect to the arguments r_i with corresponding subscript.

(γ) $G: R^n x R \to R$. It is continuous in x,t differentiable with respect to x and G_x is continuous with respect to x and t. Further $G(\varphi_1(t-a),t) < 0$ for $t \in [a - \sigma, a]$ and $G(\varphi_2(t-b),t) < 0$ for $t \in [b-\sigma, b]$. After these general assumptions, we shall state some definitions and results concerning Duboviskii-Milyutin theory. Let Y be a Banach space and $Q_i \subset Y$, $i = 1,2,\ldots k+1$. Let F be a real functional on C. We consider the following problem of optimization.

What conditions must $x_0 \in \bigcap_{i=1}^{k+1} Q_i = Q$ satisfy when x_0 gives a minimum to the functional F on $\bigcap_{i=1}^{k+1} Q_i$ i e if $F(x_0) = \min_{x \in Q} F(x)$.

Definition 1 A convex cone K with vertex at the origin is called a regular feasible cone of a set $P \subset Y$ at y_0 if for each $y \in K$, there exists an $\varepsilon_0 > 0$ and a neighbourhood of the origin U such that $U + y \subset K$ and $y' \in U + y$ implies that $y_0 + \varepsilon y' \in P$ for $0 < \varepsilon < \varepsilon_0$.

Definition 2 A convex cone K with vertex at the origin is called a tangent cone of a set $P \subset Y$ at y_0 if for each $y \in K$, there exists $\varepsilon_0 > 0$ and $r(\varepsilon)$ such that $y_0 + \varepsilon y + r(\varepsilon) = y(\varepsilon) \in Q$ for $0 < \varepsilon < \varepsilon_0$ and $r(\varepsilon) = o(\varepsilon)$.

Theorem 1 (Dubovitskii-Milyutin [5]) Suppose that functional F be continuously Frechet differentiable and the differential not zero. Let K_is be regular feasible cones of Q_i, $i = 1,2,\ldots n$, at y_0. Suppose also that K_{n+1} is a tangent cone of Q_{n+1} at y_0. Then, if y_0 is a minimal point of F in $Q = \bigcap_{i=1}^{n+1} Q_i$, there must exist

$\lambda_0 \geq 0, w_i$ $i = 1,2,\ldots n+1$, not all of them zero such that

$\lambda_0 F'(y_0) = \sum_{i=1}^{n+1} w_i$, where $F'(y_0)$ is the Frechet derivative of F at y_0

and w_is are continuous linear functionals in Y such that

$y \in K_i \Rightarrow w_i(y) \geq 0$.

Now we state a theorem of functional analysis which we shall make use of later.

Theorem 2 Let Y_1 and Y_2 be two Banach spaces and let $A \in B(Y_1, Y_2)$. Suppose the range of A, R(A) be closed. Then $R(A^*) = (N(A))^{\perp}$ where A^* denotes the adjount of A, N(A) denotes the nullset of A and the superscript '$\perp$' denotes orthogonality.

A proof of this theorem is given in Luenberger [10].

Finally, we reformulate our optimization problem in a form amenable to the theorem of Dubovitskii-Milyutin. We introduce the following operators:

$$P(x,u) = \begin{cases} x(t) - \varphi_1(t-a) & t \in [a-\sigma, a] \\ x(t) - \varphi_1(0) - \int_a^t A(s)x(s-\sigma)ds - \int_a^t f(s,x(s),x(s-\sigma),u(s))ds & (3) \\ \text{for } t \in [a,b] & (4) \end{cases}$$

$$S(x,u) = x_b - \varphi_2.$$

Let us denote $(P(x,u),s(x,u)) = T(x,u)$
It can be shown Jacobs and Kao [8], under the assumptions enumerated above, that $T:XxL_2 \to XxX_\rho$. For brevity, we shall denote XxL_2 through E.

$$Q_1 \tilde{=} \{(x,u) \in E: T(x,u) = 0\}$$
$$Q_2 \tilde{=} \{(x,u) \in E: u \in U \subset L_2\}$$
$$Q_3 \tilde{=} \{(x,u) \in E: H(x) \leq 0\}$$

Then our problem can be reformulated as

Minimize $J(x,u)$
$(x,u)\ Q_1 \cap Q_2 \cap Q_3$

In the following section we derive a necessary condition for optimality of a point x_0, formulated as a maximum principle, which is the main prupose of this article.

2 Derivation of a necessary condition.

Under the assumptions (β) one could verify that J is continuously Frechet differentiable and the differential is given as follows:

$$J'(x_0,u_0)(\bar{x},\bar{u}) = \int_a^b (D_1L(s),\bar{x}(s)) + (D_2L(s),\bar{u}(s)) \quad ds \tag{5}$$

in the righthand expression, we have used (,)to denote inner product in R^n or R^m depending on the context. Further $(\bar{x},\bar{u}) \in E$ and $D_1L(t)$

$$= \frac{\delta L(t,x_0(t),u_0(t))}{\delta x} \quad \text{and} \quad D_2L(t) = \frac{\delta L(t,x_0(t),u_0(t))}{\delta u}$$

First of all we analyse the three constraint sets Q_1, Q_2, Q_3

Q_1: Since the set Q_1 is obtained through an equality relation, it is natural to approximate Q_1 through a tangent cone. The conditions (α) imply defferentiability of the maps P and S. In view of the theorem of Liusternik [9], the tangent cone is given by

$$K_1 = \{(\bar{x},\bar{u}) \in E : T'(x_0,u_0)(\bar{x},\bar{u}) = 0\}$$

provided $T'(x_0u_0)$ maps E onto XxX_σ. In the lemma that follows, we give a sufficient condition ensuring the above requirement. The conditions given below are exactly, those, which were given by Banks and Jacobs [2] for a different mapping.

Let X(t,s) be the unique nxn matrix solution to the following integral equation

$$X(t,s) = I + X(t,s+\sigma)A(s+\sigma) + \int_s^b X(t,\alpha)A_1(\alpha)d\alpha + \int_{s+\sigma}^t X(t,\alpha)A_2(\alpha)d\alpha$$

for $\quad a_0 \leq s < t \leq b$ subject to the conditions (7)

$\quad x(t,t) = I, \; x(t,s) = 0$ for $s > t$.

Then the variation of constants formula for the equation

$$\frac{dx}{dt} = A(t)x(b-\sigma) + A_1(t)x(t) + A_2(t)x(t-\sigma) + B(t)u(t) \tag{8}$$

is given by

$$x(t,\varphi,u) = x(t,\varphi,0) + \int_a^t X(t,s)B(s)u(s)ds \qquad a \leq t \leq b \tag{9}$$

Let

$$A_1(t) = D_1f(t) = \frac{\delta f(t,x_0(t),x_0(t-\delta),u_0(t))}{\delta x},$$

$$A_2(t) = D_2f(t) = \frac{\delta f(t,x_0(\delta),x_0(t-\delta),u_0(t))}{\delta y}$$

$B(t) = \dfrac{\delta f(t,x_0(t),x_0(t-\delta),u_0(t))}{\delta u} = D_3f(t)$, where $\frac{\delta f}{\delta y}$ denotes the partial derevative of f with respect to the third argument.

Now we state some hypotheses in the notation of Banksand Jacobs [2]:

(H1) $G(a,b-\sigma) = \int_a^{b-\sigma} X(b-\sigma,s)B(s)B^*(s)X^*(b-\sigma,s)ds$ has rank n;

(H2) There exist bounded measurable matrix function Γ and Γ_2 mapping $[b_1 -\sigma,b] \to \mathcal{L}_{mn}$ such that,

$$\begin{aligned} A(t) &= B(t)\,\Gamma(t) \\ A_2(t) &= B(t)\Gamma_2(t) \end{aligned} \quad , \; b - \sigma \leq t \leq b.$$

(H3) $B^+(t)$ the generalized inverse of $B(t)$ (see [12]) is bounded on $[b-\sigma,b]$.

Lemma: If (H1)-(H3) are satisfied, then $T'(x_0,u_0)$ maps E onto XxX_σ.

Proof: Given arbitrarily chosen $(y,\lambda)\in XxX_\sigma$, we need to find $(h,u)\in XxL_2$ such that $T'(x_0,u_0)(h,u) = (y,\lambda)$ (10)

where $T'(x_0,u_0)(h,u) = [P'(x_0,u_0)(h,u),S'(x_0,u_0)(h,u)]$. (11)

But $P'(x_0,u_0)(h,u)(t) = h(t)$ for $t\in[a-\sigma,a]$ und

$$h(t)-\int_a^t A(s)h(s-\sigma)ds - \int_a^t A_1(s)h(s)ds - \int_a^t A_2(s)h(s-\sigma)ds - \int_a^t B(s)\,u(s)ds \text{ for } t\in[a,b] \quad (12)$$

and $S'(x_0,u_0)(h,u) = h_b$. (13)

So the equation (10) is equivalent to the following equations:
$y(t) = h(t)\in t\ [a-\sigma,a]$
$\dot y(t) = \dot h(t) - A(t)h(t-\sigma) - A_1(t)h(t) - A_2(t)h(t-\sigma) - B(t)u(t)$
for $t\in[a,b]$ and $h(b+t) = \lambda(t)$ for $t\in[-\sigma,0]$.

It is easily verified with the help of the variation of constants formula
(9) that $h(t,y,u) = \tilde h(t,y,0) + \int_a^t X(t,s)B(s)u(s)ds$, where $\tilde h(b,y,0)$ is a solution of the equation

$\dot y(t) = \dot h(t) - A(t)h(t-\sigma) - A_1(t)h(t) - A_2(t)h(t-\sigma)$ for $t\in|a,b|$
with $h(t) = y(t)$ for $t\in[a-\sigma,a]$.

Choose ξ as a solution of the equation

$$G(a,b-\sigma)\xi = \lambda(-\sigma) - h(b-\sigma,y,0)$$

and $u(t) = B^*(t)X^*(b-\sigma,t)\xi$. Obviously, U is bounded in $L_2[a,b-\sigma]$.

$$\text{So } h(b-\sigma,y,u) - \tilde{h}(b-\sigma,y,0) = \int_a^{b-\sigma} X(b-\sigma,s)B(s)B^*(s)X^*(b-\sigma,s)\xi ds$$

$$= G(a,b-\sigma)\xi = \lambda(-\sigma) - \tilde{h}(b-\sigma,y,0)$$

and hence $$h(b-\sigma,y,u) = \lambda(-\sigma). \qquad (14)$$

Now one needs only to show the existence of an u in $L_2[b-\sigma,b]$ such that

$$h(t) - A_1(t)h(t) - y(t) = B(t)[\Gamma(t)h(t-\sigma) + \Gamma_2(t)h(t-\sigma) + u(t)|$$

for $t \in [b-\sigma,b]$ and $h(b+t) = \lambda(t)$ for $t \in [-\sigma,0]$.

This is now obvious from the condition (H3) when one takes $h(b+t) = \lambda(t)$, $t \in [-\sigma,0]$ for $t \in [-\sigma,0]$. This is possible in view of Thus the lemma is proved. (14)

In the following (H1),(H2),(H3) are assumed to be fulfilled and hence the tangent cone is given by (6). So, taking into accownt (11),(12), (13), we have the tangent cone given by the set of all $(\bar{x},\bar{u}) \in E$ such that

$$\bar{x}(t) = \begin{cases} 0 \text{ for } a-\sigma \leq t \leq a \\ \int_a^t A(s)x(s-\sigma)ds + \int_a^t A_1(s)x(s)ds + \int_a^t A_2(s)x(s-\sigma)ds \\ + \int_a^t B(s)U(s)ds \text{ for } t \in [a,b] \end{cases}$$

and $\bar{x}_b = 0$.

Constraint Q_2:
We have U, a closed konvex set in L_2 with nonempty interior and $Q_2 = \{(x,u) \in E: u \in U\}$. Define $N = \{\lambda u \in L_2 : u+u_0 \in \overset{o}{U}, \lambda \geq 0\}$

Then it is not difficult to verify that $K_2 = XxN$ is the cone of feasible direktions for Q_2 at (x_0,u_0).

Constraint Q_3:
The constraint Q_3 on the phase co-ordinates could be written as

$Q_3 = \{(x,u) \in E : H(x) \leq 0\}$.

We assume that $G_x(x,t) \neq 0$ for all (x,t) for which $G(x,t) = 0$. This extra assumption is over and above the general assumption (γ).

Note that there exists a positive constant C such that

$$\max_{t \in [a-\sigma,b]} |x(t)| \leq C \;\|x\| \text{ where } `\| \;\|' \text{ denotes the norm}$$

in the Hilbert space X. By a method similar to the one given by Girsanov [6] we could obtain the regular feasible cone K_3 of Q_3 at (x_0,u_0).

$$K_3 = \{(\bar{x},\bar{u}) \in E : G_x(x_0(t),t)\bar{x}(t) < 0, \; t \in R \subset [a-\sigma,b]\}$$

where $R = \{ t \in [a-\sigma,b] : G(x_0(t),t) = H(x_0)\}$.

Further the non-negative functionals for this cone are of the form $-\int_{a-\sigma}^{b} (G_x(x_0(t),t)\bar{x}(t)d\mu(t)$ where μ is a non-negative Stieltjets measure with support on R. If $H(x_0) < 0$, then x_0 is an interior point of Q_3 in view of the compactness of $[a-\sigma,b]$. So K_3 must be equal to E and the measure μ should be a zero measure. Thus there is no loss of generality in stating that the non-negative continuous linear functional on K_3 is of the form

$-\int_{a-\sigma}^{b} G_x(x_0(t),t),\bar{x}(t))d\mu(t)$ where μ is a non-negative Stieltjets measure with support on $R = \{t \in [a-\sigma,b] : H(x_0) = 0\}$

After these preliminary condierations, we are in a position to obtain a maximum principle for our problem as a necessary condtion for a pair (x_0,u_0) to be optimal.

Theorem 3 (Maximum priciple)
Suppose (x_0,u_0) be minimal point for the functional J subject to the constraints (1) and (2). Suppose also that the general conditions $(\alpha),(\beta),(\gamma)$ and the regularity conditions for $T'(x_0,u_0)$ be satisfied. Let $G_x(x,t) \neq 0$ whenever $G(x,t) = 0$. Then there exist $\lambda_0 \geq 0$, a non-negative measure μ with support on R, a Θ X and Ψ satisfying equations (19) whre α_0 satisfies (20); not all of λ_0, Ψ, Θ zero such that,

$$\int_a^b [- \lambda_0(D_2L(t),u(t)) + (D_3f(t)\ \Psi\ (t),u(t))]dt \leq$$

$$\int_a^b [- \lambda_0(D_2L(t),u_0(t)) + (D_3f(t)\ \Psi\ (t),u_0(t))]dt$$

for all $u \in U$.

Proof: Suppose $(x_0,u_0) = z_0$ be an optimal pair for our problem. By the assumptions made above, K_1 is a regular tangent cone of Q_1; K_2 and K_3 are regular feasible cones of Q_2 and Q_3. So by Dubovitskii-Muliutin theorem, there exists continuous linear functionals χ_1,χ_2,χ_3 on E such that $i \in \{1,2,3\}$ and $z \in K_i$ implies that $\chi_i(z) \geq 0$.

Further there exist $\lambda_0 \geq 0$ such that the Euler equation

$$\chi_0 = \lambda_0 J'(x_0,u_0) = \chi_1 + \chi_2 + \chi_3 \tag{15}$$

is satisfied, where not all the $\{\chi_i\}$, $i = 0,1,2,3$ are zero.

Since K_1 is a subspace of E and χ_1, is non-negative on K_1, χ_1, must be an anhilator of K_1. So from the equation (15) it follows that

$$\chi_0(z) - \chi_2(z) - \chi_3(z) = 0 \text{ for all } z \in K_1 \tag{16}$$

By theorem 2 and equation (16) it is clear that ther exist $\eta \in X$ and $\Theta \in X_\sigma$ such that $\chi_1(z) = \langle \eta, P'(x_0,u_0)z \rangle + \langle \Theta, S'(x_0,u_0)z \rangle$ for all $z \in E$. Considering equation (13) the last equation could be rewritten as

$$\chi_1(z) = \langle \eta, P'(x_0,u_0)z \rangle + \langle \Theta, \bar{x}_b \rangle \text{ where } z = (\bar{x},\bar{u}) \text{ and } \Theta \in X_\sigma \tag{17}$$

We now, further that $\chi_2(z) \geq 0$ for $z \in K_2$ and χ_3 is of the form

$$- \int_R G_x(x_0(t),t)x(t)d\mu(t).$$

Continuing to analyse the Euler equation (15) let us take any $\bar{u} \in L_2$ and obtain an $\bar{x}$ satisfying the equation $P'(x_0,u_0)(\bar{x},\bar{u}) = 0$, which is always possible under our assumptions. Then we have the following reduced Euler equation:

$$\lambda_0 J'(x_0,u_0)(\bar{x},\bar{u}) = \chi_2(\bar{x},\bar{u}) - \int_{a-\sigma}^{b} (G_x(x_0(t),t),\bar{x}(t))d\mu(t)$$

$$+ (\theta(-\sigma), \bar{x}(b-\sigma)) + \int_{-\sigma}^{0} (\dot{\theta}(t), \dot{\bar{x}}(t+b))dt$$

From the structure of K_2 it is also evident that

$$\chi_2(\bar{x},\bar{u}) = \overset{*}{\chi}_2(u) \geq 0 \text{ for } u \in N.$$

Now we introduce an auxiliary function Ψ defined as follows:

$$-\Psi(t) = \begin{cases} \alpha_0 - A^*(t+\sigma)\,\Psi(t+\sigma) - \int_t^{b-\sigma} (D_1 f)\,\Psi(s)ds - \int_{t+\sigma}^{b}(D_2 f)\,\Psi(s)ds \\ \quad + \lambda_0 \int_t^{b-\sigma} D_1 L(s)ds + \int_t^{b-\sigma} G_x(s)d\mu(s) \text{ for } t \in [a,b-\sigma] \\ -\dot{\theta}(t-b) - \int_t^{b} (D_1 f)^*\,\Psi(s)ds + \lambda_0 \int_t^{b} D_1 L(s)ds \text{ for } t \in [b-\sigma,b] \end{cases} \quad (19)$$

Inview of the assumption (γ), μ has support in a compact subset of $(a,b-\sigma)$. So the existence of the solution of (19) could be obtained by a slight modification of a result of Das and Sharma (Theorem 4) [4].

Now let $\bar{u} \in L_2$ be arbitrary and $\bar{x}$ obtained by the equation $P'(x_0,u_0)(\bar{x},\bar{u}) = 0$.

Then $\frac{d\bar{x}}{dt}$ satisfies the variational equation obtained from (12) through differentiation.

We obtain the following multiplying $\frac{d\bar{x}}{dt}$ to $-\Psi$ and integrating by parts

$$- \int_a^b (\frac{d\bar{x}}{dt}, \Psi(t)dt = (\bar{x}(b-\sigma), \alpha_0) - \int_a^{b-\sigma} (\dot{\bar{x}}(t)\ A^*(b+\sigma)\ \Psi(t+\sigma))dt$$

$$- \int_a^{b-\sigma} (\bar{x}(t),(D_1 f)\ \ \Psi(t))dt - \int_a^{b-\sigma} (\bar{x}(t),D_2 f^*(t+\sigma)\ \ \Psi(t+\sigma)\ dt$$

$$+ \lambda_0 \int_a^{b-\sigma} (\bar{x}(t),D_1 L(t))dt + \int_a^{b-\sigma} (\bar{x}(t),G_x(x_0(t),t))d\mu(t) + (\bar{x}(b-\sigma),$$

$$\int_{b-\sigma}^b (D_1 f^*)\ \ \Psi(s)ds) - \lambda_0(\bar{x}(b-\sigma), \int_{b-\sigma}^b D_1 L(s)ds) - \int_{b-\sigma}^b(\dot{\bar{x}}(t),\dot{\Theta}(t-b))dt$$

$$- \int_{b-\sigma}^b (\bar{x}(t),(D_1 f^*)\ \Psi(t))dt + \lambda_0 \int_{b-\sigma}^b (\bar{x}(t),D_1 L(t))dt.$$

Choosing $\alpha_0 = - \int_{b-\sigma}^b (D_1 f\ )\ \Psi(s)ds + \lambda_0 \int_{b-\sigma}^b D_1 L(s)ds - \Theta(-\sigma)$ (20)

and considering that $x_a = 0$, we have,

$$- \int_a^b(\frac{d\bar{x}}{dt}, \Psi(t))dt = - \int_a^b (D_1 f(t)\bar{x}, \Psi(t))dt - \int_a^b(D_2 f(t)\bar{x}(t-\sigma), \Psi(t))dt$$

$$+ \lambda_0 \int_a^b(\bar{x}(t),D_1 L(t))dt - (\bar{x}(b-\sigma), \Theta(-\sigma)) - \int_{b-\sigma}^b (\dot{\bar{x}}(t), \dot{\Theta}(t-b))dt$$

$$+ \int_a^b(\bar{x}(t),G_x(x_0(t),t))d\mu(t) - \int_a^b (A(t)\dot{\bar{x}}(t-\sigma), \Psi(t))dt.$$

Or, $\lambda_0 \int_a^b(\bar{x}(t),D_1 L(t))dt + \int_a^b(\bar{x}(t),G_x(x_0(t),t))d\mu(t) - \langle \bar{x}_b,\Theta \rangle$

$$= - \int_a^b ((D_3 f)\bar{u}(t), \Psi(t))dt.$$

In obtaining the last equation, we used the variational equation satisfied by $\bar{x}$ and the fact that μ has support only in the interior of $(a,b-\sigma)$.

Finally taking into consideration the reduced Euler equation (18), we have

$$x_2^*(\bar{u}) = x_2(\bar{x},\bar{u}) = \int_a^b -[((D_3 f^* \Psi)(t),\bar{u}(t)) + \lambda_0(D_2L(t),\bar{u}(t))]\, dt \geq 0$$

For all $\bar{u} \in M$.

Or equivalently,

$$\int_a^b [-\lambda_0(D_2L(t),u(t)) + (D_3 f^* \Psi)(t)),u(t))]dt \leq \int_a^b [-\lambda_0(D_2L(t),u_0(t))$$

$$+ (D_3 f^*(t)\ \Psi(t),\ u_0(t))]dt \qquad \text{for all } u \in U.$$

If $\lambda_0 = 0$, $\Psi = 0$ and $\Theta = 0$, then from (14) and (15) it follows that $\chi_3 = 0$ and hence from (13) $\chi_2 = 0$. Thus we have

$$\lambda_0 = 0,\ \Theta = 0,\ \chi_2 = 0,\ \chi_3 = 0$$

This implies $\chi_1 = 0$, constradicting the conclusion of the theorem 1 that not all of them are zero. This concludes the proof our theorem.

Note! It is easy to verify that the results of Jacobs and Kao [8] and Banks and Jacobs [2] are particular cases of the maximum principle derived above.

References

1) H.T. Banks - Control of functional differential equations with function space boundary conditions.
Proc. Park City, Differential Equation Symposium
Academic Press New York (1972)

2) H.T. Banks and M.Q. Jacobs - An attainable set approach to Optimal control of Functional Differential Equations with Function-space Terminal conditions.
J. Differential Equations Vol. 13 (1973) pp 127-149.

3) H.T.Bank and G.A.Kent - Control of functional differential equations of retarded and neutral type to target sets in function space
SIAM J. Control vol. 10 No. 4 (1972) pp 567-593

4) P.C.Das and R.R.Sharma - Optimal control for measure delay differential equations
SIAM J.Control vol 9 (1971) pp 43-61

5) A.Ya.Dubovitskii and A.A.Milyutin - Extremum problems in the presence of constraints
Zh. Vichis. Mat. i Mat Fiz von 5 No. 3 1965 pp 395-453

6) I.V.Girsanov - Lectures on mathematical theory of extremum problems
Springer-Verlag Berlin, Heidelberg, New York (1972)

7) J.K.Hale - Functional differential equations
Springer-Verlag New York (1971)

8) M.Q.Jacobs and T.J.Kao - An optimal settling problem for time lag systems
J. Math. Anal. and Appl. 40 (1970) pp 687- 707

9) L.A.Liusternik and L.L.Sobolev - Elements of functional analysis
Trans. F. Ungar New York (1961)

10) D.G.Luenberger - Optimization by nector space methods
John Wiliy and Sons New York (1969)

11) L.W.Neustadt - An abstract variational theory with applications to a brood class of optimization problems
I A General theory.
SIAM J. control 4 (1966) pp 505-527

12) R.Penrose - A generalized inverse of matrices
Proc. Cambridge Phil. soc. 51 (1955) pp 116-229.

A Relaxation Strategy for the Modified Newton Method

Peter Deuflhard

Mathematisches Institut
der Technischen Universität

München, West Germany

§ 0. Introduction

The modified (underrelaxed, damped) Newton method is one of the most popular and efficient methods for the numerical solution of nonlinear equations (for ref. see e.g. Goldstein [8], Stoer [16]). If some initial guess x_o of the solution data is given, then iterates x_k are determined by

(0.1) $$x_{k+1} := x_k + \lambda_k \Delta x_k \;, \quad 0 < \lambda_k \leq 1$$

where Δx_k denotes the correction vector of the ordinary Newton method and λ_k denotes the *relaxation* (or damping) *factor*. Under mild assumptions global convergence of the algorithm can be assured.

In applications, λ_k is usually selected *ad hoc* by means of some monotonicity test. Moreover, in order to take advantage of the quadratic convergence of the ordinary Newton method (for iterates sufficiently close to the solution point), the first trial of λ_k (say $\lambda_k^{(o)}$) is determined from $(\lambda_{k-2}, \lambda_{k-1})$ e.g. by

(0.2) $$\lambda_{k-2} > \lambda_{k-1} \implies \lambda_k^{(o)} := \lambda_{k-1}$$
$$\lambda_{k-2} \leq \lambda_{k-1} \implies \lambda_k^{(o)} := \min\,\{1, 2\lambda_{k-1}\}$$

Those simple, but vigorous *empirical relaxation strategies* have been extremely helpful, for instance in solving numerically sensitive equations arising in multiple shooting techniques (for ref. see Keller [11], Osborne [13], Bulirsch [5], and Stoer/Bulirsch [17]). In some highly nonlinear systems of equations of that type, however, the empirical relaxation strategy may be unsatisfactory: for λ_k "too large", exponential overflow may easily occur (thus terminating the computation without result), for λ_k "too small", the method tends to be inefficient. On the other hand, upper bounds for λ_k obtained from theoretical results (see e.g. Meyer [12], Amann [1], or [7]) were both too pessimistic and in terms of computationally unavailable quantities. It is the purpose of this paper to close this gap between theory and applications.

In § 1, a computationally convenient upper bound for the λ_k is derived on a theoretical basis that is a slight refinement of results given in [7]. In § 2, details of the numerical realization of a new relaxation strategy are worked out. The method obtained is a generalization of Aitken's Δ^2- method. In addition, a simple a-priori condition for the economic alternative use of quasi-Newton corrections (in lieu of Newton corrections) is proposed (for reference see e.g. Broyden [3],[4]). In § 3, a numerical comparison of the modified Newton method with old (empirical) and new (theoretically backed) relaxation strategy is given. The gain in efficiency and reliability is demonstrated solving a well-known nonlinear least squares test problem and realistic multiple shooting examples.

§ 1. Theoretical background

The notations of this paragraph are close to [7]. Let

$$F(x) := \begin{bmatrix} f_1(\xi_1,\dots,\xi_n) \\ \cdot \\ \cdot \\ \cdot \\ f_n(\xi_1,\dots,\xi_n) \end{bmatrix} , \quad F(x^*) = 0$$

denote the system of nonlinear equations to be solved and $J(x)$ the Jacobian (n,n)-matrix (assumed to be nonsingular here). Then the modified Newton method is defined by the iteration $(k = 0,1,\dots)$:

$$x_{k+1} := x_k + \lambda_k \Delta x_k \, , \quad 0 < \lambda_k \le 1 \tag{1.1}$$

where

$$\Delta x_k := -J(x_k)^{-1} F(x_k)$$

The relaxation factors λ_k are determined by means of a *monotonicity test*

$$T(x_k + \lambda_k \Delta x_k | A) \le T(x_k | A) \tag{1.2}$$

where

$$T(x|A) := \| AF(x) \|_2^2 \equiv (AF(x))^T AF(x) \tag{1.3}$$

denotes a *level function* associated with a nonsingular scaling matrix A (say $A = I$, or $A = J(x_k)^{-1}$). At first, the theoretical results given in [7] are refined.

Theorem 1.1

I. *Let* $F(x) \in C^1(D)$ *with* $D \subset \mathbb{R}^n$, $J(x) \equiv F'(x)$

II. *Let exist* $J(x)^{-1}$ *with* $\| J(x)^{-1} \|_2 \le \beta$ *for all* $x \in D$

III. Let J(x) satisfy a Lipschitz-condition

$$\| J(x) - J(y) \|_2 \leq \gamma \| x - y \|_2 \quad \text{for } x,y \in D$$

IV. For some (n,n)-matrix A with $\text{cond}_2(A) \leq M < \infty$ *let*

$$G_k(A) := \left\{ x \in D \,\middle|\, T(x|A) \leq T(x_k|A) \right\} \subset D$$

denote a level set with $T(x_k|A) > 0$. *Let the path-connected component of* x_k *in* $G_k(A)$ *be compact.*

Then, with the notations

$$J_k \equiv J(x_k) \;,\; h_k := \beta^2 \gamma \| F(x_k) \|_2$$

the following results hold:

$$T(x_k + \lambda \Delta x_k | A) \leq t_k^2(\lambda|A) T(x_k|A) \quad \text{for } 0 \leq \lambda \leq \mu_k(A)$$

where

$$t_k(\lambda|A) := 1 - \lambda + \tfrac{1}{2}\lambda^2 h_k (1 + \lambda h_k) \; \text{cond}_2(AJ_k) \leq 1$$

$$(1.4) \quad \mu_k(A) := \min\left\{ 1, \left(\sqrt{1 + \frac{8}{\text{cond}(AJ_k)}} - 1 \right) / (2h_k) \right\}$$

Proof: The proof of Theorem 3.2 of [7] can nearly be copied with one main exception: using the notations of that paper, the estimates

$$\| J_k^{-1} \bar{J} \|_2 \leq 1 + h_k, \; \| x(\lambda,\delta) - \bar{x}(\lambda) \| \leq \delta \alpha_k h_k \lambda^2$$

can be refined to yield

$$\| J_k^{-1} \bar{J} \|_2 \leq 1 + \lambda h_k, \quad \| x(\lambda,\delta) - \bar{x}(\lambda) \| \leq \tfrac{1}{2} \delta \alpha_k h_k \lambda^2$$

Thus the bound (3.12) in [7], which is *second* order in λ, can be replaced by the *third* order bound

$$(1.5) \qquad \frac{\| AF(x_k + \lambda \Delta x_k) \|_2}{\| AF(x_k) \|_2} \leq 1 - \lambda + \tfrac{1}{2}\lambda^2 h_k (1 + \lambda h_k) \, \text{cond}_2(AJ_k)$$

Now, a straightforward calculation shows that $\mu_k(A)$ in (1.4) is selected so that

$$(1.6) \qquad t_k(0|A) = t_k(\mu_k(A)|A) = 1$$

$$t_k(\lambda|A) < 1 \quad \text{for } 0 < \lambda < \mu_k(A)$$

The rest of the proof can be omitted. ♦

The *optimal choice* of λ_k with respect to the estimate $t_k(\lambda|A)$ is

$$(1.7) \qquad \lambda_k(A) := \min\left\{1, \left(\sqrt{1+\frac{6}{\text{cond}(AJ_k)}} - 1\right) / (3h_k)\right\}$$

The *extremal properties* of natural scaling ($A = J(x_k)^{-1}$) as stated in [7] remain unchanged:

$$(1.8) \qquad \begin{aligned} &\text{a)}\ t_k(\lambda|A) \geq t_k(\lambda|J_k^{-1}) \\ &\text{b)}\ \mu_k(A) \leq \mu_k(J_k^{-1}) = \frac{1}{h_k} \\ &\text{c)}\ \lambda_k(A) \leq \lambda_k(J_k^{-1}) = \frac{1}{3}(\sqrt{7}-1)\frac{1}{h_k} \end{aligned}$$

for all nonsingular (n,n)-matrices A

The global convergence theorem associated with Theorem 1.1 is omitted here, since it is quite similar to Theorem 3.3 of [7]. Instead, the computationally unavailable quantity $\frac{1}{h_k}$ in (1.8) is replaced by an upper bound.

Lemma 1.2

Same notations and assumptions as in Theorem 1.1. Let

$$\overline{\Delta x_k} := -J(x_{k-1})^{-1}F(x_k)$$

denote the simplified Newton correction. Let $\gamma \neq 0$, $\overline{\Delta x_k} \neq \Delta x_k$ *(purely nonlinear case). Then*

$$(1.9) \qquad \frac{1}{h_k} \leq \frac{1}{\hat{h}_k} \leq \frac{\|\Delta x_{k-1}\|_2}{\|\overline{\Delta x_k} - \Delta x_k\|_2} \cdot \lambda_{k-1}$$

where $\hat{h}_k := \|\Delta x_k\|_2 \beta\gamma$

Proof: With the notations $J_k \equiv J(x_k)$, $F_k \equiv F(x_k)$ it is obtained

$$\|\overline{\Delta x_k} - \Delta x_k\|_2 = \|(J_{k-1}^{-1} - J_k^{-1})F_k\|_2 \leq \|J_{k-1}^{-1}\| \, \|J_k - J_{k-1}\| \, \|\Delta x_k\|$$

$$\leq \beta \|\Delta x_k\| \gamma \|x_k - x_{k-1}\| = \hat{h}_k \lambda_{k-1} \|\Delta x_{k-1}\|$$

Moreover, $\hat{h}_k \leq h_k$

With the assumptions above this implies (1.9). ♦

Remark: One result of the well-known *Newton-Kantorovič theorem* (see [10]) can be stated in the form

$$(1.10) \qquad \frac{1}{\hat{h}_k} > 0.5 \Rightarrow \lambda_k = 1$$

It may be noted that the right hand side of (1.9) is *invariant* to the linear transformations

$$(1.11)\qquad \begin{array}{llll} \text{(I)} & F(x) \longrightarrow & G(x) & = BF(x) \\ \text{(II)} & x \longrightarrow & y & = \alpha K x \end{array}$$

where B nonsingular, K orthogonal, $\alpha \neq 0$

§ 2. Numerical realization

The theoretical results (1.8) and (1.9) readily yield a computationally convenient relaxation strategy. Moreover, a simple condition for the economic alternative use of quasi-Newton corrections (in lieu of Newton corrections) can be derived.

a) Relaxation strategy for the modified Newton method

For the time being, let the Jacobian matrix $J(x_k)$ be nonsingular. Then, based on Lemma 1.2, the first trial of the relaxation factor λ_k (for $k \geq 1$) is determined by

$$(2.1)\qquad \lambda_k := \begin{cases} 1 & \text{if} \quad \mu_k > 0.7 \\ \mu_k & \text{if} \quad \mu_k \leq 0.7 \end{cases}$$

where

$$(2.2)\qquad \mu_k := \frac{\| \Delta x_{k-1} \|_2}{\| \overline{\Delta x_k} - \Delta x_k \|_2} \cdot \lambda_{k-1}$$

with

$$\Delta x_k := -J(x_k)^{-1} F(x_k) \, , \quad \overline{\Delta x_k} := -J(x_{k-1})^{-1} F(x_k)$$

Remark 1:
Compared with (1.8), a factor $\frac{1}{3}(\sqrt{7} - 1) \approx 0.55$ is dropped, since one cannot expect more than the order of magnitude from theoretical estimates that include the least favorable case. Moreover, with this choice, (2.1) is a formal generalization of Aitken's Δ^2-method in the linear convergence phase of the iteration (i.e. for $\lambda_k = \mu_k$). In fact, it appeared to be the most efficient choice for all examples tested so far. In the case of linear $F(x)$, one obtains $\mu_k = \infty$ and hence $\lambda_k = 1$ - independent of any choice of that factor.

Remark 2: The *Newton-Kantorovič* bound 0.5 in (1.10) is replaced by 0.7, since μ_k is just an upper bound of $\hat{h}_k^{-1}$. This choice is backed by numerical experience, too.

The *computational amount* for evaluating $\overline{\Delta x_k}$ in (2.2) is $O(n^2)$ operations - which is far less than is usually needed for one function evaluation in real life applications (e.g. in multiple shooting one "function evaluation" means the numerical solution of a series of initial value problems for ordinary differential equations!). It is negligible, if natural scaling ($A = J(x_k)^{-1}$) is used in the monotonicity test (1.2).

As for *rounding errors*, there is no trouble arising from the difference term in the denominator of μ_k, since in (2.1) μ_k is replaced by 1, if the denominator is "small".

Monotonicity tests. At each iteration step the choice λ_k due to (2.1) is monitored by means of two tests: If

(2.3)
$$T(x_k + \lambda_k \Delta x_k | A) > T(x_k | A)$$
$$\text{for } A = I \ \underline{\text{and}} \ A = J(x_k)^{-1}$$

then λ_k is reduced, say

(2.4) $$\lambda_k := \lambda_k / 2$$

Rank reduction. Let $\lambda_{\min}$ denote an input parameter (as introduced in [7]) in the range

(2.5) $$0 < \lambda_{\min} \leq 1$$

(say $\lambda_{\min} \approx 0.01$). Then, if one obtains

(2.6) $$\lambda_k < \lambda_{\min}$$

from (2.1) or (2.4), the pseudo-rank of the Jacobian matrix $J(x_k)$ is reduced and the *inverse* is replaced by the Penrose *pseudoinverse* ([14]). As a consequence, μ_k from (2.2) has to be replaced by some μ_k'. Let P and $\bar{P}$ denote two projection matrices with

$$P + \bar{P} = I \ , \ s := \text{rank } (P) = \text{rank}(J_k')$$

where

$$J_k' := \left[\begin{array}{c|c} J & j \\ \hline 0 & 0 \end{array}\right] , \ J \text{ nonsingular}, \ \bar{j} := J^{-1} j$$

Then, with the notations

$$(I_{n-s} + \bar{j}^T \bar{j}) =: LL^T \quad \text{(Cholesky decomposition)}$$

$$\Delta x_k' := -J_k'^{+} F(x_k)$$

the analogon of (2.2) for the reduced mapping

$$PF\left(x_k + J_k'^{\dagger} J_k'(x-x_k)\right)$$

is

$$(2.7) \qquad \mu_k' := \frac{\| \Delta x_{k-1} \|_2}{\left[\| \overline{\Delta x}_k - \Delta x_k' \|_2^2 - \delta_k^2 \right]^{1/2}} \cdot \lambda_{k-1}$$

where

$$\delta_k := \| L^{-1} (\bar{P}\overline{\Delta x}_k - \bar{J}^T P \overline{\Delta x}_k) \|_2 < \| \overline{\Delta x}_k - \Delta x_k' \|_2$$

Like in the comparison lemma 4.3 of [7], it is obtained

$$(2.8) \qquad \mu_k' > \mu_k \qquad (\text{if } \bar{P}F(x_k) \neq 0)$$

Proof: $\qquad J' \equiv J_k'$.

One uses the projection properties

$$(I - J'^{\dagger} J') \Delta x_k' = 0 \ , \ \Delta x_k' = J'^{\dagger} J \Delta x_k \ , (J'^{\dagger} J')^T = J'^{\dagger} J'$$

where $\Delta x_k = -J(x_k)^{-1} F(x_k)$ is the associated full rank Newton correction (assumed to be finite here without loss of generality).

These properties imply

$$\| J'^{\dagger} J' (\overline{\Delta x}_k - \Delta x_k) \|_2^2 = \| \overline{\Delta x}_k - \Delta x_k \|_2^2 - \| (I - J'^{\dagger} J')(\overline{\Delta x}_k - \Delta x_k) \|_2^2 < \| \overline{\Delta x}_k - \Delta x_k \|_2^2$$

$$\| J'^{\dagger} J' (\overline{\Delta x}_k - \Delta x_k) \|_2^2 = \| \overline{\Delta x}_k - \Delta x_k' \|_2^2 - \| (I - J'^{\dagger} J') \overline{\Delta x}_k \|_2^2$$

Hence

$$\mu_k' := \frac{\| \Delta x_{k-1} \| \lambda_{k-1}}{\| J'^{\dagger} J' (\overline{\Delta x}_k - \Delta x_k) \|_2} > \frac{\| \Delta x_{k-1} \| \lambda_{k-1}}{\| \overline{\Delta x}_k - \Delta x_k \|_2} = \mu_k$$

Upon using the pseudo-inverse representation as suggested in [7], a straightforward calculation shows

$$\delta_k = \| (I - J'^{\dagger} J') \overline{\Delta x}_k \|_2 = \| L^{-1} (\bar{P}\overline{\Delta x}_k - \bar{J}^T P \overline{\Delta x}_k) \|_2 \qquad \blacklozenge$$

Finally, it may be noted that the right hand side of (2.7) is invariant to the class (1.11) of linear transformations, if

$$PB^T B\bar{P} = 0 \ .$$

In fact, this is the same class as obtained in [7] for the invariance of $\Delta x_k'$.

b) Alternative use of quasi-Newton corrections

In [5] Bulirsch suggested to replace the ordinary Newton corrections by quasi-Newton corrections at suitably selected iterates: thus the time consuming evaluation of the Jacobian matrices can be avoided by (repeated) use of approximations (e.g. due to Broyden [3]). The set of empirically obtained selection conditions, however, appeared to be not sufficiently reliable in real life applications: in some examples, more than 5o% of computing time is saved, whereas in other examples the global convergence behavior of the algorithm is heavily disturbed, unless the number of recursive quasi-Newton approximations is restricted (by some input parameter p to be selected by the user of the algorithm). This typical behavior is illustrated by Examples II and III of § 3.

The purpose of this section is to propose a computationally convenient *selection condition* based on the results presented above. From numerical experience, the following condition would be desirable:

$$\hat{\lambda}_{k+1} \geq \lambda_k \tag{2.9}$$

where $\hat{\lambda}_{k+1}$ denotes the relaxation factor obtained at the quasi-Newton step. (2.9), however, is an *a-posteriori* condition. That is why $\hat{\lambda}_{k+1}$ is replaced by its *a-priori* estimate

$$\hat{\mu}_{k+1} := \frac{\| \Delta x_k \|_2}{\| \widehat{\Delta x}_{k+1} - \overline{\Delta x}_{k+1} \|_2} \cdot \lambda_k \tag{2.10}$$

where $\widehat{\Delta x}_{k+1}$ denotes the quasi-Newton correction. Hence, in lieu of (2.9), one requires

$$\hat{\mu}_{k+1} \geq \sigma \lambda_k \tag{2.11.a}$$

for some safety factor $\sigma > 1$

This is equivalent to

$$\| \widehat{\Delta x}_{k+1} - \overline{\Delta x}_{k+1} \|_2 \leq \frac{1}{\sigma} \| \Delta x_k \|_2 \tag{2.11.b}$$

The computational test of condition (2.11) needs no additional function evaluation. In the numerical experiments done so far, the choice of σ appeared to be not very sensitive. A reasonable range of choice turned out to be

$$3 \leq \sigma \leq 10 \tag{2.12}$$

Remark : In the case of linear $F(x)$ one obtains $\hat{\mu}_{k+1} = \infty$ - so (2.11) is satisfied independent of σ .

(2.11) can be further simplified, if the quasi-Newton corrections are computed from rank-1 approximations of the Jacobian matrix.

Lemma 2.1

Notations as before. Let J_k be nonsingular and

$$(2.13) \qquad \alpha_{k+1} := \frac{\overline{\Delta x}_{k+1}^T \Delta x_k}{\Delta x_k^T \Delta x_k} \neq 1$$

For the modified Newton method the rank-1 approximation $\hat{J}_{k+1}$ due to [3] *can be simplified to yield*

$$(2.14) \qquad \hat{J}_{k+1} = J_k + \left(F(x_{k+1}) - (1-\lambda_k)F(x_k)\right) \frac{\Delta x_k^T}{\lambda_k \Delta x_k^T \Delta x_k}$$

This implies:

a) the quasi-Newton correction is

$$(2.15) \qquad \widehat{\Delta x}_{k+1} := -\hat{J}_{k+1}^{-1} F(x_{k+1}) = \left(\overline{\Delta x}_{k+1} - (1-\lambda_k)\alpha_{k+1}\Delta x_k\right) / (1-\alpha_{k+1})$$

b) condition (2.11) is equivalent to

$$(2.16) \qquad |\alpha_{k+1}| \, \| \overline{\Delta x}_{k+1} - (1-\lambda_k)\Delta x_k \|_2 \leq \frac{1}{\sigma} |1-\alpha_{k+1}| \, \| \Delta x_k \|_2$$

Proof: a) Use the well-known theorem of Householder [9]

$$(I-uv^T)^{-1} = I + \frac{uv^T}{1-u^Tv} \quad \text{for } u^Tv \neq 1$$

b) straightforward substitution of (2.15) into (2.11) ♦

Remark: A computational convenient *sufficient* condition for (2.13) is

$$(2.17) \qquad \| \overline{\Delta x}_{k+1} \|_2 < \| \Delta x_k \|_2$$

(i.e. decrease of the naturally scaled level function). Then, by means of the Cauchy-Schwarz inequality, it is obtained from (2.17):

$$|\alpha_{k+1}| < 1$$

For Jacobian approximations $\hat{J}_{k+1}$ with

$$(2.18) \qquad \text{rank } (\hat{J}_{k+1} - J_k) > 1$$

Lemma 2.1 does not apply, but condition (2.11) may be required.

Examples of that type of approximation are rank-2 approximations (see e.g. Broyden [4]) or the approximation obtained in multiple shooting for the (sparse) total Jacobian bloc matrix.

For the numerical solution of real life problems a *scaled* version of (2.2), (2.7), (2.10) and (2.14) is recommended, so that the iteration is invariant to re-gauging of the components of the variable x. An ALGOL program containing the results of this paper employed to multiple shooting will be published in [6].

§ 3. Numerical results

The numerical experiments were run on the TR 440 of the *Leibniz-Rechenzentrum der Bayerischen Akademie der Wissenschaften.* The computations were performed in FORTRAN single precision with a 38 bit mantissa.

Example I : Nonlinear Least Squares Test Problem (due to [2])
The problem is to minimize

$$T(x|I) := \sum_{i=1}^{10} \left(f(\xi_1,\xi_2,p_i) - f(1,10,p_i)\right)^2$$

for $\qquad f(\xi_1,\xi_2,p) := e^{-\xi_1 p} - e^{-\xi_2 p}\,,\quad p_i = 0.1(0.1)1.0$

In Table *A*, the number of function evaluations necessary to reduce $T(x|I)$ to less than 10^{-5} is listed for different starting points. For comparison, the results published by Brown/Dennis [2] for the Levenberg-Marquardt method are arranged together with results given in [7] for a modified Newton method using an empirical relaxation strategy. The analytic expression of the Jacobian was used (counted for 2 function evaluations). A comparison of row 2 and row 3 of Table *A* shows a reduction of 10% - 20% in computational amount. This effect was found to be typical for *non-sensitive* examples (here the condition number of the Jacobian matrix at the solution point was ≈ 2.0).

Table *A* Number of function evaluations required to reduce $T(x|I)$ to less than 10^{-5}

Starting point x_0 and $T(x_0|I)$

	(0,0) 3.06	(0,20) 2.09	(5,0) 19.6	(5,20) 1.81	(2.5,10) 0.808	Sum
Levenberg-Marquardt method [2]	22	25	25	31	16	119
old modified Gauss-Newton method [7]	19	14	21	14	13	81
new modified Gauss-Newton method	16	10	17	12	10	65

Example II : Re-entry Problem (two-point boundary value problem, 7 differential equations, 9 nodes)

The underlying physical model describes the re-entry of an Apollo 4 - type satellite into a parking orbit around Earth. For reference see [17] (7.3.7.1),(7.3.7.6) and (7.3.7.7). The sensitivity of the physical problem goes with the well-known numerical sensitivity of the mathematical problem (condition number of the Jacobian matrix at the solution point : $\approx 10^6$).
In Table *B* four different multiple shooting algorithms (involving the modified Newton method) are compared. "Old" and "new" mean the old and new relaxation strategy, "I" means Jacobian approximation by numerical differentiation at each iterate (counted for 7 trajectories), "II" means Jacobian approximation by rank-1 corrections at selected iterates. Eight experiments with eight starting values of the relaxation factor $\lambda_o^{(o)}$ were run (common initial data, common prescribed relative precision 10^{-4}). Rows "old I" and "new I" clearly show a 10% - 30% efficiency gain in the non-critical cases, but a 50% gain in the critical case $\lambda_o^{(o)} = 1/4$. Rows "old II" and "new II" indicate the additional increase in reliability when using condition (2.11) (here with $\sigma = 3$) for selecting the quasi-Newton steps. It may be said that in "new I" and "new II" the computational amount turned out to be nearly independent of the choice $\lambda_o^{(o)}$. This effect was found to be typical for a wide class of problems tested so far.

Table *B* Number of trajectories needed to solve the re-entry two-point boundary value problem

$\lambda_o^{(o)} \rightarrow$	1	$\frac{1}{2}$	$\frac{1}{4}$	$\frac{1}{8}$	$\frac{1}{16}$	$\frac{1}{32}$	$\frac{1}{64}$	$\frac{1}{128}$	average amount
old I	99	98	231*)	114	114	122	130	146	132
new I	90	89	119*)	106	106	98	106	98	102
old II	77	76	over-flow!	114	62	67	75	84	79 +)
new II	53	52	92	60	39	47	48	54	56

*) rank reduction necessary at some iterate

+) average amount skipping the "overflow" result

Example III : Ginzburg-Landau Equations
(singular two-point boundary value problem, 4 differential equations, 14 nodes)

The problem arises in the physical theory of superconductivity. The mathematical model leads to the following singular two-point boundary value problem:

$$f'' = -\frac{f'}{x} + f\left(f^2 - 1 + (a - \frac{1}{x})^2\right)$$

$$a'' = -\frac{a'}{x} + \frac{a}{x^2} + f^2(a - \frac{1}{x})$$

$$a(0) = f(0) = f'(10) = 0\ ,\ a(10) = 0.5$$

Details of the numerical treatment of this problem by multiple shooting techniques are given by Rentrop [15] (in the case of *singular* b.v.p. some analytical preparation is necessary, here for $x \to 0$). He found that there are at least 3 independent solutions of the problem differing only in the fourth significant digit at $x = 0$ - compare Fig. 1. Hence, the level set $G_k(A)$ defined in assumption IV. of Theorem 1.1 consists of at least 3 connected components (for x_k sufficiently close to one of the solution points).

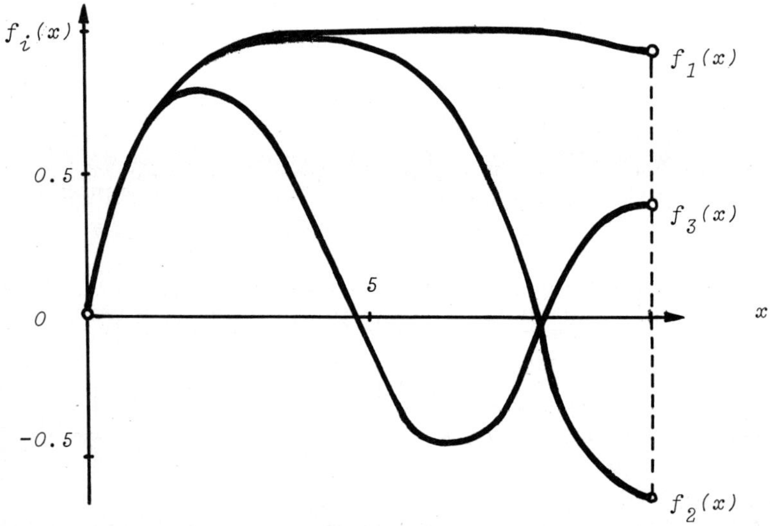

Three solutions of the Ginzburg-Landau Equations (due to [15])

\- Fig. 1 -

The use of quasi-Newton corrections appeared to be sensitive when solving this problem. Let p denote the input parameter of maximum permitted number of successive quasi Newton steps (as already mentioned in § 2.b). A usual standard choice would be $p = 4$ (number of differential equations). Table C shows a comparison of the old and new algorithm for common initial data.

Table *C* Computational amount for solving the Ginzburg-Landau singular two-point boundary value problem

	Number of iterations	Number of trajectories
old II ($p=4$) (standard version)	no convergence (after 14 iterations)	no convergence
old II ($p=2, \lambda_o^{(o)}=0.5$) (optimized version)	*7*	*19*
new II ($\lambda_o^{(o)}=0.01$) (standard version)	*7*	*19*

It may be noted that the first relaxation estimate obtained in "new II" from (2.1) with $\lambda_o^{(o)} = 0.01$ turned out to be $\lambda_1^{(o)} = 0.518$ - which is close to the optimal value $\lambda_o^{(o)} = 0.5$ in row II. Thus the *standard* version of the *new* algorithm mainly reproduces the version obtained by *optimizing* the set of input parameters of the *old* algorithm (with respect to this particular example).

Conclusion

By means of the results presented above, the efficiency and - what is even more important in applications - the reliability of the modified Newton method is increased. The results apply to general systems of equations and nonlinear least squares problems. The progress is significant, if the systems to be solved are highly nonlinear and numerically sensitive.

Acknowledgement

The author wishes to thank Prof. R. Bulirsch for encouraging this work and P. Rentrop for preparing example III. In particular, the author is indebted to Mrs. D. Jahn for her careful and patient typing of the present manuscript.

References

[1] Amann, H.:
Über die näherungsweise Lösung nichtlinearer Integralgleichungen.
Numer. Math. 19, 29 - 45 (1972)

[2] Brown, K.M., Dennis jr., J.E.:
Derivative-free analogues of the Levenberg-Marquardt and Gauss algorithms for nonlinear least squares approximation.
Numer. Math. 18, 289 - 297 (1972)

[3] Broyden, C.G.:
The convergence of single-rank quasi-Newton methods.
Math. Comp. 24, 365 - 382 (1970)

[4] Broyden, C.G.:
The convergence of a class of double-rank minimization algorithms.
J.Inst. Math. Applic. 6, 76 - 90 (1970)

[5] Bulirsch, R.:
Die Mehrzielmethode zur numerischen Lösung von nichtlinearen Randwertproblemen und Aufgaben der optimalen Steuerung.
Lecture "Flugbahnoptimierung" of the Carl-Cranz-Gesellschaft e.V., Oct. 1971

[6] Bulirsch,R., Stoer,J., Deuflhard, P.:
Numerical solution of nonlinear two-point boundary value problems I
to be published in Numer. Math., Handbook Series Approximation

[7] Deuflhard, P.:
A Modified Newton Method for the Solution of Ill-conditioned Systems of Nonlinear Equations with Application to Multiple Shooting.
Numer. Math. 22, 289 - 315 (1974)

[8] Goldstein, A.A.:
Cauchy's Methode der Minimierung.
Numer. Math. 4, 146 - 150 (1962)

[9] Householder, A.S.:
Principles of Numerical Analysis.
New York: Mc Graw-Hill, 1953

[10] Kantorovič, L., Akilow, G.:
Functional Analysis in Normed Spaces.
Moscow: Fizmatgiz, 1959

[11] Keller, H.B.:
Numerical methods for two-point boundary value problems.
London: Blaisdell, 1968

[12] Meyer, G.H.:
On solving nonlinear equations with a one-parameter operator imbedding.
University of Maryland, Comp. Sc. C.: Techn.Rep. 67-50

[13] Osborne, M.R.:
On shooting methods for boundary value problems.
J. Math. Anal. Appl. 27, 417 - 433 (1969)

[14] Penrose, R.:
A Generalized Inverse for Matrices.
Proc. Cambridge Philos. Soc. 51, 406 - 413 (1955)

[15] Rentrop, P.:
Numerical Solution of the Singular Ginzburg-Landau Equation.
to be published in Computing

[16] Stoer, J.:
Einführung in die Mumerische Mathemathik I.
Heidelberger Taschenbuch 105 , Berlin-Heidelberg-New York: Springer, 1972

[17] Stoer, J., Bulirsch, R.:
Einführung in die Numerische Mathematik II.
Heidelberger Taschenbuch 114, Berlin-Heidelberg-New York: Springer, 1973

Optimale Steuerungen für Gleitflugbahnen Maximaler Reichweite beim Eintritt in Planetenatmosphären

E.D. Dickmanns

Übersicht

Es wird eine analytische Näherungslösung unter der Annahme quasistationären Gleitens für Bahnen größter Seitenreichweite abgeleitet und mit numerischen Ergebnissen verglichen. Der Einfluß einer Zustandsraumbegrenzung durch kinetische Aufheizung auf optimale Bahnen wird anhand numerischer Ergebnisse dargelegt. Die maximal mögliche Landefläche bei gegebenen Eintrittsbedingungen und die zugehörigen Steuerverläufe für oszillierende Bahnen werden für ein Space-Shuttle-ähnliches Eintrittsfahrzeug anhand sehr genauer numerischer Lösungen mit der Mehrzielmethode diskutiert.

Einleitung

Zukünftige Raumtransportersysteme (wie z.B. der amerikanische Space-Shuttle) werden die Atmosphäre zur zweckmäßigen Gestaltung ihrer Flugbahn ausnutzen. Bisher waren die Trägerraketen Verlustgeräte, und zu bergende Nutzlasten erforderten eigene Wiedereintrittskapseln, die meist als reine Widerstandskörper ausgebildet waren. Sie konnten deshalb nur an einem Ort der Erde niedergehen, der sich zum Zeitpunkt der Landung gerade mit der Erddrehung durch die raumfeste Umlaufebene dieses Raumflugkörpers hindurchbewegte. Dies machte die bekannten aufwendigen Bergungsmanöver auf Land oder See erforderlich.

Bei den amerikanischen Gemini- und Apollo-Kapseln wurde bereits eine kleine (in der Größe nicht steuerbare) Kraftkomponente senkrecht zur aerodynamischen Anströmung erzeugt, um den gewünschten Landeort auch bei auftretenden Bahnstörungen genauer erreichen zu können. Dies wurde durch eine Steuerung der Kraftrichtung, des aerodynamischen Auftriebsquerneigungswinkels, erreicht. Mit den zukünftigen Raumtransporter-Oberstufen, die eine große Kraftkomponente senkrecht zur Anströmung durch entsprechende aerodynamische Formgebung und Winkelanstellung erreichen und die deshalb wie ein Segelflugzeug zur Erde niedergleiten, können dreidimensionale Bahnen mit erheblichen seitlichen Reichweiten aus der Satellitenbahnebene heraus erflogen werden. Dies gestattet eine häufigere Rückkehrmöglichkeit aus einer gegebenen Umlaufbahn an einen vorgegebenen (entsprechend ausgerüsteten) Lande- und Startplatz ohne nachfolgende langwierige Verhole-Aktivitäten.

Den von einem gegebenen Satz von Eintrittsbedingungen aus erreichbaren Bereich von Landepunkten auf der Planetenoberfläche nennt man in Anspielung an die Form dieser Fläche *Fußabdruck* (engl. footprint). Gesucht werden nun für ein Fahrzeug mit gegebener aerodynamischer Charakteristik *die* Verläufe der Steuerfunktionen Anstellwinkel $\alpha(t)$ und Auftriebsquerneigungswinkel $\mu(t)$, die diese Fläche maximieren. Da aber die Reichweite in der Umlaufebene durch den Zeitpunkt des Bremsimpulses beliebig angepaßt werden kann, interessiert vor allem die seitliche Reichweite als zu optimierende Größe.

Der wesentliche Kern des realen Problems bleibt erhalten, wenn man folgende vereinfachende Annahmen trifft:

A_1: der Planet ruht und ist kugelförmig,

A_2: seine Atmosphäre ist nur höhenabhängig, zeitlich konstant und in Ruhe,

A_3: die Luftdichte nimmt exponentiell mit der Höhe ab.

Durch die ersten beiden Annahmen sind der aerodynamische und der inertiale Geschwindigkeitsvektor identisch, wodurch die Gleichungen wesentlich vereinfacht werden. Die letzte Annahme kommt analytischen Näherungslösungen entgegen; für die numerischen Rechnungen bringt sie keine Vorteile.

Die folgenden Ausführungen vervollständigen die Ergebnisse aus [1]. Für die analytische Näherungslösung wird eine neue, einsichtigere Herleitung gegeben, und für das aufheizungsbeschränkte Problem werden genaue numerische Ergebnisse von PESCH [2] mitgeteilt. Diese wurden mit der Mehrzielmethode nach BULIRSCH [3] und DEUFLHARD [4] erhalten und umfassen eine Variation eines Oberflächentemperaturparameters; es ergeben sich Flugbahnen mit bis zu vier aufheizungsbeschränkten Teilbögen.

Mathematisches Modell

Das Fahrzeug wird als Punktmasse aufgefaßt; seine Winkellagen sind damit momentan einstellbar und stellen die Steuerfunktionen dar. Es wird angenommen, daß eine Schiebewinkelregelung den Geschwindigkeitsvektor der aerodynamischen Anströmung stets in der Symmetrieebene des Fahrzeugs hält, so daß zwei Steuerfunktionen vorhanden sind: Der Anstellwinkel α in der Symmetrieebene zwischen Anströmung und Flugkörper-Referenzlinie (meist die Nullauftriebsrichtung), der die Größe der Luftkraftbeiwerte festlegt, und der Auftriebsquerneigungswinkel μ, gemessen aus der Vertikalebene heraus und senkrecht zur Anströmung, der den Auftrieb in eine horizontale bahndrehende und eine vertikale Komponente zerlegt.

Als Koordinatensystem (Bild 1) wird ein flugbahnorientiertes rechtshändiges Achsenkreuz gewählt, dessen x-Achse stets in Richtung des Geschwindigkeitsvekors zeigt und dessen Ursprung im Fahrzeugschwerpunkt liegt. Die z-Achse liegt in der Vertikalebene, positiv nach unten. Die Position des Fahrzeugs wird relativ zum Anfangszeitpunkt und zur dort gegebenen Flugrichtung angegeben. In der Anfangsflugrichtung zählt der Reichweitenwinkel Θ, normal dazu die Seitenreichweite Λ. Die Höhe über Grund h' vervollständigt die Positionsbeschreibung. Der Geschwindigkeitsvektor bildet mit der Horizontalebene den Bahnneigungswinkel γ, positiv für zunehmende Höhe, und seine Horizontalkomponente bildet mit dem Parallelkreis zur Anfangsflugebene den Azimutwinkel χ in der Horizontalebene, positiv für zunehmende Seitenreichweite Λ.

Hieraus ergeben sich die bekannten Differentialgleichungen

$$
\begin{aligned}
\dot{V} &= -g \sin\gamma - W/m \quad , \\
\dot{\chi} &= -\frac{V}{r} \cos\gamma \cos\chi \tan\Lambda + \frac{A}{mV} \frac{\sin\mu}{\cos\gamma} \quad , \\
\dot{\gamma} &= \left(\frac{V}{r} - \frac{g}{V}\right) \cos\gamma + \frac{A}{mV} \cos\mu \quad , \\
\dot{\Theta} &= V \cos\gamma \cos\chi/(r \cos\Lambda) \quad , \\
\dot{\Lambda} &= V \cos\gamma (\sin\chi)/r \quad , \\
\dot{r} &= \dot{h} = V \sin\gamma \quad ,
\end{aligned}
\qquad (1)
$$

mit

$$
\begin{bmatrix} A \\ W \end{bmatrix} = \begin{bmatrix} C_A(\alpha) \\ C_W(\alpha) \end{bmatrix} S \frac{\rho}{2} V^2 \; ; \qquad \begin{matrix} \text{Auftrieb} \\ \text{Widerstand} \end{matrix}
$$

$$
\begin{aligned}
r &= R + h' \; ; \qquad & R &= \text{Planetenradius} \\
g &= G/r^2 \; ; \qquad & G &= \text{Planetenmasse} \cdot \text{Gravitationskonstante.}
\end{aligned}
\qquad (2)
$$

Unter gewissen Voraussetzungen, die hier als gegeben betrachtet werden, läßt sich die Anstellwinkelabhängigkeit eliminieren durch einen Übergang auf den Auftriebsbeiwert als Steuerfunktion. Der Widerstandsbeiwert läßt sich dann näherungsweise beschreiben durch ein Gesetz der Form

$$C_W = C_{Wo} + k\,C_A^n \,, \qquad (3)$$

wobei C_{Wo}, k und n die aerodynamischen Eigenschaften des Fahrzeugs charakterisieren.

Auf der Eintrittsflugbahn ist der erlaubte Zustandsraum für das Fahrzeug beschränkt, um Fahrzeugbeschädigungen durch zu starke kinetische Aufheizung oder zu hohen Staudruck zu verhindern. Für Bahnen größter Reichweite ist die Aufheizungsbeschränkung ein wesentlicher Faktor bei der Bahnauslegung. Im folgenden wird eine strahlungsgekühlte Außenhaut als kritisches Element betrachtet. Unter der Annahme, daß die Zeitkonstante zur Aufheizung der Außenhaut klein ist im Vergleich zur zeitlichen Änderung der Größen: Geschwindigkeit, Höhe und Anstellwinkel, kann eine quasistationäre Näherung als gültig betrachtet werden. Dann gilt als Zustandsraumbeschränkung (vgl. [1])

$$C = C_A - C_{AH}(V,h') \leq 0 \,. \qquad (4)$$

Die experimentell und aus aufwendigen numerischen Rechnungen gewonnene Grenzfunktion $C_{AH}(V,h')$ wurde durch folgende Funktion approximiert

$$C_{AH} = B_i\,H_i + \Delta C_{AH} \,, \qquad i = 1 \ldots 5 \qquad (5)$$

mit

$$B_i = g_{i,j}\,\hat{h}^{(j-1)} \,, \qquad \hat{h} = h'[\mathrm{km}]/50 - 1 \,,$$

der Koeffizientenmatrix

$$g_{i,j} = \begin{bmatrix} 0.110717 & 0.834519 & 1.213679 & -1.060833 \\ -0.672677 & 2.734170 & -0.864369 & -12.100000 \\ 0.812241 & 2.337815 & 10.316280 & 22.974860 \\ -3.151267 & -13.621310 & -40.485500 & -57.833330 \\ 2.368095 & 19.073400 & 69.869050 & 127.777778 \end{bmatrix}$$

und

$$H_1 = a^2 \;; \qquad H_2 = a - H_1 \;;$$

$$H_3 = 1 - a - H_2 \;; \qquad H_4 = 1/a - 2 + a - H_3 \;;$$

$$H_5 = 1/a^2 - 3/a + 3 - a - H_4 \;;$$

$$a = b\,h'/V\;[\mathrm{km,s}] \;; \qquad b = 0.095[\mathrm{s}^{-1}] \,.$$

Der additive Term ΔC_{AH} in Gl. (5) gestattet, auf einfache Weise eine Verschärfung bzw. Abschwächung der Aufheizungsbegrenzung näherungsweise zu untersuchen. $\Delta C_{AH} = 0.03$ entspricht dabei einer Grenztemperaturanhebung um ~200 °F (~110 °C) gegenüber dem Referenzwert von 2 000 °F (1 093 °C), für den der Koeffizientensatz gilt.

Die Gleichungen (1) bis (5) liegen den numerischen Rechnungen zugrunde (s.u.). Als Fahrzeugdaten wurden folgende Werte gewählt (sie entsprechen in etwa der amerikanischen Space-Shuttle-Oberstufe):

$$C_{Wo} = 0.04 \quad ; \quad k = 1.0 \quad ; \quad n = 1.86 \quad ; \quad m/S = 250[kg/m^2] \quad . \tag{6}$$

Dies liefert eine beste inverse Gleitzahl von $E_m = (C_A/C_W)_{max} = 2.22$ bei $C_{AE} = 0.192$; der Wert E_m charakterisiert die aerodynamische Güte für Gleitflugbahnen maximaler Reichweite.

Analytische Näherungslösung

Analytische Lösungen gestatten ein leichteres Verständnis grundsätzlicher Zusammenhänge als numerische Ergebnisse. Die oben angegebenen Differentialgleichungen lassen jedoch selbst bei Einführung eines exponentiellen Dichteverlaufs (A_3):

$$\rho = \rho_o \, e^{-\beta' h'} \tag{7}$$

(mit $\rho_o \simeq 1.54[kg/m^3]$ und $\beta' \simeq 1/6.9[km^{-1}]$ für die Erde) keine solchen Lösungen als erreichbar erscheinen. Deshalb werden folgende zusätzliche Annahmen getroffen:

A_4: kleine Bahnneigungswinkel, so daß $\cos\gamma \simeq 1$, $\sin\gamma \simeq \gamma$;

A_5: der (inertiale) 1. Term in der $\dot{\chi}$-Gleichung ist gegenüber dem 2. (aerodynamischen) vernachlässigbar bzw. mit $\tan\Lambda \simeq \Lambda$ zu approximieren;

A_6: das vertikale Gleichgewicht ist immer gut erfüllt ($\dot{\gamma} \simeq 0$ ist die Summe zweier vergleichsweise großer, entgegengesetzt gleicher Terme); dies erfordert nicht, daß γ konstant ist, sondern sich vergleichsweise langsam ändert (im folgenden kurz als QSG-Bedingung, *q*uasi*s*tationäres *G*leiten, bezeichnet);

A_7: die spezifische dimensionsbefreite mechanische Gesamtenergie des Fahrzeugs

$$e = V^2/(2\,g\,r) + h'/r \tag{8}$$

wird über weite Teile der Bahn im wesentlichen durch den ersten Term repräsentiert.

Führt man nun die dimensionsbefreite Zeit

$$\tau = t \sqrt{g/r} \, , \tag{9}$$

den Geschwindigkeits-(Energie-)Parameter

$$v = V^2/(r\ g) \tag{10}$$

und die dimensionslose Höhe h = h'/r sowie den kombinierten Fahrzeug/Planetenparameter

$$p = 0.5\ \rho_0\ r\ S/m \tag{11}$$

ein, so erhält man folgendes reduzierte Gleichungssystem unter Ausnutzung der Annahmen A_3 bis A_8 und mit der Schreibweise $(\mathring{\ }) = d/d\tau(\ )$ sowie $v = 2(e - h)$ aus (8), (10):

$$\mathring{e} = -\sqrt{v}\, p \exp(-\beta\ h)\ C_W\ v \, , \tag{12}$$

$$\mathring{\chi} = \sqrt{v}\,(p \exp(-\beta\ h)\ C_A \sin\mu - \underline{\Lambda \cos\chi})^* \, , \tag{13}$$

$$\mathring{\Theta} = \sqrt{v}\, \cos\chi/\cos\Lambda \, , \tag{14}$$

$$\mathring{\Lambda} = \sqrt{v}\, \sin\chi \, , \tag{15}$$

$$\mathring{\gamma} \simeq 0 = \sqrt{v}\,[p \exp(-\beta\ h)\ C_A \cos\mu - (1 - v)/v] \quad (QSG) \, . \tag{16}$$

Gl. (14) ist von den übrigen entkoppelt, da Θ nicht in der rechten Seite der Differentialgleichungen auftritt. (16) in (12) und (13) eingeführt, liefert als Variationsgleichungssystem für die QSG-Bedingung bei offener Reichweite in Anfangsflugrichtung ($\lambda_\Theta = 0$)

$$\mathring{e} = -\sqrt{v}\,\frac{1 - v}{E \cos\mu} \quad ; \quad \mathring{\lambda}_e = \frac{\sqrt{v}\,\lambda_e}{E \cos\mu}\left(1 - \frac{\lambda_\chi^2\ E^2}{\lambda_e^2\ v^3}\right) \ ; \tag{17}$$

$$\mathring{\chi} = \sqrt{v}\,\frac{(1 - v)}{v} \tan\mu \ ; \quad \mathring{\lambda}_\chi = -\sqrt{v}\,\lambda_\Lambda \cos\chi \ ; \tag{18}$$

$$\mathring{\Lambda} = \sqrt{v}\,\sin\chi \quad ; \quad \mathring{\lambda}_\Lambda = \sqrt{v}\,\lambda_\chi \cos\chi^* \, . \tag{19}$$

Die Hamiltonsche Funktion lautet dann

$$H = \sqrt{v}\left\{- \lambda_e \frac{1 - v}{E \cos\mu} + \lambda_\chi \frac{1 - v}{v} \tan\mu + \lambda_\Lambda \sin\chi\right\} \, . \tag{20}$$

Gemäß dem Maximumprinzip [5] ist H bezüglich C_A und μ zu minimieren,

*) Der zweite Term in Gl.(13) wird nur zur Ableitung der λ_Λ-Differentialgleichung berücksichtigt, für die weiteren Näherungslösungen jedoch vernachlässigt.

damit die Zielfunktion ihr Maximum annimmt. Für $\lambda_e < 0$, $v < 1$ und $\cos\mu > 0$ verlangt dies, daß E maximal gewählt wird:

$$E_{opt} = (C_A/C_W)_{max} \quad . \tag{21}$$

H wird minimal bezüglich μ für

$$\sin\mu_{opt} = \frac{\lambda_\chi E}{\lambda_e v} \quad . \tag{22}$$

Gemäß Gl. (8) steckt in jedem v der Gleichungen (12) bis (20) neben der Energie e auch die Höhe h als Variable. Die Differentialgleichung (12) für die Energie wurde aus einer Addition der Differentialgleichungen für v/2 und h gewonnen, wobei sich die $\sin\gamma$-Terme aufhoben. Die Höhe h tritt nur noch auf der rechten Seite der Differentialgleichungen auf und ist damit in dieser Formulierung zu einer Steuerfunktion geworden. Man kann nun die Hamiltonsche Funktion Gl. (20) auch bezüglich h extremieren und erhält

$$\sin\mu^* = \frac{\lambda_e v^2}{E \lambda_\chi} \quad . \tag{23}$$

Dies ist nur dann mit Gl. (22) kompatibel, wenn gilt

$$\frac{\lambda_{\chi,G}^2 E^2}{\lambda_{e,G}^2 v^3} = 1 \quad , \quad \text{d.h.} \quad \sin\mu_G = \sqrt{v} \quad . \tag{24}$$

Die Lösung erscheint zunächst unsinnig, da der Verlauf der optimalen Steuerung nicht mehr von irgendwelchen Randwerten (z.B. λ_χ und λ_e in Gl. (22)) abhängen soll. Andererseits erfüllt diese Bahn sowohl bezüglich des μ-Verlaufs als auch des Höhenverlaufs die notwendigen Bedingungen der Optimalität. Wegen der Einfachheit des Steuergesetzes: $\sin\mu_G$ = Verhältnis von momentaner Geschwindigkeit zu örtlicher Satellitengeschwindigkeit lassen sich leicht einige Integrale der Bewegungsgleichungen erhalten (vgl. [1]); sie werden im folgenden mit Grundlösung (Index G) bezeichnet:

$$\begin{aligned} \lambda_{eG} &= \text{const} \quad , \\ \chi_G &= E(\sqrt{v_o} - \sqrt{v}) \quad , \\ \lambda_{\chi,G} &= \frac{\lambda_{e,G}}{E} v^{3/2} \\ \tau_G &= E\left[\arctan\sqrt{\frac{v_o}{1 - v_o}} - \arctan\sqrt{\frac{v}{1 - v}}\right] \quad , \end{aligned} \tag{25}$$

$$\left.\begin{aligned} \sin\gamma_G &= -\frac{1}{E}\, q(v)\, \sqrt{1 - v} \\ \mathring{\mu}_G &= -\frac{1}{E}\left\{1 - q(v)\right\} \end{aligned}\right\} \text{mit } q(v) = \frac{2 - v}{2 + v[\beta(1-v) - 1]} .$$

Da in dieser dimensionsbefreiten Form für Eintrittsgeschwindigkeiten entsprechend der Satellitengeschwindigkeit stets $v \lesssim 1.$ gilt, andererseits die atmosphärische Skalenhöhe sehr große Werte hat (z.B. $\beta \simeq 930$ für die Erde), folgt, daß mit Ausnahme der Bereiche $v \simeq 1$ und $v \simeq 0$ für die Funktion $q(v)$ gilt: $q << 1$. Aus der letzten Gleichung von (25) folgt damit $\mathring{\mu}_G \simeq -1/E$, d.h. der Auftriebsquerneigungswinkel sollte mit guter Näherung eine lineare Funktion der Zeit sein. Diese Aussage wird von den numerischen Rechnungen für das vollständige Gleichungssystem bestätigt.

Für seitliche Reichweitenmaximierung $\phi = \Lambda(\tau_f)$ mit offener Endzeit und freiem Azimutwinkel am rechten Rand folgt mit

$$H(\tau_f) = 0 \; ; \quad \lambda_\chi(\tau_f) = 0 \; ; \quad \lambda_\Lambda(\tau_f) = -1 \tag{26}$$

aus Gl. (22)

$$\mu_{opt}(\tau_f) = 0 \tag{27}$$

und hiermit aus Gl. (20)

$$\sin\chi(\tau_f) = -\frac{\lambda_e}{E}(1 - v) . \tag{28}$$

Aus der physikalischen Deutung von $\lambda_e = -\partial\Lambda(\tau_f)/\partial e$ und E als Umsetzungsfaktor von Energiehöhe in Reichweite läßt sich folgern, daß $-\lambda_e/E \simeq 1$ sein muß. Da $v(\tau_f) << 1$ ist (bei der Machzahl 3 in der Erdatmosphäre ist $v \simeq 0.01$), ergibt sich für den Azimutwinkel am Ende einer Bahn maximaler seitlicher Reichweite

$$\chi(\tau_f) \simeq \arcsin(1 - v(\tau_f)) \simeq \pi/2 . \qquad (29$$

Hiermit läßt sich aus der zweiten Gleichung von (25) schließen (mit $v_o \simeq 1$ und $v_f << 1$), daß die Grundlösung für Fahrzeuge mit der Gleitzahl $E \simeq \pi/2$ oder etwas größer am ehesten eine gute Näherungslösung sein kann. Der Zufall will es, daß das erste für die Erdatmosphäre in Entwicklung stehende Fahrzeug dieser Art, die Space-Shuttle-Oberstufe, in diesem Bereich liegen wird.

Das Ergebnis λ_{eG} = const (Gl. (25), erste Zeile) besagt, daß sich die Grundlösung dadurch auszeichnet, daß eine gegebene Energieänderung an jedem Punkt der Bahn die gleiche Reichweitenänderung $\Delta\Lambda$ hervorrufen würde.

Einige der Grundlösungsergebnisse sind in Bild 2 dargestellt.

Aus den Gleichungen (18) und (19) für die adjungierten Variablen λ_χ und λ_Λ und den Randbedingungen (26) ergibt sich nach Integration durch Separation der Variablen

$$\lambda_\Lambda = - \sqrt{1 - \lambda_\chi^2} \quad . \tag{30}$$

Diese Beziehung ist, wie sich inzwischen aus numerischen Rechnungen ergab, auch eine gute Näherung für oszillierende Bahnen, bei denen die QSG-Bedingung nicht gilt (vgl. auch [1]).

Numerische Ergebnisse

Den numerischen Rechnungen liegt das Gleichungssystem (1) bis (5), ergänzt um die Differentialgleichungen für die adjungierten Variablen und die Nebenbedingungen, wie sie in [1], Kapitel 4.3, gegeben wurden, zugrunde. Die Ergebnisse in [1] waren mit einem dem Gradientenverfahren im Funktionenraum ähnlichen Algorithmus [6] bei vergleichsweise geringem Rechenaufwand erzielt worden. Sie können nur als erste Näherung betrachtet werden, da die Iteration abgebrochen wurde, als der Zuwachs der Zielfunktion pro Iteration klein wurde.

In der Zwischenzeit hat H.J. PESCH [2] unter Verwendung eines Digitalrechnerprogramms nach der Mehrzielmethode nach BULIRSCH [3] und DEUFLHARD [4] numerische Ergebnisse von hoher Genauigkeit erzielt, die im folgenden dargelegt werden. Das Wiedereintrittsproblem der Reichweitenmaximierung, vor allem mit Aufheizungsbeschränkungen, ist numerisch äußerst delikat. Die Kondition der Iterationsmatrix stieg bis auf 10^{19} an; dies vermittelt einen Eindruck von der Empfindlichkeit der Lösungen. Für ingenieurmäßige Routinerechnungen würde eine andere Formulierung des Problems, die diese Schwierigkeiten vermeidet, sehr begrüßt werden.

Die neuen numerischen Ergebnisse beziehen sich auf Anfangsbedingungen des Wiedereintritts, die nicht dem Zustand quasistationären Gleitens entsprechen, sondern infolge eines zu steilen Anfangsbahnneigungswinkels zu schwingenden Bahnen führen. Die Randbedingungen sind in Tab. 1 zusammengestellt. Als Zielfunktion gilt eine gewichtete Summe der Reichweiten

$$\phi = \Lambda + P\,\Theta \;, \tag{31}$$

die für P = 0 den Fall maximaler seitlicher Reichweite und für $P \gg 1$ den Fall der Reichweite in Eintrittsrichtung umfaßt. Die Berechnung der möglichen Landefläche für ein Fahrzeug mit der Charakteristik nach Gl. (6) wird zunächst ohne Aufheizungsbegrenzung durchgeführt. Dann wird für Bahnen größter Seitenreichweite der Einfluß verschieden starker Aufheizungsbegrenzungen untersucht.

Maximal mögliche Landefläche ohne Aufheizungsbegrenzung

In [1] war ein *footprint* für ein Fahrzeug mit $E_m = 1.4$ unter QSG-Bedingungen angegeben worden. Bild 3 zeigt Ergebnisse für ein Fahrzeug mit $E_m = 2.22$, wobei die QSG-Bedingung nicht erfüllt ist. Die Vergrößerung von E_m um knapp 60% bringt etwa eine Verdopplung der Seitenreichweite und für größere Werte von P zusammen mit dem steileren Eintrittswinkel ein noch stärkeres Anwachsen der Reichweite in Anfangsflugrichtung auf fast 3/4 des Kugelumfangs. Für diese letzteren Bahnen nimmt der Auftriebsbeiwert (Bild 3b) einen stark oszillierenden Verlauf an, um bei den anfänglich kleinen Auftriebsquerneigungswinkeln μ

Tabelle 1: Randbedingungen für Bahnen maximaler Reichweite beim Eintritt in die Erdatmosphäre.

Zustandsvariable: Anfangsbedingung	Zustandsvariable: Endbedingung (offene Endzeit: $H(t_f) = 0$)	adjungierte Variable†: Endbedingung
$V_o = 7.85$ km/s	*$V_f \simeq 1.1$ km/s	*λ_{Vf} = unbekannt oder vorgegeben
$\chi_o = 0$	χ_f offen	$\lambda_{\chi f} = 0$
$\gamma_o = -1.25^o$	*$\gamma_f \simeq -2.7^o$	*$\lambda_{\gamma f}$ = unbekannt oder vorgegeben
$\Theta_o = 0$	Θ_f = Elemente der Zielfunktion	$\lambda_\Theta = -P$, $0 \le P \le 1$
$\Lambda_o = 0$	Λ_f = Elemente der Zielfunktion	$\lambda_{\Lambda f} = -1$
$h_o = 95$ km	$h_f = 30$ km	λ_{hf} = unbekannt

*Diese Bedingungen ergeben sich aus der Endhöhe und dem dort geforderten QSG-Zustand mit $C_A \simeq C_{AE}$ aus Näherungsbeziehungen (vgl. [1]).

†Anfangsbedingungen alle unbekannt.

(Bild 3a) durch eine Anpassung der vertikalen Kraftkomponente das schwingende Verhalten günstig zu beeinflussen.

Für P = 0 ist der Auftriebsbeiwert fast konstant bei $C_A = C_{AE} = 0.192$, und der Auftriebsquerneigungswinkel μ hat eine fast konstante Abnahmerate (nach einem anfänglich konstanten Wert, während sich das Fahrzeug weit oberhalb der QSG-Höhe befindet), wie die Grundlösung das andeutet. Allerdings ist die Abnahmerate etwa 12% größer, als die letzte Gleichung von (25) angibt. Bahnstücke mit der gleichen Änderungsgeschwindigkeit für μ treten bis zu Werten für P von etwa 0.5 auf; für wachsende P werden diese Bahnstücke eingeleitet von immer ausgedehnteren Anfangsphasen mit zunehmendem Bahnneigungswinkel, die bei immer kleineren, für $\Theta_f > \pi$ sogar bei negativen Anfangswerten beginnen. Um möglichst weit nach links zu kommen, muß man anfangs also nach rechts steuern, was für diese sphärischen Bahnen physikalisch einleuchtet. Für P = 0 führt die erste Bahnschwingung auf $h_{min} \approx 53$ km und $h_{max} \approx 65$ km, während für P = 1 die entsprechenden Werte $h_{min} \approx 67$ km und $h_{max} \approx 100$ km sind; dieser starke Sprung tritt auf trotz der erheblichen Reduktion des Auftriebsbeiwertes im ersten Minimum und einer vorhergehenden, abbremsenden C_A-Erhöhung auf 0.23. Bei P = 1 wird im ersten Eintauchvorgang die Aufheizungsbegrenzung nicht berührt, während für P = 0 ihre Verletzung hier am größten ist (Bild 4a).

Maximale Seitenreichweite mit Aufheizungsbegrenzung

Auf Zustandsraumbegrenzungen, die eine Steuerfunktion wie im vorliegenden Fall explizit enthalten (vgl. Gl. (4)) kann diese Steuerfunktion nicht mehr frei gewählt werden, sondern es muß gelten

$$\frac{\partial C}{\partial x}\,\delta x + \frac{\partial C}{\partial u}\,\delta u = 0 \tag{32}$$

oder

$$\delta u = -\left[\frac{\partial C}{\partial u}\right]^{-1} \frac{\partial C}{\partial x}\,\delta x \quad ,$$ wobei x der Zustandsvektor und u der Steuervektor ist.

Hiermit schreiben sich die linearen Störungsgleichungen des Differentialgleichungssystems $\dot{x} = f(x,u)$:

$$\delta\dot{x} = \left(\frac{\partial f}{\partial x} - \frac{\partial f}{\partial u}\left[\frac{\partial C}{\partial u}\right]^{-1}\frac{\partial C}{\partial x}\right)\delta x \quad ,$$

und die adjungierten Differentialgleichungen auf der Zustandsraumbegrenzung erhalten additive Terme (unterstrichen):

$$\dot{\lambda} = -\left[\frac{\partial f}{\partial x}\right]^T \lambda + \underline{\left[\frac{\partial f}{\partial u}\left(\frac{\partial C}{\partial u}\right)^{-1}\frac{\partial C}{\partial x}\right]^T \lambda} \ . \tag{33}$$

Da in der Zustandsraumbegrenzung (4) als einzige Steuerfunktion der Auftriebsbeiwert C_A und als Zustandsvariable die Geschwindigkeit V und die Höhe h enthalten sind, folgt als Änderung auf einem beschränktem Bahnbogen im vorliegenden Fall

$$\begin{aligned} \dot{\lambda}_{Vb} &= \dot{\lambda}_{Vub} + K\ \partial C_{AH}/\partial V \ , \\ \dot{\lambda}_{hb} &= \dot{\lambda}_{hub} + K\ \partial C_{AH}/\partial h \end{aligned} \tag{34}$$

mit

$$K = \left(\frac{S\rho V}{2m}\left[\frac{\lambda_\chi^{\ 2}}{\cos^2\gamma} + \lambda_\gamma^{2}\right]^{1/2} + V\lambda_V \ \ k\ n\ C_{AH}^{\ (n-1)}\right) \ .$$

Die numerischen Rechnungen laufen bei den Schießverfahren (auch Verfahren der benachbarten Extremalen genannt), zu denen die Mehrzielmethode gehört, folgendermaßen ab: Die nicht bekannten Anfangswerte für die Integration der Zustands- und adjungierten Variablen werden zunächst geschätzt. Von diesen Werten aus werden die Differentialgleichungen integriert, wobei die Steuerfunktionen nach dem Maximumprinzip aus x und λ bestimmt werden. Geht nun im Laufe der Integration eine Schaltfunktion durch Null - hier die Zustandsraumbeschränkung Gl. (4) - so wird der Zeitpunkt des Nulldurchgangs ermittelt und eine logische Variable gesetzt, die das Gleichungssystem entsprechend den neuen Bedingungen ändert. Hiermit ist eine flexible Handhabung von Unstetigkeitsstellen möglich. Mit Hilfe von Variationen der Anfangswerte wird dann die Iterationsmatrix bestimmt, aus der zusammen mit den Fehlern in den Randbedingungen verbesserte Schätzwerte für die unbekannten Randwerte zu Beginn der Integration abgeleitet werden.

Bild 4 zeigt eine Folge von Verläufen des Auftriebsbeiwertes C_A, wie sie sich bei zunehmend schärferer Aufheizungsbeschränkung als optimal ergeben. $\Delta C_{AH} = 0.06$ entspricht etwa einer Grenztemperatur der Außenhaut des Fahrzeugs von 2 400 $^{\circ}$F (≃ 1 320 $^{\circ}$C), $\Delta C_{AH} = 0.03$ einer solchen von 2 200 $^{\circ}$F (≃ 1 200 $^{\circ}$C) und $\Delta C_{AH} = 0.008$ etwa 2 050 $^{\circ}$F (≃ 1 120 $^{\circ}$C). Jedes Teilbild b) bis k) enthält zwei Kurven: eine durchgezogene, die die aktive Steuerung darstellt und eine gestrichelte, die auf unbeschränkten Teilbögen die bezüglich der Aufheizungsbegrenzung mögliche Größe des Auftriebsbeiwertes und auf beschränkten

Teilbögen die nach dem Maximumprinzip berechneten Werte darstellt. Bild 4a zeigt die Verletzung der Aufheizungsbegrenzung für $\Delta C_{AH} = 0$ für die optimale Bahn ohne Zustandsraumbegrenzung. Es gibt zwei getrennt liegende Phasen mit zu großer Erhitzung während der ersten beiden Schwingungen und eine größere zusammenhängende während der folgenden drei Schwingungen.

Bei $\Delta C_{AH} = 0.072$ ergibt sich *ein* aufheizungsbeschränkter Bogen im ersten Höhenminimum (Bild 4b). Bei $\Delta C_{AH} = 0.066$ treten drei (Bild 4c) und bei $\Delta C_{AH} = 0.054$ vier aufheizungsbeschränkte Bahnstücke auf (Bild 4d). Mit weiterer Verschärfung der Temperaturbegrenzung verschmelzen zunächst die letzten beiden Teilbögen (Bild 4e und f) und dehnen sich aus, bis bei $\Delta C_{AH} = 0.035$ auch der zweite Bogen mit einbezogen wird (Bild 4g) und sich ein Flug von über 12 Minuten auf der Begrenzung ergibt. Bei $\Delta C_{AH} = 0.008$ bleibt nur mehr *eine* aufheizungsbeschränkte Flugphase von etwa 23 Minuten aus 40 Minuten Gesamtflugzeit übrig. Nach Verlassen der Begrenzung ist der Auftriebswert etwa konstant bei $C_A = C_{AE}$. Für kleinere Werte von ΔC_{AH} wurden keine Rechnungen versucht, da die Kondition der Iterationsmatrix von 10^9 bei 0.072 auf 10^{19} bei $\Delta C_{AH} = 0.008$ angewachsen war und Rechnungen mit mehr als doppelter Genauigkeit aus Rechenzeitgründen nicht vertretbar erschienen.

Bild 5 zeigt einen Vergleich von Steuerfunktionen für aufheizungsbeschränkte und unbeschränkte Bahnen. Mit zunehmender Stärke der Aufheizungsbegrenzung (kleinere ΔC_{AH}) nimmt anfangs der Auftriebsquerneigungswinkel ab, um die die Sinkgeschwindigkeit verkleinernde Komponente des Auftriebs zu vergrößern; dies führt zu einem weichen Übergang auf einen nichtschwingenden aufheizungsbeschränkten Teilbogen für $\Delta C_{AH} = 0.008$ (Bild 6a und b). Während des ersten Höhenminimums wächst der Auftriebsquerneigungswinkel rasch an, um eine zu starke Beschleunigung nach oben zu verhindern (Bild 5a). Auf dem aufheizungsbeschränkten Teilstück nimmt der Auftriebsquerneigungswinkel schneller ab als im unbeschränkten Fall, was zu einer Verringerung des Azimutwinkels am Ende dieser Phase führt (Bild 6c). Dies wird kompensiert durch einen zweiten Anstieg des Auftriebsquerneigungswinkels vor Verlassen der Zustandsraumbegrenzung. Zusammen mit dem Auftriebsbeiwertverlauf (Bild 5b) führt dies zu einem schwingenden Flug nach Verlassen der Zustandsraumbegrenzung (Bild 5c und 6a, $t \gtrsim 1600$ [s]). Der glättende Effekt auf die Flugbahn durch die Aufheizungsbegrenzung, während diese wirksam ist, kann aus den Bildern 5c und 6a ersehen werden.

Die Flugzeit für die beschränkte Bahn bis zur gleichen Endflughöhe stieg leicht an (84 sec $\hat{=}$ 3.5%). Die Reichweite in Anfangsflugrichtung Θ stieg um 8%, während die zu maximierende Seitenreichweite sich um 1.5% verringerte (Bild 6c). Der Einfluß auf die adjungierten Variablen λ_V, λ_χ und λ_Λ ist relativ gering. Bemerkenswert ist die Tatsache, daß die Beziehung Gl. (30) eine gute Näherung sowohl für die beschränkte als auch die unbeschränkte Bahn ist (Bild 6d). Stark beeinflußt durch die Zustandsraumbegrenzung werden die adjungierten Variablen zum Bahnneigungswinkel λ_γ und der Höhe λ_h (Bild 6e). Vor allem vor und während des beschränkten Teilbogens treten große Abweichungen auf. Da λ_χ sich nur wenig ändert, wird die Anpassung der optimalen Steuerung ($\mu = \lambda_\chi/(\lambda_\gamma \cos\gamma)$) für den Auftriebsquerneigungswinkel durch λ_γ vollzogen.

Literatur

[1] Dickmanns,E.D.: Maximum Range Threedimensional Lifting Planetary Entry. NASA TR R-387, 1972.

[2] Pesch,H.J.: Numerische Berechnung optimaler Steuerungen mit Hilfe der Mehrzielmethode, dokumentiert am Problem der optimalen Rückführung eines Raumgleiters unter Berücksichtigung von Aufheizungsbegrenzungen. Diplomarbeit, Math. Institut der Universität Köln, 1973.

[3] Stoer,J., Bulirsch,R.: Einführung in die numerische Mathematik II. Springer Verlag, Heidelberger Taschenbücher 114, 1973 (Kap. 7.3.5: Die Mehrzielmethode, S. 170-191).

[4] Deuflhard,P.: Ein Newton-Verfahren bei fastsingulärer Funktionalmatrix zur Lösung von nichtlinearen Randwertaufgaben mit der Mehrzielmethode. Dissertation, Math. Institut der Universität Köln, 1972.

[5] Pontryagin,L.S., Boltyansky,V.G., Gamkrelidse,R.V., Mishchenko,E.F.: Mathematische Théorie optimaler Prozesse. Oldenbourg Verlag, 1964.

[6] Dickmanns,E.D.: Optimierung von Flugbahnen durch iterative Anwendung des Maximumprinzips. WGLR-Jahrbuch, 1967, S. 272-301, 1968.

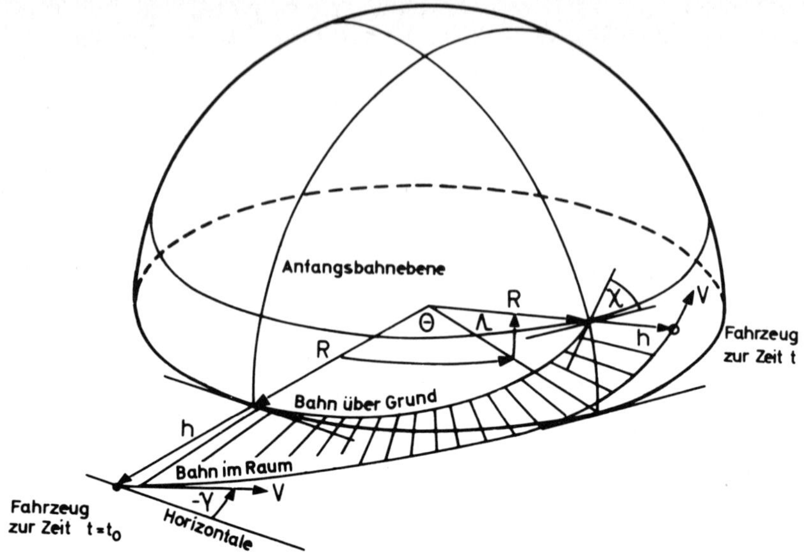

Bild 1: Koordinatensystem.

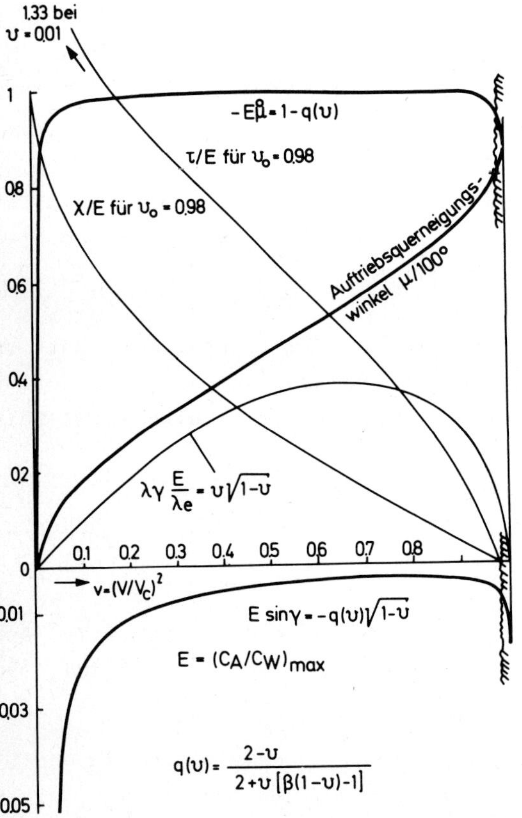

Bild 2:
Analytische Näherungslösung für maximale seitliche Reichweite unter der Bedingung quasistationären Gleitens ("Grundlösung").

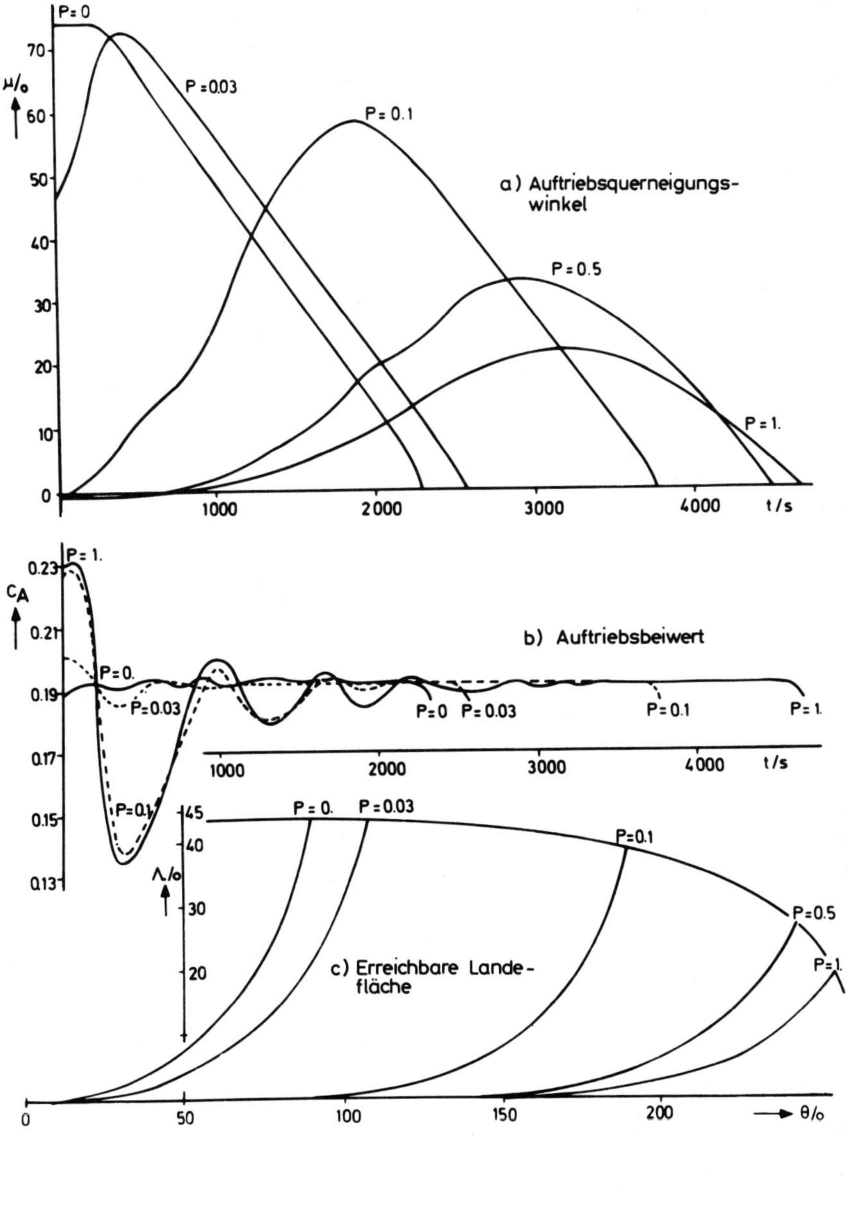

Bild 3: Erreichbare Landefläche und zugehörige Steuerfunktionen für ein Fahrzeug mit E_m = 2.22 ohne Aufheizungsbegrenzung.

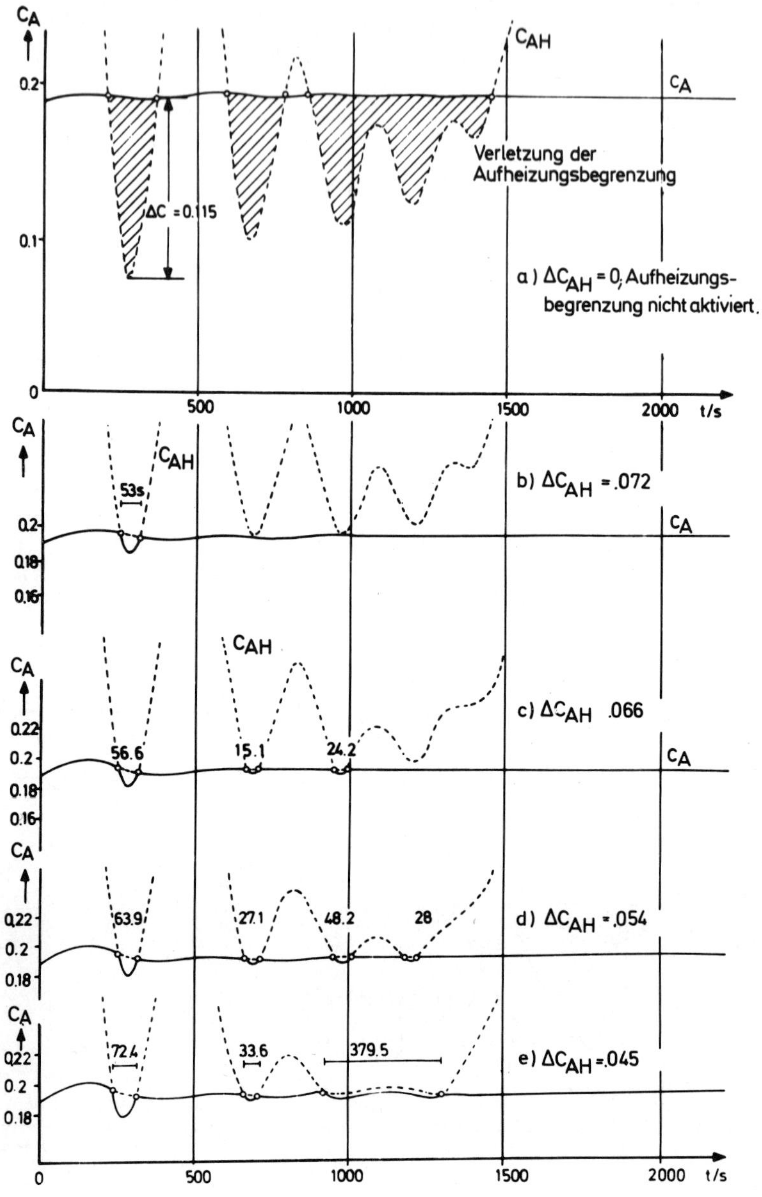

Bild 4a)-e): Verlauf des Auftriebsbeiwertes und der Begrenzungsfunktion C_{AH} für verschieden stark aufheizungsbeschränkte Bahnen größter Seitenreichweite $0.072 \geq \Delta C_{AH} \geq 0.008$.

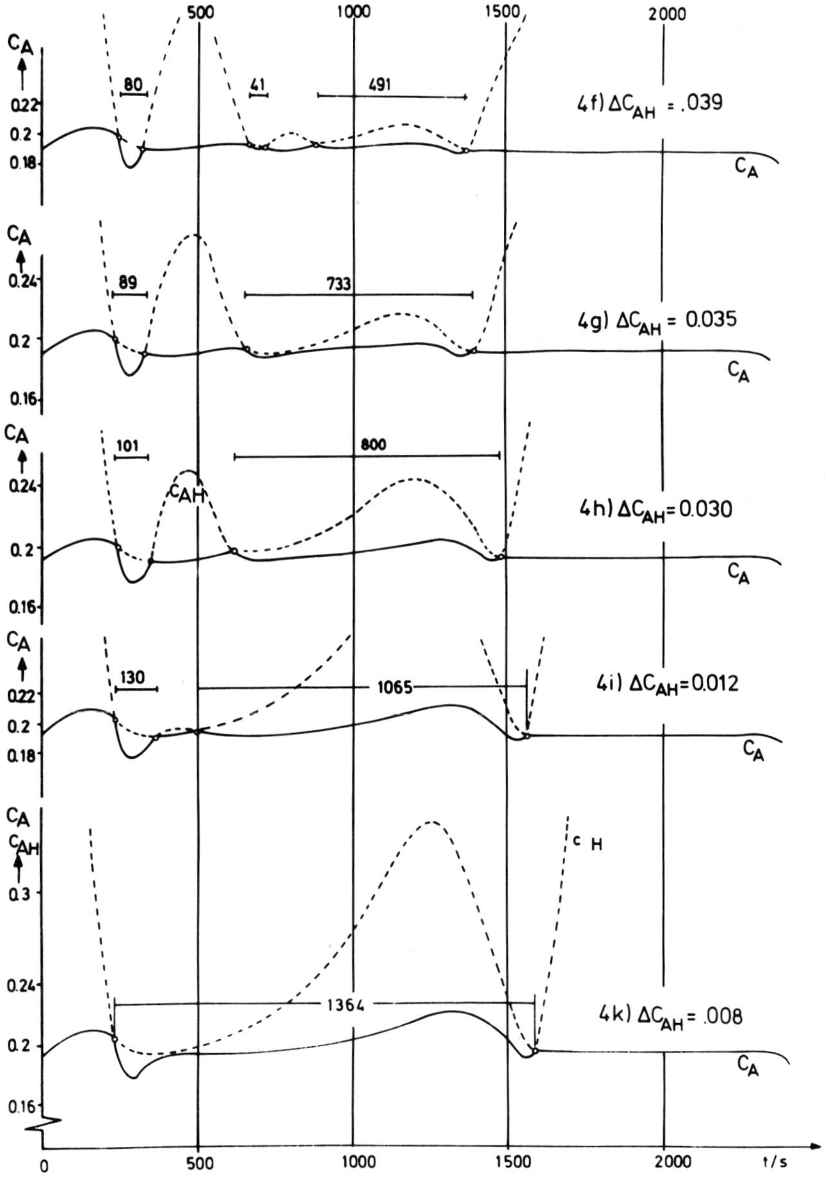

Bild 4f)-k): (Fortsetzung).

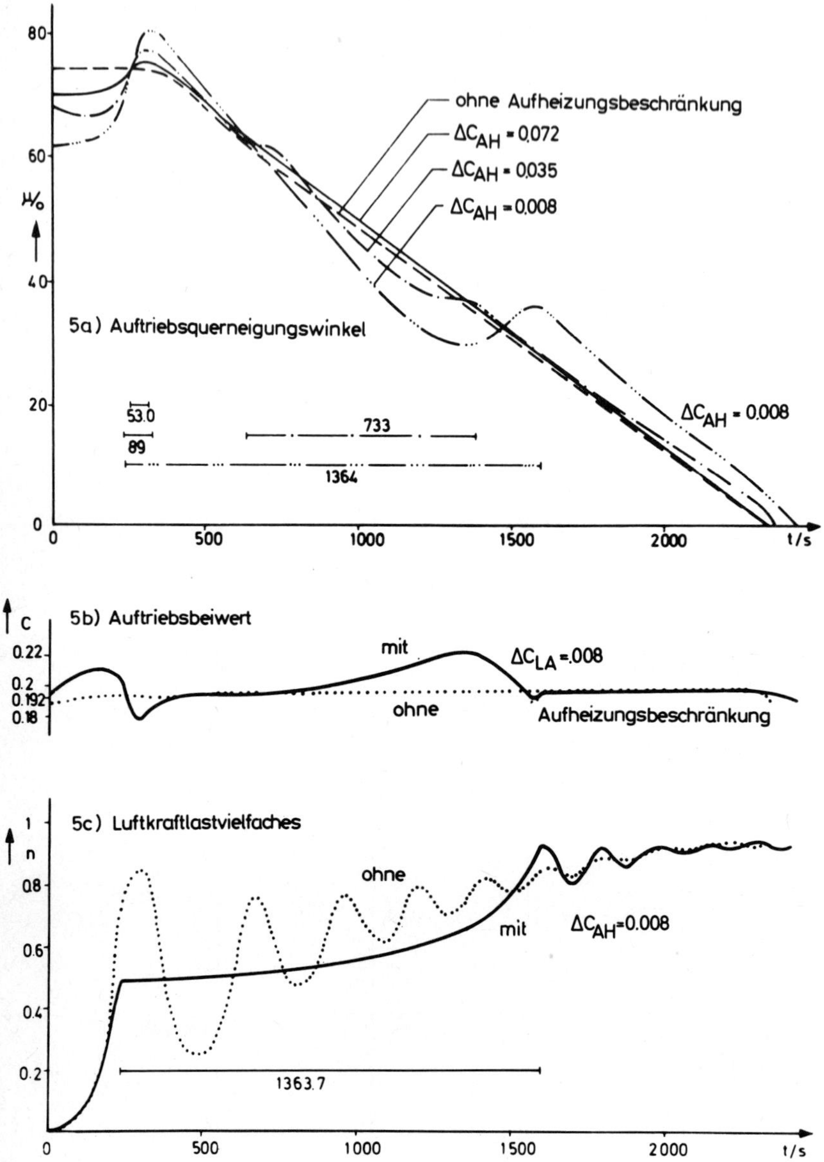

Bild 5: Vergleich optimaler Steuerungen für Bahnen größter Seitenreichweite mit und ohne Aufheizungsbeschränkungen.

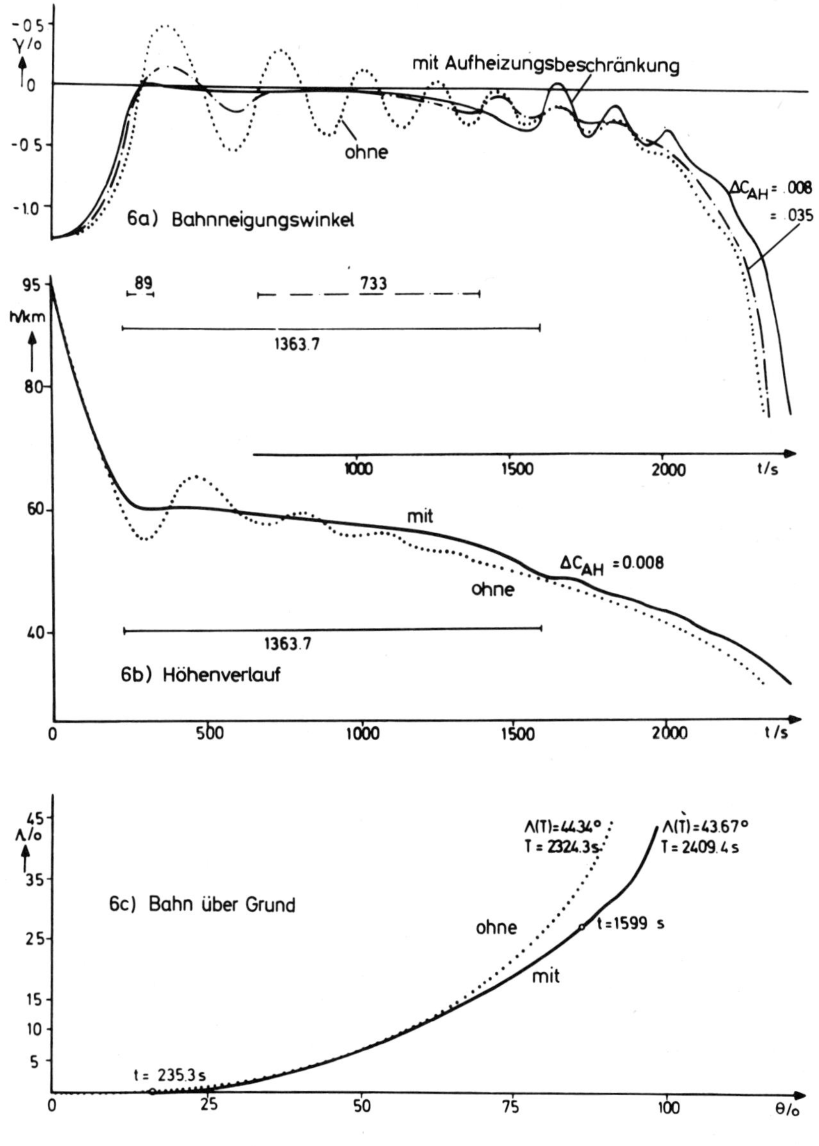

Bild 6a)-c): Bahnen größter Seitenreichweite mit und ohne Aufheizungsbeschränkung.

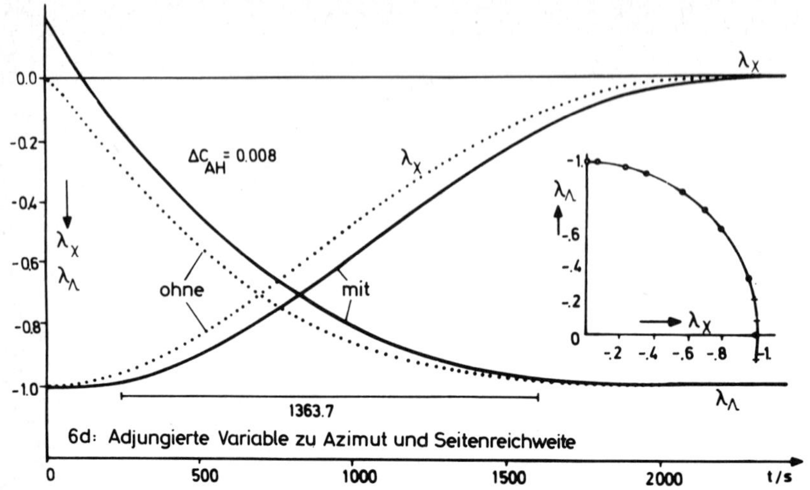
6d: Adjungierte Variable zu Azimut und Seitenreichweite
$\Delta C_{AH} = 0.008$
ohne
mit
1363.7
t/s

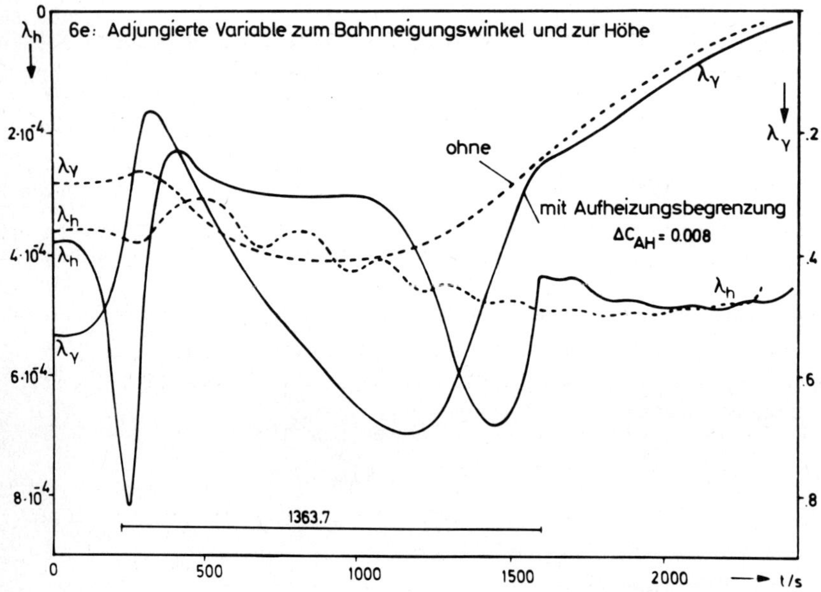
6e: Adjungierte Variable zum Bahnneigungswinkel und zur Höhe
ohne
mit Aufheizungsbegrenzung
$\Delta C_{AH} = 0.008$
1363.7
t/s

Bild 6d)-e): (Fortsetzung).

On an Optimization Problem Related to Minimal Surfaces with Obstacles

Ulrich Eckhardt

1. Statement of the Problem. In very different applications of mathematics one has to consider the following problem:

Find the minimum of

$$\varphi(x) = \sum_{j=1}^{n} w_j \cdot \sqrt{q_j(x)},$$

where

$$q_j(x) = x^T C_j x + c_j^T x + \gamma_j, \quad j = 1,\cdots,n$$

are quadratic functionals $q_j: R^d \longrightarrow R$. C_j, c_j and γ_j are choosen in such a way that $q_j(x) \geq 0$ for all $x \varepsilon R^d$ and all $j = 1,\cdots, n$. Moreover, C_j are positive definite matrices. The weights w_j are positive for all $j = 1,\cdots, n$.

Problems of this type arise in various contexts. In economics one often has to minimize sums of distances (location problem, generalized Weber problem). This problem was first stated by Fermat and solved by Weiszfeld [17] theoretically and numerically (see Katz [7]). Kuhn [8] gives a brief historical sketch.

A quite different application was described by Thomas [16] in connection with a system of springs.

The most extensively investigated problem of the above type is the discretized problem of Plateau. Given a region $B \subset R^2$ with boundary Γ and a function g on Γ, all with "nice" properties, one is looking for a function u on $B \cup \Gamma$ with

$$\int_B \sqrt{1 + u_x^2 + u_y^2}\, dx\, dy \longrightarrow \text{Minimum}$$

such that $u = g$ on Γ. By discretization we obviously get our problem. Here we have a large amount of literature, e.g. Wilson [18], Concus [2], Meis [11], Meis and Törnig [12], Mittelmann [13], Hinata, Shimasaki and Kiyono [6].

2. The Restricted Problem. For appropriately choosen matrices A, B and vectors a, b we define the polyhedral set

$$P = \{x \varepsilon R^d \mid Ax \leq a,\ Bx = b\}$$

and formulate the restricted problem:

Find $\min_{x \varepsilon P} \varphi(x)$. (A)

Inequality constraints arise quite naturally in the practical applications described above. Especially in the minimal surface case one can impose as an additional restriction that the solution u lies above a given "obstacle" $\psi(x, y)$ in $B \cup \Gamma$ (see Nitsche, [14], [15]).

By virtue of the definiteness of the matrices C_j we can easily prove the following

Theorem 1: (A) has a solution if $P \neq \emptyset$.

This solution needs not to be unique. The case of (A) having more than one solution, however, can only occur under very restrictive and numerically well detectable conditions.

Theorem 2: If (A) has more than one solution, then the following is true:

(i) For each $j = 1,\cdots, n$ there is a $x_j \in R^d$ with $q_j(x_j) = 0$,

(ii) There exists $y \in R^d$, $x_o \in R^d$ and numbers μ_j such that $x_j = x_o + \mu_j \cdot y$ for $j = 1,\cdots, n$ (i.e. all x_j are collinear).

Sketch of Proof: It is clear that $\varphi(x)$ is a convex function. Writing $q_j(x) = (x - x_j)^T C_j (x - x_j) + \gamma_j'$, $j = 1,\cdots, n$, it becomes evident that φ is strictly convex whenever one of the γ_j' is positive. For $\gamma_j' = 0$, $\sqrt{q_j(x)}$ is not strictly convex only on straight lines containing x_j. From this the proof of the Theorem is clear.

Note that for the more general $\varphi(x)$ essentially the same conditions for being not strictly convex are true as in the special case of Weiszfeld.

3. Weiszfeld's Method. In order to formulate Weiszfeld's algorithm we assume that $q_j(x) > 0$ for all $x \in P$. Later on we will see that this assumption is in fact no restriction.

For given numbers $y_j > 0$, $j = 1,\cdots, n$, we state the following problem

$$\min_{x \in P} \sum_{j=1}^{n} w_j \cdot q_j(x)/y_j. \qquad (A(y))$$

Now Weiszfeld's method looks like this:

0. Choose $y_j^{(o)} > 0$ for $j = 1,\cdots, n$; $r := 0$,
1. $x^{(r+1)}$ is the solution of $(A(y^{(r)}))$,
2. $y_j^{(r+1)} = \sqrt{q_j(x^{(r+1)})}$,
3. Apply an appropriate termination test. If it fails, put $r := r + 1$ and go to 1.

The convergence properties of this variant of Weiszfeld's method are characterized by the following two Theorems:

Theorem 3: $\varphi(x^{(r+1)}) \leqslant \varphi(x^{(r)})$.

Theorem 4: Let x^* be any solution of (A). Then there exists a number $\gamma > 0$ such that

$$0 \leqslant \varphi(x^{(r)}) - \varphi(x^*) \leqslant \gamma \cdot \sqrt{\varphi^2(x^{(r)}) - \varphi^2(x^{(r+1)})}.$$

Theorems 3 and 4 together prove the convergence of the method. Since γ can be calculated explicitly, Theorem 4 gives an a posteriori estimate of the error. Due to the square root, however, this estimate is actually not very good.

Proof of Theorems 3 and 4: In R^n we consider the following set

$$K = \{(\eta_1, \cdots, \eta_n) \in R^n \mid \eta_j \geqslant \sqrt{q_j(x)},\ j = 1, \cdots, n \text{ for any } x \in P\}.$$

Obviously, K is a convex set. Furthermore, we state the following problems

$$\min_{\eta \in K} \sum w_j \cdot \eta_j \tag{A'}$$

and

$$\min_{\eta \in K} \sum w_j \cdot \eta_j^2 / y_j. \tag{A'(y)}$$

It is clear that for every solution η to (A') (or (A'(y))) we can find a solution x to (A) (or (A(y))) such that $\eta_j = \sqrt{q_j(x)}$.

Let $y = (y_1, \cdots, y_n) \in K$ be fixed and consider the convex set of all η with

$$\sum w_j \cdot \eta_j^2 / y_j \leqslant \sum w_j \cdot y_j.$$

By convexity we have [1], [10]:

$$\sum w_j \cdot \eta_j^2 / y_j \geqslant \sum w_j \cdot y_j + 2 \cdot \sum w_j \cdot (\eta_j - y_j) =$$

$$= 2 \cdot \sum w_j \cdot \eta_j - \sum w_j \cdot y_j$$

for all η contained in this set. Consequently

$$\sum w_j \cdot y_j \geqslant \sum w_j \cdot \eta_j$$

for all such η. This yields the assertion of Theorem 3.

In order to prove Theorem 4 we fix $y \in K$ and consider the linear transformation $T: R^n \longrightarrow R^n$ defined by

$$T\,\eta = (\eta_j \cdot \sqrt{w_j} / \sqrt{y_j})_{j=1}^n.$$

Denoting by y^* a solution of (A') we minimize $\|\alpha(\mu)\|^2$ where

$$\alpha(\mu) = Ty + \mu \cdot (Ty^* - Ty) \in TK \quad \text{for } 0 \leqslant \mu \leqslant 1.$$

Assuming $\|\alpha(\mu_0)\| = \min \|\alpha(\mu)\|$ we maximize the linear functional $\sum w_j \cdot \eta_j$ for all η with $\|T\eta\|^2 \leqslant \|\alpha(\mu_0)\|^2$. This maximization can also be performed explicitly and the maximum value of this functional will provide an upper bound for $\sum w_j \cdot \tilde{\eta}_j$ where $\tilde{\eta}$ is the solution of (A'(y)).

Assuming $\|Ty - Ty^*\|^2 \leqslant C_0$ we can prove the inequality of Theorem 4 with $\gamma^2 = \frac{1}{C_0} \cdot \min_{x \in P} \varphi(x)$.

At the begin of this paragraph we stated the assumption that $q_j(x) > 0$ for all x and all j. It is very easy to see that this assumption is indeed not necessary if one modifies the method such that the case $q_j(x) = 0$ is taken into account.

Recalling the definition of the x_j from Theorem 2 we see that the assumption will be violated whenever one of the iterates comes close to one of the x_j. If this is the case, we simply disregard the corresponding $q_j(x)$ for the next iterations until eventually the iterates will move away from x_j. If this does not happen, x_j is obviously the solution of the problem or very close to the solution (see [17] and [8]).

4. Numerical Considerations. In performing the numerical calculations we note that the problems $(A(y))$ are ordinary quadratic programming problems which can be solved by well-known methods (see [1], [4], [9]).

Especially in the case of the discretized Plateau problem it seems to be advisable to choose $y_j^{(o)} = 1$ for all j. Then one is actually solving the discretized Dirichlet problem and, as well-known (see [3]), there are close relationships between both problems. So we can hope that $x^{(1)}$ will be a "good" solution of problem (A). Physically speaking, we replace the soap film of the minimal surface problem by a membrane. In our numerical experiments, the quadratic programming problem $(A(y^{(o)}))$ was solved by Lemke's method for quadratic programming ([4], [9]).

Having obtained an approximate solution $x^{(r)}$, we solve $(A(y^{(r)}))$ to get $x^{(r+1)}$. If $x^{(r)}$ was already a "good" solution to (A), $x^{(r+1)}$ will be not very different from $x^{(r)}$. Consequently one solves the quadratic programming problems $(A(y^{(r)}))$ for $r = 1, 2, \dots$ by means of an iterative method [5] to take advantage of the good starting solution $x^{(r)}$.

A number of computations was performed, mainly to provide test problems for the computer programs for quadratic programming written at the Central Institute for Applied Mathematics of the Nuclear Research Center Jülich. All programs were run at the IBM 370-168 of the Institute. As an example we consider the set

$$B = \{(x, y) \mid -1 \leqslant x, y \leqslant 1\}$$

and the function

$$g(x, y) = \begin{cases} 0 & \text{for } y = \pm 1, \ -1 \leqslant x \leqslant 1 \\ \sqrt{1 - y^2} & \text{for } x = \pm 1, \ -1 \leqslant y \leqslant 1. \end{cases}$$

The obstacle is defined by

$$(x, y) = \sqrt{r^2 - x^2 - y^2}$$

whereever the radicand is nonnegative.

This problem was discretized by introducing piecewise linear functions on rectangular triangles defined by the grid points $x_j = \frac{1}{10} \cdot j$, $y_j = \frac{1}{10} \cdot j$, $j = -10\ (1)\ 10$. Symmetries of the problem were used during the calculations. The pictures show the resulting minimal surfaces for $r = 0$ and $r = 1$.

5. Acknowledgements. The author is very much indebted to Mr. H. T. Zimmermann for providing a program for Lemke's method and to Mrs. L. Lange for coding the programs for the method described here.

The author also wants to express his thanks to all those participants of the Oberwolfach meeting who gave him very interesting and helpful comments. He never experienced such a stimulating audience.

References

1. Collatz, L. and W. Wetterling: Optimierungsaufgaben. 2nd ed. Heidelberger Taschenbücher, Vol. 15. Berlin, Heidelberg, New York: Springer-Verlag 1971.
2. Concus, P.: Numerical solution of the minimal surface equation. Math. Comput. 21, 340 - 350 (1967).
3. Courant, R.: Dirichlet's Principle, Conformal Mapping, and Minimal Surfaces. Pure and Applied Mathematics, Vol. III. New York, London: Interscience Publishers 1950.
4. Eaves, B. C.: On quadratic programming. Management Sci. 17, 698 - 711 (1970/71).
5. Eckhardt, U.: Quadratic programming by successive overrelaxation. KFA Jülich Research Report, Jül-1064-MA, April 1974.
6. Hinata, M., M. Shimasaki and T. Kiyono: Numerical solution of Plateau's problem by a finite element method. Math. Comput. 28, 45 - 60 (1974).
7. Katz, I. N.: On the convergence of a numerical scheme for solving some locational equilibrium problems. SIAM J. Appl. Math. 17, 1224 - 1231 (1969).
8. Kuhn, H. W.: A note on Fermat's problem. Math. Programming 4, 98 - 107 (1973).
9. Lemke, C. E.: On complementary pivot theory. In: G. B. Dantzig and A. F. Veinott, eds.: Mathematics of the Decision Sciences, Part I. Lectures in Applied Mathematics, Vol. 11. pp. 95 - 114. Providence: American Mathematical Society 1968.
10. Mangasarian, O. L.: Nonlinear Programming. New York etc.: McGraw-Hill Book Company 1969.

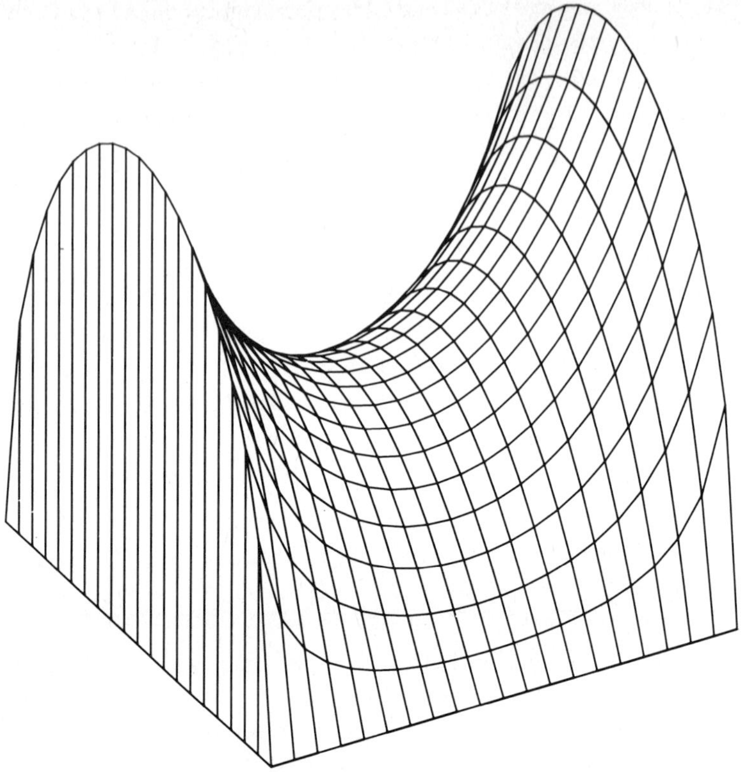

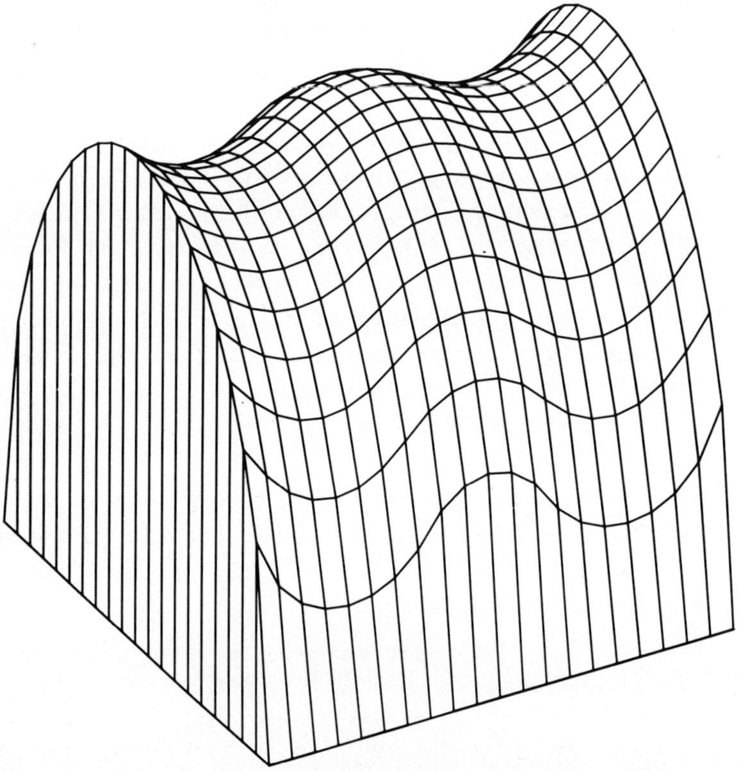

11. Meis, Th.: Zur Diskretisierung nichtlinearer elliptischer Differentialgleichungen. Computing 7, 344 - 352 (1971).

12. Meis, Th. and W. Törnig: Diskretisierungen des Dirichletproblems nichtlinearer elliptischer Differentialgleichungen. In: B. Brosowski and E. Martensen, eds.: Methoden und Verfahren der mathematischen Physik, Band 8, pp. 47 - 72. Mannheim/Wien/Zürich: Bibliographisches Institut, B.I.-Wissenschaftsverlag 1973.

13. Mittelmann, H. D.: Die Approximation der Lösungen gemischter Randwertprobleme quasilinearer elliptischer Differentialgleichungen. Technische Hochschule Darmstadt, Fachbereich Mathematik, Preprint Nr. 114, Januar 1974.

14. Nitsche, J. C. C.: Variational problems with inequalities as boundary conditions, or, How to fashion a cheap hat for Giacometti's brother. Arch. Rat. Mech. Anal. 35, 83 - 113 (1969).

15. Nitsche, J. C. C.: Minimal surfaces with partially free boundary. Annales Academiae Scientiarum Fennicae Ser. A. I. Mathematica. No. 483 (1971).

16. Thomas, J.: Zur Statik eines gewissen Federsystems im E_n. Math. Nachr. 23, 185 - 195 (1961).

17. Weiszfeld, E.: Sur le point par lequel la somme des distances de n points donnés est minimum. Tôhoku Mathematics J. 43, 355 - 386 (1937).

18. Wilson, W. L., Jr.: On discrete Dirichlet and Plateau problems. Numer. Math. 3, 359 - 373 (1961).

Ulrich Eckhardt
KFA Jülich
Z A M
D517 Jülich
Germany

Optimal Control of One-Dimensional Linear Parabolic Differential Equations

Klaus Glashoff

In this paper we consider a class of control problems which for example arise in connection with heat diffusion processes. The control function operates through the boundary- or initial-conditions or through the inhomogeneous term in the parabolic equation. The purpose of controlling such an initial-boundary-value control problem can be to minimize the difference (measured in a certain norm) between the temperature $y(x,T)$ of a thin rod at a given time T and for a given function $z(x)$ ($0 \leqq x \leqq 1$).

Problems of this type as well as their possible technological applications have been extensively discussed in the book of BUTKOVSKIY (1969). We also refer to the important paper by Yu. V. YEGOROV (1963).

The present paper gives a short and unifying existence and characterization theory for a class of problems including the ' problems of best approximation ' considered by Yu. V. YEGOROV. Our results on the ' refined ' bang-bang principle seem to be new even for the simplest case (L_2 - minimum-norm problem).

Numerous further references can be found in the survey article by ROBINSON (1971). See also BUTKOVSKI, A. I. EGOROV and LURIE

(1968) but compare carefully their statements on Yu. V. YEGOROV's results with his original paper and with the present article. Other approaches to the bang-bang principle for certain parabolic problems are given by FATTORINI (1964), LIONS (1968), WECK (1975).

Without mentioning all the simple modifications and extensions of the following results we want to point out that a generalization to parabolic equations in higher dimensions is nontrivial but possible.

In the following Section 1 we define the three control problems to be treated in this paper. In Section 2 we consider briefly a general linear minimum-norm control problem for which we give simple existence and characterization results. In the final Section 3 we return to the parabolic problems and show that the assumptions of the general theorems of Section 2 are met for the control problems defined in Section 1.

1. Three control problems

1.A) Boundary control. Let $y(u;x,t)$ denote the solution of the parabolic initial-boundary-value problem

$$y_t(x,t) - y_{xx}(x,t) + q(x)y(x,t) = 0\,, \qquad 0<x<1,\ 0<t\leqq T \tag{1}$$

$$\left.\begin{aligned} y(1,t) + \alpha y_x(1,t) &= u(t) \qquad (2)\\ y_x(0,t) &= 0 \qquad (3)\end{aligned}\right\}\ 0<t\leqq T$$

$$y(x,0) = 0\,, \qquad 0\leqq x\leqq 1. \tag{4}$$

Here $T>0$ and $\alpha>0$ are fixed parameters and $q(x)\geqq 0$ is a fixed continuously differentiable real function on the interval $[0,1]$.

We assume for a moment that for each function u in a certain function space $\mathcal{L}[0,T]$ there exists a unique solution (in a sense to be specified) to the initial-boundary-value problem stated above. Suppose further that for each $u\in\mathcal{L}[0,T]$ we have (as a function in x)

$$y(u;\cdot,T)\in\mathcal{G}[0,1]\,,$$

where $\mathcal{G}[0,1]$ is a certain normed function space on $[0,1]$ with norm $\|\cdot\|$. Then it is senseful to consider the following control problem: For a given $z \in \mathcal{G}[0,1]$,

$$\text{Minimize } \|y(u;\cdot,T) - z(\cdot)\|$$
$$\text{under the constraint } |u(t)| \leqq 1 \quad (t \in [0,T]),\ u \in \mathcal{L}[0,T] .$$

We proceed to an exact definition of the boundary control problem to be discussed.

It is a well known fact that - under suitable smoothness- and compatibility - requirements on the function $u(t)$ - there exists a unique 'classical' solution $y(u;x,t)$ of the equations (1) - (4) which is represented by

$$y(u;x,t) = \sum_{k=1}^{\infty} A_k \mu_k^2 \int_0^t u(\tau)\exp(-\mu_k^2(t-\tau))\,d\tau\ v_k(x) \tag{5}$$

where $\{v_k\}_{k\geqq 1}$ and $\{\mu_k^2\}_{k\geqq 1}$ are respectively the normed eigenfunctions and the eigenvalues of the boundary value problem

$$-V''(x) + q(x)V(x) = \lambda V(x), \qquad x \in (0,1)$$
$$V'(0) = 0$$
$$V(1) + \alpha V'(1) = 0$$

and

$$A_k = \frac{1}{\alpha}\mu_k^{-2}\ v_k(1) \qquad \text{for all } k \geqq 1.$$

Remark 1.1. In the case of the heat equation ($q \equiv 0$) the μ_k are the positive solutions of the equation

$$\mu \tan \mu = 1/\alpha ,$$

and

$$v_k(x) = (2\mu_k/(\sin\mu_k\cos\mu_k + \mu_k))^{1/2} \cos\mu_k x ,$$

$$A_k = \int_0^1 v_k(x)\,dx = (\mu_k \sin\mu_k\cos\mu_k + \mu_k^2)^{-1/2}\sqrt{2}\,\sin\mu_k . \tag{6}$$

In the following theorem we take the series (5) as definition for generalized solutions of the equations (1) - (4). The operator A to be defined in the theorem assigns to each boundary control $u(t)$ a (generalized) end temperature distribution $y(u;x,T)$.

<u>Theorem 1.1.</u> By

$$(Au)(x) = \sum_{k=1}^{\infty} A_k \mu_k^2 \int_0^T u(\tau)\exp(-\mu_k^2(T-\tau))\,d\tau \; v_k(x), \quad x \in [0,1] \tag{7}$$

for each $u \in L_p[0,T]$ ($2 < p \leqq \infty$) a continuous function Au is defined:

$$(Au)(\cdot) = y(u;\cdot,T) \in C[0,1] \; .$$

The linear operator

$$A : \; L_p[0,T] \longrightarrow C[0,1]$$

is bounded.

An easy <u>proof</u> for this is given in GLASHOFF and GUSTAFSON (1975) where it is shown that the series in (7) is uniformly convergent on $[0,1]$ for each $u \in L_p[0,T]$, $2 < p \leqq \infty$. The main tool for proving this are the inequalities

$$(k-1)\pi \leqq \mu_k \leqq (k+1)\pi \tag{8}$$

and

$$|A_k \mu_k^2 v_k(x)| \leqq C, \qquad x \in [0,1], \; k \geqq 1, \tag{9}$$

which can be established using the asymptotic expressions for μ_k and $v_k(x)$ in BIEBERBACH (1956), pp. 218, 220.

<u>Remark 1.2.</u> It follows that A is continuous regarded as a map into any space $L_r[0,1]$, $1 \leqq r \leqq \infty$, with the norm $\| \; \|_r$. We shall make no notational difference between the norm in $L_\infty[0,1]$ and in $C[0,1]$.

Now we are able to <u>define</u> the optimal control problem described by the equations (1) - (4):

(P1): Minimize $\| Au - z \|_r$ ($z \in L_r[0,1]$, $1 \leqq r \leqq \infty$)
under the constraint
$u \in Q_1 = \{ u \in L_p[0,T] \; / \; |u(t)| \leqq 1 \;\text{ a.e.} \}$

where p is a fixed real number satisfying

$$2 < p < \infty \; .$$

<u>Remark 1.3.</u> This choice of p is of a rather technical nature as the set Q_1 is trivially independent of p for $1 \leqq p \leqq \infty$. We require $2 < p$ because it is then possible to define A on the space $L_p[0,T]$. The inequality $p < \infty$ implies the reflexivity of $L_p[0,T]$ which simplifies the proofs of the theorems in Section 2 and Section 3.

1.B) Distributed control. Here a control function $v(t)$ appears in the inhomogeneous term of the differential equation:

$$y_t(x,t) - y_{xx}(x,t) + q(x)y(x,t) = g(x)v(t), \quad 0 < x < 1, \; 0 < t \leqq T \quad (10)$$

with homogeneous initial-boundary-conditions

$$\left.\begin{aligned} y(1,t) + \alpha y_x(1,t) &= 0 \quad (11)\\ y_x(0,t) &= 0 \quad (12)\end{aligned}\right\} \quad 0 < t \leqq T$$

$$y(x,0) = 0 \qquad 0 \leqq x \leqq 1. \quad (13)$$

Here $g(x)$ is a fixed given L_2 - function on $[0,1]$ and there are constraints on $v(t)$:

$$-1 \leqq v(t) \leqq 1 \qquad \text{for } t \in [0,T] .$$

The solution $y(v;x,t)$ of (10) - (13) is represented by the following series:

$$y(v;x,t) = \sum_{k=1}^{\infty} g_k \int_0^t v(\tau)\exp(-\mu_k^2(t-\tau))d\tau \; v_k(x)$$

where for all $k \geqq 1$

$$g_k = \int_0^1 g(x)v_k(x)dx \quad . \quad (14)$$

The same arguments as in the proof of Theorem 1.1 lead directly to the following theorem.

Theorem 1.2. By

$$(Bv)(x) = \sum_{k=1}^{\infty} g_k \int_0^T v(\tau) \exp(-\mu_k^2(T-\tau))d\tau \; v_k(x), \quad x \in [0,1] \quad (15)$$

for each $v \in L_2[0,T]$ a continuous function Bv is defined:

$$(Bv)(\cdot) = y(v;\cdot, T) \in C[0,1] .$$

The linear operator

$$B : L_2[0,T] \longrightarrow C[0,1]$$

is bounded.

We define the distributed control problem as

(P2): Minimize $\|Bv - z\|_r$ $\quad (z \in L_r[0,1], \; 1 \leqq r \leqq \infty)$
under the constraint
$v \in Q_2 = \{ v \in L_2[0,T] \; / \; |v(t)| \leqq 1 \;\; \text{a.e.} \}$

1.C) Initial control. In this case the control $w(x)$, $0 \leqq x \leqq 1$ operates through the initial conditions:

$$y_t(x,t) - y_{xx}(x,t) + q(x)y(x,t) = 0, \quad 0 < x < 1, \quad 0 < t \leqq T \tag{16}$$

$$\left.\begin{aligned} y(1,t) + \alpha y_x(1,t) &= 0 \quad (17) \\ y_x(0,t) &= 0 \quad (18) \end{aligned}\right\} \quad 0 < t \leqq T$$

$$y(x,0) = w(x), \quad 0 \leqq x \leqq 1 . \tag{19}$$

It is desired to minimize a similar functional as in 1.A) and 1.B) under the constraints

$$-1 \leqq w(x) \leqq 1 \qquad \text{for} \quad x \in [0,1] .$$

For $w \in L_2[0,1]$ the solution of (16) - (19) is given by

$$y(w;x,t) = \sum_{k=1}^{\infty} \exp(-\mu_k^2 t) \int_0^1 w(s)v_k(s)ds \; v_k(x),$$

Theorem 1.3. By

$$(Cw)(x) = \sum_{k=1}^{\infty} \exp(-\mu_k^2 T) \int_0^1 w(s)v_k(s)ds \; v_k(x), \quad x \in [0,1] \tag{20}$$

for each $w \in L_2[0,1]$ a continuous function Cw is defined:

$$(Cw)(\cdot) = y(w;\cdot,T) \in C[0,1] .$$

The linear operator

$$C : L_2[0,1] \longrightarrow C[0,1]$$

is bounded.

We define the initial control problem described by (16) - (19) as follows:

(P3): Minimize $\|Cw - z\|_r$, $(z \in L_r[0,1], 1 \leqq r \leqq \infty)$
under the constraint
$w \in Q_3 = \{ w \in L_2[0,1] \;/\; |w| \leqq 1 \quad \text{a.e.} \}$

Remark 1.4. One can also consider other Sturm- Liouville boundary conditions in (2)-(3), (11)-(12) and (17)-(18). The case of inhomogeneous initial conditions in (4) and (13) and of inhomogeneous boundary conditions in (3), (11)-(12) and (17)-(18) may be treated as well.

2. Existence, unicity and characterization

2.A) Existence of optimal controls

The control problems (P1), (P2) and (P3) in the preceding section have the following common structure:

$$\text{(P):} \qquad \begin{array}{l} \text{Minimize } \phi(u) := \| Su - z \| \\ \text{under the constraint } u \in Q \end{array}$$

where S is a bounded linear transformation mapping $E = L_p[a,b]$ ($1 < p < \infty$) into a normed space F (with norm $\|\cdot\|$);

$$Q = \{ u \in L_p[a,b] \ / \ |u| \leq 1 \quad \text{a.e.} \}$$

and z is a given vector in F.

Remark 2.1. In problem (P1) we have $a = 0$, $b = T$, $2 < p < \infty$, in (P2) $a = 0$, $b = T$ and $p = 2$, in (P3) $a = 0$, $b = 1$ and $p = 2$. In (P1), (P2), (P3) F is a space $L_r[0,1]$, $1 \leq r \leq \infty$.

Theorem 2.1. There is a solution to each of the problems (P1), (P2) and (P3) for any r, $1 \leq r \leq \infty$.

Proof. The functional ϕ is convex and continuous (Theorems 1.1, 1.2, 1.3) on the reflexive space E; this implies that ϕ is weakly lower semicontinuous (DEMYANOV and RUBINOV (1970), p. 10). As a bounded subset of the reflexive space E, Q is weakly relatively sequentially compact and as a convex and closed subset Q is weakly closed (DEMYANOV and RUBINOV, p. 7). The result follows at once if we use the well known theorem of Weierstraß.

Remark 2.2. One can show that the sets $A(Q_1)$, $B(Q_2)$ and $C(Q_3)$ are compact in $L_r[0,1]$, $1 \leq r \leq \infty$. This also implies Theorem 2.1. GLASHOFF and GUSTAFSON (1975) proved the compactness of A - it is easy to show that B and C are compact operators too.

2.B) Principle of maximum and bang-bang solutions

We are now going to prove a characterization theorem for the general problem formulated in the preceding section.

Let $$S : E = L_p[a,b] \longrightarrow F \qquad (1 < p < \infty)$$

be a continuous linear transformation mapping E into the normed space F (with norm $\|\cdot\|$).

As usual F' denotes the topological dual of F normed by

$$\|l\|_{F'} = \sup_{\substack{\|y\|=1 \\ y \in F}} |l(y)| \quad .$$

Further
$$S' : F' \longrightarrow E' = L_q[a,b] , \qquad 1/p + 1/q = 1, \qquad (21)$$

is the operator adjoint to S which means

$$l(Su) = (S'l,u)$$

for all $u \in E$, $l \in F'$ (we define $(\lambda,u) = \int_0^1 \lambda(s)u(s)ds$ for $\lambda \in L_q[0,T]$ and $u \in L_p[0,T]$).

Theorem 2.2. (Principle of maximum) Let $\hat{u}$ be a solution of (P). Then there is a functional $\hat{l} \in F'$ such that

$$(S'\hat{l},\hat{u}) = \max_{u \in Q} (S'\hat{l},u) \quad . \qquad (22)$$

If $\|S\hat{u} - z\| > 0$, then for each $\hat{l} \in F'$ with (22) we have

$$\hat{l} \neq 0_{F'} \quad .$$

This theorem is a corollary to the well-known ' duality theorem ' for minimum-norm problems (KÖTHE (1966), p. 348).

Definition 2.1. The operator S is called F - controllable if its range Im(S) is dense in F.

For the following definition we remind the reader that

$$Im(S') \subset L_1[a,b] \qquad \text{because of} \quad (21).$$

Definition 2.2. The operator S is called F - normal, if for each $\lambda \in \mathrm{Im}(S'), \lambda \neq 0_{L_q[a,b]}$, the set of zeroes of λ has measure zero.

Theorem 2.3. Let $z \in F$ be given and suppose that

(i) $\inf\limits_{u \in Q} \| Su - z \| > 0$ (i.e. z is ' not reachable ')

(ii) S is F - controllable and F - normal.

Then there exists a unique solution $\hat{u}$ of (P) for which

$$|\hat{u}(t)| = 1 \qquad \text{a.e. on } [a,b] .$$

Proof. If $\hat{u}$ is optimal there is a

$$\lambda = S'\hat{1} \in L_q[a,b] \quad , \quad \hat{1} \neq 0_{F'} \tag{23}$$

such that

$$\max_{u \in Q} (\lambda, u) = \max_{\|u\|_\infty \leq 1} \int_0^1 \lambda(s) u(s) ds = \|\lambda\|_1 = (\lambda, \hat{u}). \tag{24}$$

So if we define as usual

$$\mathrm{sgn}\, \beta = \beta/|\beta| \ (\beta \in \mathbb{R}, \ \beta \neq 0), \ \mathrm{sgn}\, 0 = 0,$$

we know that $\hat{u}$ is uniquely defined by (24) in the following way:

$$\hat{u}(s) = \mathrm{sgn}\, \lambda(s)$$

on the complement of the set of zeroes of λ in $[a,b]$. By virtue of the F - normality of S we only have to show that λ is not identically zero on $[a,b]$. But $\lambda = 0_{L_q[a,b]}$

implies

$$0 = (\lambda, u) = (S'\hat{1}, u) = \hat{1}(Su) \qquad \text{for all } u \in L_p[a,b],$$

which means (because of the F - controllability of S)

$$\hat{1}(x) = 0 \qquad \text{for all } x \in X$$

where $X = \mathrm{Im}(S)$ is dense in F. But this implies $\hat{1} = 0_{F'}$, in contradiction to (23).

Corollary 2.1. Under the assumptions of the preceding theorem there is an integrable function λ with the following properties:

(a) $\lambda \in \mathrm{Im}(S') \setminus 0_{L_q[a,b]}$.

(b) The set of zeroes of λ on $[a,b]$ has measure zero.

(c) $\hat{u}(s) = \mathrm{sgn}\, \lambda(s)$, $s \in [a,b]$, where $\hat{u}$ is the unique solution of (P).

Further properties of the set of zeroes of λ lead directly to a sharper characterization of the ' jumps ' of $\hat{u}$. We investigate this in the following section for the operators A, B and C defined in Section 1.

Remark 2.3. If an operator S is F - controllable and F - normal for $F = C[0,1]$ then the same holds for $F = L_r[0,1]$, where

$$1 \leqq r < \infty$$

because $C[0,1]$ is a dense subspace of each of these spaces.

Definition 2.3. ' Controllable ' means $C[0,1]$ - controllable and ' normal ' means $C[0,1]$ - normal.

3. Normality, controllability and the refined bang-bang principle

In this section we show that the requirements of Theorem 2.3 (normality and controllability) are met by the transformations A, B and C in the Problems (P1), (P2) and (P3) (Secton 1). We obtain even stronger results on the cardinality and distribution of the set of zeroes of $\lambda \in Im(A')$ and $Im(B')$ and $Im(C')$ respectively which, by Corollary 2.1, lead directly to a sharper characterization (refined bang-bang priciple) of the optimal controls.

3.A) Normality

Proposition 3.1. The operator A defined in Theorem 1.1 is normal.

Proof. We show that any $\lambda \in Im(A')$ has an analytic continuation into the open half-plane

$$G_T = \{ \zeta \in \mathbb{C} \ / \ \mathcal{Re}\, \zeta < T \ \} .$$

Let $l \in C[0,1]'$ be given. We have $\lambda = A'l \in L_q[0,T]$ where $1<q<2$ because of $2<p<\infty$. As the series in (7) is uniformly convergent, we have for each $u \in L_p[0,T]$

$$l(Au) = \sum_{k=1}^{\infty} A_k \mu_k^2 l(v_k)(e_k,u) \tag{26}$$

where we defined $e_k(t) = \exp(-\mu_k^2(T-t))$. Now it is not difficult to show by means of (8) and (9) that there is a constant $\tilde{C}$ such that for all sufficiently large n we have

$$\left|\sum_{k=1}^{n} A_k \mu_k^2 l(v_k) e_k(t)\right| \leqq \tilde{C}(T-t)^{-1/2}$$

and by the well-known Lebesgue theorem we may interchange summation and integration sign in (26):

$$l(Au) = \sum_{k=1}^{\infty} A_k \mu_k^2 l(v_k)(e_k,u) = \left(\sum_{k=1}^{\infty} A_k \mu_k^2 l(v_k) e_k, u\right) = (A'l,u) .$$

We see that the analytical continuation of λ is given by

$$\lambda(\zeta) = \sum_{k=1}^{\infty} A_k \mu_k^2 l(v_k) \exp(-\mu_k^2(T-\zeta)) , \quad \zeta \in G_T , \tag{27}$$

because for each $\delta > 0$ this series has a convergent majorant

$$C \sum_{k=1}^{\infty} \exp(-\mu_k^2 \delta)$$

in the closed plane $\bar{G}_{T-\delta} = \{\zeta \in \mathbb{C} \ / \ \mathrm{Re}\,\zeta \leqq T-\delta\}$.

Thus if $\lambda \not\equiv 0$, as an analytical function it has only finitely many zeroes on each interval $[0,T-\delta], \delta > 0$. This implies that the set of zeroes of λ on $[0,T)$ is at most countable and has no other accumulation point than T. This proves the normality of A.

Remark 3.1. It is not known whether T is really an accumulation point of the zeroes of λ or whether there are only finitely many zeroes of λ on $[0,T]$.

Definition 3.1. An operator $S : L_p[a,b] \longrightarrow F$ $(1<p<\infty)$ is called F - countably normal resp. F - finitely normal if each

$$\lambda \in \mathrm{Im}(S') \setminus 0_{L_q[a,b]}$$

has at most countably many zeroes resp. finitely many zeroes . ' Countably normal ' means $C[0,1]$ - countably normal and ' finitely normal' means $C[0,1]$ - finitely normal .

Remark 3.2. We recall that any $\eta \in L_p[a,b]$ is an equivalence class of functions differing at most on subsets $N \subset [a,b]$ of measure zero. So if we speak of a certain property of η this means : there is a function $\tilde{\eta}$ in the equivalence class denoted by η which has this property.

Corollary 3.1. The operator A is countably normal; the zeroes of $\lambda \in \mathrm{Im}(A') \setminus 0_{L_q[0,T]}$ accumulate at most at the right endpoint of $[0,T]$.

Proposition 3.2. The statement of the preceding corollary (with $p = q = 2$) is also true for the operator B defined in Theorem 1.2.

(The proof is almost identical with that of Proposition 3.1 and is omitted).

Corollary 3.2. If for all $k \geqq 1$

$$|g_k| = |\int_0^1 g(x) v_k(x) dx| \leqq \eta\, e^{-\gamma \mu_k^2} \tag{28}$$

holds with $\eta, \gamma > 0$, then B is finitely normal.

Proof. For each $l \in C[0,1]'$, $\lambda = B'l \in L_2[0,T]$ we have

$$\lambda(t) = \sum_{k=1}^{\infty} g_k\, l(v_k) \exp(-\mu_k^2 (T-t)).$$

Because of (28) λ can be analytically continued into the open half-plane

$$G_{T+\gamma} = \{ \zeta \in \mathbb{C} \,/\, \mathcal{Re}\, \zeta < T + \gamma \}$$

and we see that λ can only have finitely many zeroes on $[0,T]$.

Example: (28) is certainly true if g is a finite linear combination of the v_k's.

We turn to the ' initial operator ' C defined in Theorem 1.3.

Assumption 3.1. $q \equiv 0$ in (16).

In this case (heat equation) the v_k's are given by (6).

Proposition 3.3. The operator C given in Theorem 1.3 is finitely normal.

Proof. For $l \in C[0,1]'$, $\lambda = C'l \in L_2[0,1]$ we have

$$\lambda(\xi) = \sum_{k=1}^{\infty} e^{-\mu_k^2 T} \alpha_k \, l(\cos \mu_k(\cdot)) \cos \mu_k \xi \quad , \quad \xi \in [0,1] \quad ,$$

where
$$\alpha_k = 2\mu_k / (\sin \mu_k \cos \mu_k + \mu_k) \; .$$

There is a $\varkappa > 0$ such that for all $k \geqq 1$ we have

$$|\alpha_k \cos \mu_k \xi| \leqq \varkappa \qquad \text{for all } \xi \in [0,1] \; .$$

Now there is an analytical continuation of λ into the whole complex plane $\mathbb{C}$: for $\zeta \in \mathbb{C}$, $|\zeta| \leqq R$ ($R > 0$ arbitrary) we have

$$\Big| \sum_{k=1}^{\infty} \exp(-\mu_k^2 T) \alpha_k \, l(\cos \mu_k(\cdot)) \cos \mu_k \zeta \Big|$$

$$\leqq \|l\|_{C[0,1]'} \sum_{k=1}^{\infty} \exp(-\mu_k^2 T) |\cos \mu_k \zeta|$$

and

$$\exp(-\mu_k^2 T) |\cos \mu_k \zeta| = \exp(-\mu_k^2 T) \, \Big| \tfrac{1}{2}(\exp(i\mu_k \zeta) + \exp(-i\mu_k \zeta)) \Big|$$

$$\leqq \exp(-\mu_k^2 T) \exp \mu_k |\zeta| \leqq \exp(-\mu_k^2 (T - R/\mu_k))$$

and the last term is not greater than $\exp(-\mu_k^2 T/2)$ for sufficiently large $k \geqq k(R)$. This shows that the series (29) is majorized by an absolutely convergent series in the region $|\zeta| \leqq R$ for any $R > 0$. The analytical continuation of λ to the whole complex plane is given by

$$\lambda(\zeta) = \sum_{k=1}^{\infty} \exp(-\mu_k^2 T) \alpha_k \, l(\cos \mu_k(\cdot)) \cos \mu_k \zeta \quad , \; \zeta \in \mathbb{C}. \tag{29}$$

If λ is not identically zero then there are only finitely many zeroes of λ on $[0,1]$.

Remark 3.3. The F - normality properties of A, B and C which we proved for $F = C[0,1]$ hold as well for

$$F = L_r[0,1] \quad , \qquad 1 \leqq r < \infty \; .$$

The same is true for the F - controllability of A,B and C which we are going to prove now. Compare Remark 2.3 !

3.B) Controllability

There are a lot of results concerning the $L_2[0,1]$ - controllability of A, B and C (and, in fact, of many other operators related to control problems with evolution equations), see FATTORINI (1968), LIONS (1968). We remind the reader that in our terminology 'controllable' means $C[0,1]$ - controllable.

Proposition 3.4. The operator A defined in Theorem 1.1 is controllable.

Proof. We recall that the orthogonal of the range of A is equal to the kernel of A':

$$(\mathrm{Im}(A))^{o} = \mathrm{Ker}(A').$$

We show that the kernel of A' consists only of the nullvector $0_{C[0,1]'}$. Suppose that

$$A'l = 0_{L_q[0,T]}$$

for $l \in C[0,1]'$. By analyticity of $\lambda = A'l$ in G_T (see (27)) it follows that

$$0 = \lambda(\zeta) = \sum_{k=1}^{\infty} A_k \mu_k^2 \, l(v_k) \exp(-\mu_k^2(T-\zeta))$$

for all $\zeta \in G_T$. But this implies

$$A_k \mu_k^2 \, l(v_k) = 0 \qquad \text{for all } k \geqq 1$$

(TSUJIOKA (1970), GLASHOFF and KRABS (1975)). Using the definition of the A_k's it is easy to show that

$$A_k \neq 0 \qquad \text{for all } k \geqq 1$$

and because of $\mu_k > 0$, $k \geqq 1$, we get

$$l(v_k) = 0 \qquad \text{for all } k \geqq 1.$$

Now the linear space V spanned by $v_1, v_2, v_3, \dots$ is dense in $H^1[0,1]$ (the space of absolutely continuous real functions on $[0,1]$ with square-integrable first derivative) (MIKHLIN (1971), p. 143) and thus V is dense in $C[0,1]$ too. This implies that $l = 0_{C[0,1]'}$ and by the Hahn - Banach theorem Im(A) is dense in $C[0,1]$.

Proposition 3.5. The operator B defined in Theorem 1.2 is controllable if

$$g_k = \int_0^1 g(x)v_k(x)dx \neq 0 \quad \text{for all } k \geqq 1.$$

(The proof is almost identical with that of Proposition 3.4 and is omitted).

Proposition 3.6. The operator C defined in Theorem 1.3 is controllable.

Proof. Because V (the space spanned by v_1, v_2, ... is dense in C[0,1] it is sufficient to show that for each v_k ($k \geqq 1$) there is a w_k in $L_2[0,1]$ such that

$$Cw_k = v_k \quad .$$

But this is certainly possible: take simply

$$w_k(x) = \exp(\mu_k^2 T)\ v_k(x), \quad x \in [0,1]\ , \ k \geqq 1 \ .$$

3.C) Summary of results

Now it is easy to combine the results of 3.A), 3.B) with Theorem 2.3 and Corollary 2.1.

In view of Remark 2.3 we assume for the remaining part of the paper that for the space F we either take $L_r[0,1]$, $1 \leqq r < \infty$, or C[0,1] . By 3.A), 3.B) the operators A, B and C are then F - normal (or even F - countably normal or F - finitely normal) and F - controllable.

Assumption 3.2. For problems (P1), (P2), (P3) we choose

either $\quad 1 \leqq r < \infty \quad$ and $\quad z \in L_r[0,1]$

or $\quad r = \infty \quad$ and $\quad z \in C[0,1]$.

For each of the problems (P1), (P2), (P3) we shall prove that the optimal solution has one of the following properties (see Remark 3.2).

Property 3.1. $u \in Q$ is called ' bang-bang ', if

$$|u(s)| = 1 \qquad \text{a.e. on } [a,b].$$

Property 3.2. $u \in Q$ is called ' countably bang-bang with a cluster point of jumps at b ' if there is a strictly increasing sequence $\{\alpha_k\}_{k \geqq 0}$ of real numbers satisfying

$$\alpha_0 = a \ , \quad \lim_{k \to \infty} \alpha_k = b$$

such that

$$u(s) = \varepsilon(-1)^k \qquad \text{for } \alpha_{k-1} \leqq s < \alpha_k, \quad k = 1, 2, \ldots$$

with $\varepsilon = +1$ or -1 .

Property 3.3. $u \in Q$ is called ' finitely bang-bang' if there is a partition

$$a = \alpha_0 < \alpha_1 < \alpha_2 < \ldots < \alpha_\varsigma = b$$

of $[a,b]$ such that

$$u(s) = \varepsilon(-1)^k \qquad \text{for } \alpha_{k-1} \leqq s < \alpha_k, \quad k = 1,..,\varsigma$$

with a certain $\varsigma \geqq 1$ and $\varepsilon = +1$ or -1.

Without further comments we state the three main theorems which follow directly from the preceding theorems and propositions.

Theorem 3.1. (Boundary control problem (P1)).
Assume $\inf\limits_{u \in Q_1} \| Au - z \|_r > 0$. Here r and z are chosen according to Assumption 3.2. Then there is a unique solution $\hat{u}$ of (P1) which is finitely bang-bang or countably bang-bang with a cluster points of jumps at T.

Theorem 3.2. (Distributed control problem (P2)).
Assume $\inf\limits_{v \in Q_2} \| Bv - z \|_r > 0$. Here r and z are chosen according to Assumption 3.2. If for all $k \geqq 1$

$$g_k = \int_0^1 g(x) v_k(x) dx \neq 0 \quad ,$$

then there is a unique solution $\hat{v}$ of (P2) which is finitely bang-bang or countably bang-bang with a cluster point of jumps at T.
If, in addition,

$$|g_k| \leqq \eta e^{-\gamma \mu_k^2} \qquad \text{for all } k \geqq 1$$

with $\eta, \gamma > 0$, then $\hat{v}$ is finitely bang-bang.

Theorem 3.3. (Initial control problem (P3)).
Assume $\inf_{w \in Q_3} \| Cw - z \|_r > 0$. Here r and z are chosen according to Assumption 3.2. Then there is an optimal solution $\hat{w}$ of (P3). Under Assumption 3.1 ($q \equiv 0$ in (16)) $\hat{w}$ is unique and finitely bang-bang.

Remark 3.4. We do not know whether in (P1) $\hat{u}$ is countably bang-bang or finitely bang-bang. This may depend on the function z.

Remark 3.5. In Assumption 3.2 the condition

' if $r = \infty$ then $z \in C[0,1]$ '

cannot be replaced by the requirement $z \in L_\infty[0,1]$. This is shown by the following simple counter-example given by Prof. N. Weck .
We define $z \in L_\infty[0,1]$ by

$$z(x) = \begin{cases} +1 & 0 \leqq x < 1/2 \\ -1 & 1/2 \leqq x \leqq 1 \end{cases} \quad .$$

It is easy to see that

$$\| Au - z \|_\infty \geqq 1$$

for all $u \in Q_1$ (because Au is a continuous function on [0,1]). But for $\tilde{u} \equiv 0$ we have

$$\| Au - z \|_\infty = \| z \|_\infty = 1 \ ;$$

this implies that $\tilde{u}$ is a solution of (P1). $\tilde{u}$ is not bang-bang. (We remark that in this example any $u \in Q_1$ solves (P1)).

Observe that the existence theorem in 2.A) is valid for the case $r = \infty$, $z \in L_\infty[0,1]$ too.

Remark 3.6. All the results of this paper also hold in the following situation. Let F be a Banch space. We replace the functional

$\phi(u) .= \| Su - z \|$ in (P), Section 2.A) by

$$\psi(u) .= f(Su)$$

where f is a convex, bounded and continuous real function on F. Obviously Theorem 2.1 (existence) does not change. This is also true for the ' principle of maximum ' (Theorem 2.2) (see for example PSCHENITSCHNY (1972), pp. 52, 54). We have to replace the condition

$$\inf_{u \in Q} \| Su - z \| = \| S\hat{u} - z \| > 0$$

by the following requirement: ' $\hat{u}$ is no solution of the unconstrained problem $\min \psi(u)$, $u \in E$ ' in Theorem 2.2 - Theorem 3.3.

References

Bieberbach, L. Einführung in die Theorie der Differentialgleichungen im reellen Gebiet. Springer Verlag, Berlin - Göttingen - Heidelberg, 1956.

Butkovskiy, A.G. Distributed control systems, Elsevier, New Y., 1969.

Butkovskiy, A.G., A. Egorov and K. Lurie Optimal control of distributed parameter systems / A survey of Soviet publications, SIAM J. Control, Vol. 6, pp. 437 - 476, 1968.

Demyanov, V.F. and A.M. Rubinov Approximate methods in optimization problems. American Elsevier Publishing Comp., Inc., New York, 197o.

Fattorini, H.O. Time optimal control of solutions of operational differential equations. SIAM J. Control, Ser. A.2, pp. 54-59, 1964.

Fattorini, H.O. Boundary control systems. SIAM J. Control, Vol. 6, pp. 349 - 385, 1968.

Glashoff, K. and S. A. Gustafson Numerical treatment of a parabolic boundary-value control problem, to appear in J. Optimization Theory and Applications, 1975.

Glashoff, K. and W. Krabs Dualität und Bang-Bang- Prinzip bei einem parabolischen Rand- Kontrollproblem. To appear in 'Bonner Mathem. Schriften', 1975.

Köthe, G. Topologische lineare Räume I, Springer Verlag, Berlin-Heidelberg - New York, 1966.

Lions, J.J. Contrôle Optimale de Systèmes Gouvernés par des Équations aux Dérivées partielles, Dunod, Paris, 1968.

Mikhlin, S.G. The numerical performance of variational methods, Wolters -Nordhoff Publishing Company, Groningen, The Netherlands, 1971.

Pschenitschny, B.N. Notwendige Optimalitätsbedingungen, R. Oldenbourg Verlag, München - Wien, 1972.

Robinson, A.C. A survey of optimal control of distributed - parameter systems, Automatica, Vol. 7, pp. 371 - 388, 1971.

Tsujioka, K. Remarks on controllability of second order evolution equations in Hilbert spaces, SIAM J. Control, Vol. 8 No. 1, 1970.

Weck, N. Über das Prinzip der eindeutigen analytischen Fortsetzbarkeit in der Kontrolltheorie, this volume, 1975.

Yegorov, Yu. V. Some problems in the theory of optimal control, USSR Comp. Math. Vol. 3, No. 5, 1963.

Klaus Glashoff
TH Darmstadt
Fachbereich Mathematik
D - 61 DARMSTADT
Schloßgartenstr. 7
W. - Germany

On the Numerical Treatment of a Multi-Dimensional Parabolic Boundary-Value Control Problem

Sven-Åke Gustafson, Royal Institute of Technology, Stockholm*.

Summary. The purpose of this paper is to show that the methods given in [Glasshoff-Gustafson] can be extended to much more general control problems. We describe how to carry out the numerical calculations in a general context.

1. Formulation of a general boundary-value control problem.

Consider the parabolic boundary-value problem

$$\frac{\partial y(x,t)}{\partial t} = Ly(x,t) \quad x \in S \ , \ 0 \le t \le T \ , \tag{1}$$

$$y(x,0) = 0 \ , \ x \in S \ , \tag{2}$$

$$\frac{\partial y(x,t)}{\partial n} + \alpha y(x,t) = u_0(x)u(t) \ , \ x \in \partial S \ , \ 0 \le t \le T \ , \tag{3}$$

where: S is a compact subset of R^k , $k < \infty$,

∂S is the boundary of S,

$\frac{\partial}{\partial n}$ the normal derivative at ∂S,

L an elliptic operator,

α, T positive numbers and

u_0 a fixed positive function, continuous on S.

We shall assume that S is sufficiently regular and that appropriate conventions are made with respect to Condition (3) at such points where the normal direction of S is not uniquely defined, in order to guarantee that the boundary-value problem (1), (2), (3) has a unique solution y for all continuous functions u. Then it is known, that y is continuous in the interior of S for all t in $[0,T]$.

In this paper we treat the slightly more general case: The interval $[0,T]$ may be partitioned, $0 = \tau_0 < \tau_1 < \ldots < \tau_\ell = T$ (ℓ is a finite number) and u is continuous in the interior of each sub-interval but may have a jump at τ_j, $j = 0,1,\ldots,\ell$. Then the controlproblem (1), (2), (3) has a unique solution y which is continuous in the interior. We note already here, that if we can integrate (1), (2), (3) numerically for continuous u, then we can handle the more general situation also:

---

* This research was financially supported by NSF under Grant GK - 31833

First we put $u(\tau_0) = \lim_{\tau\to\tau_0+0} u(\tau)$ and $u(\tau_1) = \lim_{\tau\to\tau_1-0} u(\tau)$ and integrate from τ_0 to τ_1. Hence we obtain $y(x,\tau_1)$. Next we put $u(\tau_1) = \lim_{\tau\to\tau_1+0} u(\tau)$ and $u(\tau_2) = \lim_{\tau\to\tau_2-0} u(\tau)$. Using $y(x,\tau_1)$ as initial value we integrate from τ_1 to τ_2. In the same way we advance to τ_3, τ_4 ... and reach T. We observe, that this procedure is applicable for the control problem treated in [Glasshoff-Gustafson].

Let now $u_1, u_2, \ldots, u_n$ be functions on $[0,T]$ which are continuous on each subinterval (τ_{j-1},τ_j), $j = 1,2,\ldots \ell$ and define $w_r(x)$ as $y_r(x,T)$ where y_r is the solution of the problem (1), (2), (3) with u_r replacing u. Then w_r is a continuous function on S. Our task is to determine a function

$$u = \sum_{r=1}^{n} \alpha_r u_r$$

with $0 \le \alpha_r \le 1$, $r = 1,2,\ldots,n$ such that the solution y for $t = T$ approximates a given continuous function z as well as possible in the uniform norm. This problem can be written

Program D: Compute $\inf_{\alpha} \alpha_{n+1}$

subject to

(4a) $$\left| \sum_{r=1}^{n} \alpha_r w_r(x) - z(x) \right| \le \alpha_{n+1}, \quad x \in S,$$

(4b) $$0 \le \alpha_r \le 1, \quad r = 1,2,\ldots,n,$$

(4c) $$\alpha_{n+1} \ge 0.$$

$\alpha_{n+1} = 0$ is possible, if and only if z is a linear combination of $u_1,u_2,\ldots,u_n$. We assume that this case does not occur. Since

$\left\| \sum_{r=1}^{n} \alpha_r w_r - z \right\|$ is a continuous function of $\alpha = \alpha_1,\alpha_2,\ldots,\alpha_n$ and α is confined to the compact subset of R^n defined by (4b), Program D assumes its inf-value.

2. Application of the duality theory of semi-infinite programming.

Program D is a constrained approximation problem. It may be written as a semi-infinite program and the numerical treatment may be based on the results in [Gustafson-Kortanek]. The optimal solution may be computed using the general codes in [Fahlander] provided subroutines for the determination of $w_1, w_2, \ldots, w_n$ are available. We give here a summary of the underlying theory. Thus arguing as in [Glasshoff-Gustafson] we find that Program D has the same optimal value as the nonlinear task

Program P

Compute $\sup \sum_{i=1}^{n} \rho_i\, z(x^i) - \frac{1}{2} \sum_{r=1}^{n} (|m_r| - m_r)$

over all real numbers $\rho_1, \rho_2, \ldots \rho_n$; $m_1, m_2, \ldots, m_n$ and all $x^i \in S$ such that

(5) $\sum_{i=1}^{n} \rho_i\, w_r(x^i) + m_r = 0, \; r = 1,2,\ldots,n$

(6) $\sum_{i=1}^{n} |\rho_i| = 1$

We note that Conditions (5) and (6) are consistent. Take namely $\rho_1 = 1$ and $\rho_i = 0$, $i = 2,3,\ldots,n$. Then select x^i arbitrarily in S and put $m_r = - w_r(x^1)$, $r = 1,2,\ldots,n$.

Program P assumes its optimal value since ρ is confined to a compact subset of R^n due to (6). $x^i \in S$ which is assumed compact. Therefore m_r must meet the condition $|m_r| \le ||w_r||$. Since the preference function of Program P is continuous we conclude that the optimal value is assumed. We observe that some of the ρ_i:s may be zero for an optimal solution. If $\rho_i = 0$, then x^i is arbitrary.

Using the theory in [Gustafson] we show that Programs P and D have a common optimal value. Hence, if we evaluate the preference function of Program P for any numbers ρ_i, x^i and m_r meeting (5) and (6) we get a lower bound for the optimal value of Program P.

Applying the theory on complementary slackness we can state the following necessary conditions for optimality: (Compare [Glasshoff-Gustafson]).

(7) $m_r \alpha_r (1-\alpha_r) = 0, \; r = 1,2,\ldots,n.$

(8) $\rho_i (|\sum_{r=1}^{n} \alpha_r w_r(x^i) - z(x^i)| - \alpha_{n+1}) = 0, \; i = 1,2,\ldots,n.$

Let $\alpha_1, \alpha_2, \ldots, \alpha_{n+1}$ be an optimal solution of Program D and put

$$\psi(x) = |\sum_{r=1}^{n} \alpha_r w_r(x) - z(x)|$$

Since $\psi(x) \le \alpha_{n+1}$ on S we conclude:

(9) ψ has a local maximum at x^i, if $\rho_i \neq 0$, $i = 1,2,\ldots,n$.

ψ gives rise to further relations: if x^i is in the interior of S and ψ is continuously differentiable at x^i, then

(9a) $\nabla\psi(x^i) = 0$

If x^i is in ∂S, S has a tangent plane at x^i with normal vector n^i, and ψ is continuously differentiable at x^i, then

(9b) $\nabla\psi(x^i) \, || \, n^i$

where $||$ means "parallel with".

The relations (7), (8), (9), (9a), (9b) can also be derived using Lagrange multipliers on Program P.

If we combine (5), (6), (7), (8), (9a), (9b), then we get a nonlinear system from whose solutions an optimal solution of Program D may be constructed. Since (5) through (9) only are necessary conditions one must verify that an accepted solution also meets (4a) and (4b). But if these lastmentioned conditions also are met then the solution found is optimal.

In the next Section we shall discuss the complications which arise when w_r are not known in analytic form but must be computed numerically, e.g. by a difference method.

Solving discretized versions of the semi-infinite program.

If the partition of the time-interval $[0,T]$ is equidistant, i.e. $\tau_r = \Delta T \cdot r$ with $\Delta T = T/n$, then a major simplification is possible in the solution of the control problem in [Glashoff-Gustafson]. We observe that in this case $u_r(t) = 0$, $0 \le t \le \tau_{r-1}$ and $u_r(t) = u_1(t - (r-1)\Delta T)$, $\tau_{r-1} \le t \le T$. Hence, if y^* is the solution of the boundary-value problem (1), (2) and (3) with u_1 replacing u, we find

$$w_r(x) = y^*(x, \Delta T + T - r\Delta T), \; r = 1,2,\ldots,n.$$

Therefore we need to solve the problem (1), (2), (3) only once in order to obtain

$w_r(x)$, $r = 1,2,\ldots,n$. Accordingly, the labor required for obtaining a table of $w_1, w_2, \ldots, w_n$ is independent of n for difference methods.

We shall assume that we use a uniform grid with the step-size h in all space directions. Then after performing the numerical integration we get a table of $w_r(x^j)$, $j = 1,2,\ldots,N$ and approximate Program D with the discretized task

Program D_0

Compute $\min \alpha_{n+1}$

subject to $\left|\sum_{r=1}^{n} \alpha_r w_r(x^j) - z(x^j)\right| \le \alpha_{n+1}$, $j = 1,2,\ldots,N$

and $0 \le \alpha_r \le 1$, $r = 1,2,\ldots,n$

This is a linear program which may be solved by means of standard computer codes. We recommend the stable code by [Bartels-Stoer-Zenger].

It is now possible to construct an initial approximate solution to the system (5), (6), (7), (8), (9a), (9b) as described in [Glasshoff-Gustafson]. Before discussing how to improve upon this approximate solution we make the following observations

1) The computed values $w_r(x^j)$ are not exact but affected by a truncation error which depends on the numerical method used to integrate the differential equation.

2) The grid defines a partition of S into hypercubes with side h. If w_r, $r = 1,2,\ldots,n$ is linear within each hypercube, then an optimal solution of Program D_0 is also an optimal solution of Program D. Hence, if it is known that w_r has continuous partial derivations of the second order in each compact subset of the interior of S, then we conclude that approximating Program D with Program D_0 means to perturb w_r at interior points with $O(h^2)$ (when $h \to 0$). Thus it is only worthwhile to use Newton-Raphson's method on the system (5), (6), (7), (8), (9a), (9b), if the necessary values of w_r and its derivatives are evaluated by means of at least quadratic interpolation and the truncation error in the integration method is of order $o(h^2)$ when $h \to 0$.

To meet the latter condition is possible either by selecting a high order difference method or choosing a lower order method and use Richardson-extrapolation. Assume that the latter course of action is taken. We consider the following situation: We apply a difference method such that the local error has a power expansion with the leading term $O(h)$ in the interior of S. According to [Stetter, 1965] one can show, that under certain general conditions the global error has a power expansion of the same type for each fixed compact subset of the interior of S. We apply our method using two grids G_1 and G_2 with mesh-sizes h and $2h$ respectively and such that $G_1 \cap G_2 = G_2$. Next we perform a Richardson-extrapolation and get values of w_r at G_2 with errors of the magnitude $O(h^2)$. But on the part of G_1 that does not belong to G_2 the errors are $O(h)$. As suggested by [Lindberg] for ordinary differential equations we interpolate in the corrections to determine the proper corrections for the parts of G_1 which are not in G_2. In this case linear interpolation is sufficient. It is possible to continue in the same way using 3 grids with mesh-sizes $2h$, h and $h/2$ in order to remove more powers in the error expansion. Then the corrections must be interpolated using formulas of higher order. However, one must observe the "curse of dimensionality": Since S is a subset of R^k the number of points in a grid of mesh-size h increases as h^{-k} when $h \to 0$. Therefore an alternative might be to use deferred corrections to improve a computed solution. See e.g. [Pereyra].

References:

Bartels, R.H., Stoer, J., Zenger, C.H.: A realization of the simplex method based on triangular decompositions, Grundlehren der mathematischen Wissenschaften in Einzeldarstellungen, 186, Springer, 1971.

Fahlander, K.: Computer programs for semi-infinite optimization, Techn. Rep TRITA-NA-7312, Dept of Numerical Analysis, Royal Institute of Technology, Stockholm, Sweden.

Glasshoff, K., Gustafson, S.-Å.: Numerical treatment of a parabolic boundary-value problem, Techn Rep TRITA-NA-7409, Dept of Numerical Analysis, Royal Institute of Technology, Stockholm, Sweden.

Gustafson, S.-Å., in: C.A. Hall and G.B. Byrnes (eds): Numerical solution of systems of nonlinear algebraic equations, Academic Press, New York 1973, 63-99.

Gustafson, S.-Å., Kortanek, K.O.: NRLQ 20, 477-504 (1973).

Lindberg, B.: SIAM J. Numer. Anal. 9, 662-668 (1972).

Pereyra, V.: Numer. Math. 8, 376-391 (1966).

Stetter, H.J.: Numer. Math. 7, 18-31 (1965).

Penalty-Methoden für Kontrollprobleme und Open-Loop-Differential-Spiele

Joachim Hartung (*)

Das Prinzip der wohl auf COURANT [3] zurückgehenden Penalty-Methode ist folgendes: Eine Aufgabe mit Nebenbedingungen wird äquivalent zu einer Folge von Aufgaben mit weniger oder gar keinen Nebenbedingungen.

In dieser Arbeit wird gezeigt, daß diese Methode auch anwendbar ist bei Differential-Spielen, in denen die Strategien (Kontrollen, Steuerungen) nicht vom Zustand des Systems abhängig sind (Open-Loop-Strategien). Dabei zeigt sich, daß die Methode selbst ohne Konvexitätsbedingungen für die auftretenden Funktionen auskommt und diese im wesentlichen nur gebraucht werden, um die Existenz von Sattelpunkten zu sichern. Für (nicht unbedingt konvexe) Minimum-Kontrollprobleme werden dann einige - im Vergleich zu Spielen weitergehende - Konvergenzeigenschaften der Penalty-Methode gezeigt.

Sei der 'Kontrollraum' oder 'Raum der Steuerungen' U des Spielers I eine Teilmenge vom Hilbertraum $L_2^m[0,1]$ der m-dimensionalen reellen vektorwertigen Funktionen, die definiert, meßbar und quadrat-integrierbar auf $[0,1] \subset \mathbb{R}$ sind, wobei das Skalarprodukt in natürlicher Weise definiert ist durch

$$\langle u_1, u_2 \rangle = \int_0^1 \sum_{i=1}^m u_{i1}(t) \cdot u_{i2}(t)\, dt \;, \quad \text{für} \quad u_1, u_2 \in L_2^m[0,1] \;.$$

Entsprechend sei der Kontrollraum V von Spieler II eine Teilmenge von

(*) Institut für Angewandte Mathematik der Universität Bonn
Abteilung für Wahrscheinlichkeitstheorie und Statistik,
Sonderforschungsbereich 72
(53) Bonn, Wegelerstraße 6

$L_2^n[0,1]$. ($U \neq \emptyset$, $V \neq \emptyset$). Der 'Zustandsraum' oder 'Trajektorienraum' X des Systems, das wir betrachten werden, sei eine Teilmenge des Raumes der q-dimensionalen auf $[0,1]$ definierten reellen vektorwertigen Funktionen.

Das dynamische Verhalten des Systems, oder wie man auch sagen kann, des Spieles, ist bestimmt durch ein System von gewöhnlichen Differentialgleichungen

$$\dot{x} = \frac{dx}{dt} = f(t,x,u,v) \text{ , wobei}$$

$$f: [0,1] \times X \times U \times V \to \mathbb{R}^q \text{ .}$$

Definieren wir nun ein 'Kostenfunktional' oder eine 'Auszahlung'

$$\tilde{p}: X \times U \times V \to \mathbb{R} \text{ ,}$$

so erhalten wir das Open-Loop-Differentialspiel

$$(\tilde{G}) \qquad (U,\ V,\ \tilde{p}(x,u,v)) \text{ ,}$$

mit den Zustandsgleichungen

$$(1) \qquad \dot{x} = f(t,x,u,v)$$

und den Anfangsbedingungen

$$(2) \qquad x(0) = x_o \text{ , } \quad x_o \in \mathbb{R}^q \text{ ,}$$

wobei Spieler I mit $u \in U$ die Auszahlung $\tilde{p}$ maximiert, und Spieler II $\tilde{p}$ mit $v \in V$ minimiert, unter Beachtung von (1) und (2).

Wir nehmen an, das System (1), (2) hat eine eindeutige Lösung

$$(3) \qquad x(t) = x_o + \int_o^t f(s,\ x(s),\ u(s),\ v(s))\ ds \text{ .}$$

(Hinreichende Bedingungen hierfür sind z.B. in FRIEDMAN [8] angegeben.) x(t) ist die 'Trajektorie in Bezug auf u(t) und v(t)'. Hiermit definieren wir eine neue Auszahlung

$$(4) \qquad p: \begin{array}{ccc} U \times V & \to & \mathbb{R} \\ (u,v) & \to & \tilde{p}(x,u,v) \end{array} \text{ , wobei x durch (3) gegeben ist.}$$

Dann wird das Spiel $(\tilde{G})$ zum Spiel

(G) $(U, V, p(u,v))$.

Eine Strategie $\hat{u} \in U$ ($\hat{v} \in V$) ist 'optimal' für den Spieler I (II), wenn ein $v_o \in V$ ($u_o \in U$) existiert, so daß $(\hat{u},v_o)$ ($(u_o,\hat{v})$) ein Sattelpunkt von p ist, falls einer existiert, d.h.

$$p(\hat{u},v_o) = \max_{u \in U} \min_{v \in V} p(u,v) = \min_{v \in V} \max_{u \in U} p(u,v) = p(u_o,\hat{v})$$

Sattelpunkte heißen 'Lösungen' des Spiels und $W(G) := \sup_u \inf_v p(u,v)$ heißt 'Wert' des Spiels, falls $\sup_u \inf_v p = \inf_v \sup_u p$.

Müssen die Strategien u,v nun zusätzlichen Bedingungen genügen,

$$u \in U_o,\ v \in V_o\ ,\ \text{mit}\ U_o \subset U\ ,\ V_o \subset V\ ,$$

z.B.

(5) $U_o := \{u \in U \mid h(u) \leq 0\}$, für eine Funktion $h: U \to \mathbb{R}^\nu$,

so erhalten wir ein 'Spiel mit Nebenbedingungen'

(G_o) $(U_o, V_o, p(u,v))$.

Ist die Struktur der Mengen U, V einfacher als die von U_o, V_o, und ist es so leichter, Sattelpunkte auf $U \times V$ zu bestimmen als auf $U_o \times V_o$, so schlagen wir zur Lösung von (G_o) eine sequentielle Methode vor, die "lediglich" das Aufsuchen von Sattelpunkten auf $U \times V$ verlangt.

Die 'Penalty-Funktionen'

$$P_U: U \to \mathbb{R}\ ,\quad P_V: V \to \mathbb{R}$$

haben die Eigenschaft

$$P_U(u) = \begin{cases} 0 & \text{für } u \in U_o \\ >0 & \text{sonst} \end{cases}\ ,\quad P_V(v) = \begin{cases} 0 & \text{für } v \in V_o \\ >0 & \text{sonst} \end{cases}\ .$$

z.B., wenn U_o durch (5) gegeben ist,

$$P_U(u) = \sum_{i=1}^{\nu} \max\,(0, h_i(u))^{1+\delta}\ ,\quad \delta \geq 0\ .$$

Für $r_n \in \mathbb{R}$, $r_n > 0$, $n \in \mathbb{N}$, definieren wir

$$\text{(6)} \qquad p_n : \begin{array}{lcl} U \times V & \to & \mathbb{R} \\ (u,v) & \to & p(u,v) + r_n(P_V(v) - P_U(u)) \end{array}$$

und die 'Penalty-Spiele'

(G_n) $\qquad (U, V, p_n)$.

Die <u>Penalty Methode</u> ist dann wie folgt: Wähle eine positive reelle Folge $\{r_n\}_{n\in\mathbb{N}}$ mit $r_n \to +\infty$, für $n \to \infty$, und löse die Spiele (G_n), $n \in \mathbb{N}$.

Unter geeigneten Bedingungen konvergieren dann optimale Strategien von (G_n) gegen optimale Strategien von (G_0), und entsprechend die zugehörigen Werte.

Sei von jetzt ab die Folge $\{r_n\}$ fest gewählt, so daß $r_n > 0$ und $r_n \to \infty$ für $n \to \infty$.

<u>Satz 1.</u>

Seien entweder in der starken oder in der schwachen Topologie folgende Bedingungen erfüllt:

U und V sind abgeschlossen.

(7) $P_U(u)$, $P_V(v)$ sind halbstetig nach unten.

(8) $p(u,v)$ ist halbstetig nach oben (unten) in u (v).

(9) $\bigvee_{a\in\mathbb{R}} \bigvee_{\tilde{u}\in U_0} \bigwedge_{v\in V} \quad a \leq p(\tilde{u},v)$

$\bigvee_{b\in\mathbb{R}} \bigvee_{\tilde{v}\in V_0} \bigwedge_{u\in U} \quad b \geq p(u,\tilde{v})$

(10) (G_n) hat eine Lösung (u_n,v_n), $n \in \mathbb{N}$, (nicht unbedingt eindeutig).

(11) Die Folge $\{(u_n,v_n)\}_{n\in\mathbb{N}}$ hat einen Häufungspunkt $(\hat{u},\hat{v})$, so daß für eine Unterfolge $\{(u_{n_k},v_{n_k})\}$ gilt:

$$\hat{u} = \lim_{k\to\infty} u_{n_k}$$

$$\hat{v} = \lim_{k\to\infty} v_{n_k}$$

Dann gilt:

(12) $$\lim_{n\to\infty} \frac{W(G_n)}{r_n} = 0$$

(13) $$\lim_{n\to\infty} P_U(u_n) = 0 \, , \qquad \lim_{n\to\infty} P_V(v_n) = 0$$

(14) (G_o) hat einen Wert, und es gilt $\lim_{k\to\infty} W(G_{n_k}) = W(G_o)$.

(15) $(\hat{u},\hat{v})$ ist eine Lösung des Spiels (G_o).

<u>Beweis:</u>

Es ist

$$W_n := W(G_n) = p_n(u_n,v_n) = p(u_n,v_n) + r_n(P_V(v_n)-P_U(u_n))$$

und es gilt

$$\bigwedge_{u\varepsilon U} \bigwedge_{v\varepsilon V} p_n(u,v_n) \leq W_n \leq p_n(u_n,v) \quad .$$

Da $U_o \subset U$, $V_o \subset V$ und P_U und P_V auf U_o bzw. V_o identisch verschwinden, ergibt sich

(16) $$\bigwedge_{u\varepsilon U_o} \bigwedge_{v\varepsilon V_o} p(u,v_n) + r_n P_V(v_n) \leq W_n \leq p(u_n,v) - r_n P_U(u_n) \; .$$

Wegen $P_V \geq 0$, $P_U \geq 0$ folgt daraus

(17) $$\bigwedge_{u\varepsilon U_o} \bigwedge_{v\varepsilon V_o} p(u,v_n) \leq W_n \leq p(u_n,v) \; ,$$

und mit (9)

(18) $$a \leq p(\tilde{u},v_n) \leq W_n \leq p(u_n,\tilde{v}) \leq b \; ,$$

also $\frac{W_n}{r_n} \xrightarrow[n\to\infty]{} 0$, womit (12) bewiesen ist.

Hiermit gilt dann wegen (9) und (16)

$$\frac{a}{r_n} + P_V(v_n) \leq \frac{W_n}{r_n} \leq \frac{b}{r_n} - P_U(u_n)$$

und

(19) $$0 \leq \overline{\lim_{n\to\infty}} P_V(v_n) \leq \overline{\lim_{n\to\infty}} \frac{W_n}{r_n} = 0 = \underline{\lim_{n\to\infty}} \frac{W_n}{r_n} \leq \underline{\lim_{n\to\infty}} (-P_U(u_n))$$

$$= - \overline{\lim_{n\to\infty}} P_U(u_n) \leq 0 \; ,$$

woraus dann (13) folgt.

Mit (7) erhält man aus (19)

$$P_V(\hat{v}) \leq \varliminf_{k\to\infty} P_V(v_{n_k}) \leq 0 \;,$$

$$P_U(\hat{u}) \leq \varliminf_{k\to\infty} P_U(u_{n_k}) \leq 0 \;.$$

$\hat{u}$, $\hat{v}$ sind also zulässige Strategien im Spiel (G_o), d.h. $\hat{u} \varepsilon U_o$, $\hat{v} \varepsilon V_o$.

Aus (17) folgt

$$(20) \qquad \sup_{u\varepsilon U_o} p(u,v_n) \leq W_n \leq \inf_{v\varepsilon V_o} p(u_n,v)$$

Die Funktionen $v \to \sup_{u\varepsilon U_o} p(u,v)$ und $u \to \inf_{v\varepsilon V_o} p(u,v)$ sind halbstetig nach unten bzw. nach oben, so daß

$$(21) \qquad \sup_{u\varepsilon U_o} p(u,\hat{v}) \leq \varliminf_{k\to\infty} \sup_{u\varepsilon U_o} p(u,v_{n_k}) \leq \varliminf_{k\to\infty} W_{n_k}$$

$$\leq \varlimsup_{k\to\infty} W_{n_k} \leq \varlimsup_{k\to\infty} \inf_{v\varepsilon V_o} p(u_{n_k},v) \leq \inf_{v\varepsilon V_o} p(\hat{u},v)$$

Wegen $\hat{u} \varepsilon U_o$, $\hat{v} \varepsilon V_o$ erhalten wir hieraus

$$p(\hat{u},\hat{v}) \leq \sup_{u\varepsilon U_o} p(u,\hat{v}) \leq \varliminf_{k\to\infty} W_{n_k}$$

$$\leq \varlimsup_{k\to\infty} W_{n_k} \leq \inf_{v\varepsilon V_o} p(\hat{u},v) \leq p(\hat{u},\hat{v}) \;.$$

Das heißt aber, $(\hat{u},\hat{v})$ ist ein Sattelpunkt von p auf $U_o \times V_o$ und $W_{n_k} \xrightarrow[k\to\infty]{} W(G_o)$. (w.z.b.w.)

<u>Satz 2.</u>

Seien die Voraussetzungen (7), (8) und (10) von Satz 1. erfüllt und

U und V kompakt (in der gleichen Topologie, in der (7) und (8) erfüllt sind) .

Dann gilt neben (13):

(22) Jeder Häufungspunkt (es existiert mindestens einer) der Folge $\{u_n\}$ $(\{v_n\})$ ist optimale Strategie des Spielers I (II) im Spiel (G_o), und

(23) $\lim\limits_{n\to\infty} W(G_n) = W(G_o)$.

Beweis:

Da U und V kompakt sind, ist (9) wegen (8) für alle $\tilde{u} \in U$ und $\tilde{v} \in V$ erfüllt.

Die Folge $\{u_n\}$ der optimalen Strategien des Spielers I enthält, da U kompakt ist, mindestens einen Häufungspunkt und eine dagegen konvergierende Unterfolge $\{u_{n_k}\}$. Die zugehörige Folge $\{v_{n_k}\}$ von optimalen Strategien des Spielers II braucht natürlich nicht zu konvergieren. Da V aber kompakt ist, enthält sie eine konvergente Teilfolge $\{v_{n_{k_i}}\}$. Die entsprechende Folge $\{u_{n_{k_i}}\}$ ist selbstverständlich konvergent, und mit

$$\hat{u} := \lim_{i\to\infty} u_{n_{k_i}} , \qquad \hat{v} := \lim_{i\to\infty} v_{n_{k_i}}$$

ist (11) dann erfüllt. Wegen (15) ist $\hat{u}$ dann optimale Strategie in (G_o). Entsprechendes gilt für Häufungspunkte von $\{v_n\}$.

Seien jetzt S und T die Mengen der Häufungspunkte von $\{u_n\}$ und $\{v_n\}$ jeweils, $\mathcal{S}$ und $\mathcal{T}$ folgende Indexmengen

$$\mathcal{S} := \{s \mid \{u_{n_{s_i}}\} \subset \{u_n\}, \text{ es existiert } \lim_{i\to\infty} u_{n_{s_i}} =: u_s,\ u_s \in S\},$$

$$\mathcal{T} := \{t \mid \{v_{n_{t_j}}\} \subset \{v_n\}, \text{ es existiert } \lim_{j\to\infty} v_{n_{t_j}} =: v_t,\ v_t \in T\}.$$

Wir erhalten aus (20) aufgrund der Kompaktheit von U, V

$$\begin{aligned}
(24) \qquad -\infty &< \inf_{v\in T} \sup_{u\in U_o} p(u,v) \le \inf_{t\in\mathcal{T}} \sup_{u\in U_o} p(u,v_t) \\
&\le \inf_{t\in\mathcal{T}} \underline{\lim}_{j\to\infty} \sup_{u\in U_o} p(u,v_{n_{t_j}}) \\
&\le \underline{\lim}_{n\to\infty} \sup_{u\in U_o} p(u,v_n) \le \underline{\lim}_{n\to\infty} W_n \\
&\le \overline{\lim}_{n\to\infty} W_n \le \overline{\lim}_{n\to\infty} \inf_{v\in V_o} p(u_n,v) \\
&\le \sup_{s\in\mathcal{S}} \overline{\lim}_{i\to\infty} \inf_{v\in V_o} p(u_{n_{s_i}},v)
\end{aligned}$$

$$\leq \sup_{s\in\gamma}\ \inf_{v\in V_o}\ p(u_s,v)$$

$$\leq \sup_{u\in S}\ \inf_{v\in V_o}\ p(u,v) \quad < \ +\infty .$$

Nach (22) gilt natürlich

$$S \subset U_o \ , \quad T \subset V_o \ ,$$

und (24) ergibt dann

$$(25) \qquad -\infty \ < \ \inf_{v\in V_o}\ \sup_{u\in U_o}\ p(u,v) \ \leq \ \inf_{v\in T}\ \sup_{u\in U_o}\ p(u,v)$$

$$\leq \ \underline{\lim}_{n\to\infty}\ W_n \ \leq \ \overline{\lim}_{n\to\infty}\ W_n$$

$$\leq \ \sup_{u\in S}\ \inf_{v\in V_o}\ p(u,v) \ \leq \ \sup_{u\in U_o}\ \inf_{v\in V_o}\ p(u,v) \ < \ +\infty \ .$$

Andererseits gilt die triviale Ungleichung

$$\sup_{u\in U_o}\ \inf_{v\in V_o}\ p(u,v) \ \leq \ \inf_{v\in V_o}\ \sup_{u\in U_o}\ p(u,v) \ .$$

Also haben wir in (25) überall die Gleichheit und somit

$\lim_{n\to\infty} W_n = W(G_o)$. (w.z.b.w.)

Sind U und V nicht beschränkt, so heißt $p(u,v)$ 'koerziv' in $(\tilde{u},\tilde{v}) \in U \times V$, wenn

$$p(\tilde{u},v) - p(u,\tilde{v}) \ \to \ +\infty \ , \qquad \text{für } \|u\| + \|v\| \to \ +\infty \ .$$

<u>Satz 3.</u>

Seien U_o und V_o nicht leer,

U und V stark abgeschlossen und konvex,

$P_U(u)$, $P_V(v)$ stark halbstetig nach unten und konvex,

$p(u,v)$ stark halbstetig nach oben (unten) und konkav (konvex) in u (v),

U und V beschränkt oder p koerziv in einem $(\tilde{u},\tilde{v}) \in U_o\times V_o$.

Dann gilt:

(i) (G_n) hat eine Lösung (u_n, v_n), $n \in \mathbb{N}$.

(ii) Jede Teilfolge von $\{u_n\}$, bzw. $\{v_n\}$ hat schwache Häufungspunkte und diese sind optimale Strategien für (G_o).

(iii) $P_U(u_n) \to 0$, $P_V(v_n) \to 0$, für $n \to \infty$.

(iv) $W(G_n) \to W(G_o)$, $n \to \infty$.

Beweis:

Sind U und V beschränkt, so sind sie schwach kompakt. $p_n(u,v)$ ist schwach halbstetig nach oben (unten) und konkav (konvex) in u (v). Nach dem VON NEUMANN [20] - KY FAN [14] Sattelpunkttheorem hat (G_n) dann eine Lösung. Somit ist Satz 2. in der schwachen Topologie anwendbar und liefert obige Behauptungen.

Seien jetzt U und V nicht beschränkt. Die Koerzivität von p in $(\tilde{u},\tilde{v})$ bewirkt, daß die Niveau-Mengen

$$\{u \in U \mid p(u,\tilde{v}) \geq a\}, \; a \in \mathbb{R},$$

und

$$\{v \in V \mid p(\tilde{u},v) \leq b\}, \; b \in \mathbb{R},$$

beschränkt sind. Mit obigen Voraussetzungen sind sie auch stark abgeschlossen und konvex, also schwach kompakt.

Es existieren dann:

$$\max_{u \in U} p(u,\tilde{v}) \;, \quad \min_{v \in V} p(\tilde{u},v) \;.$$

(Bedingung (9) ist hiermit erfüllt.)

Aus

$$\{p_n(u,\tilde{v}) \geq a \implies p(u,\tilde{v}) \geq a\}$$

und

$$\{p_n(\tilde{u},v) \leq b \implies p(\tilde{u},v) \leq b\}$$

folgt, daß auch die Niveau-Mengen

$$\{u \in U \mid p_n(u,\tilde{v}) \geq a\} \;, \; a \in \mathbb{R} \;,$$

$$\{v \in V \mid p_n(\tilde{u},v) \leq b\} \;, \; b \in \mathbb{R} \;,$$

schwach kompakt sind.

Nach Sattelpunkttheoremen z.B. von MOREAU [17], ROCKAFELLAR [18] folgt dann (i).

Wegen

$$p(u_n,\hat{\tilde{v}}) \leq \max_{u\in U} p(u,\tilde{v}) < +\infty$$

$$p(\tilde{u},v_n) \geq \min_{v\in V} p(\tilde{u},v) > -\infty$$

gibt es schwach kompakte Mengen U_1, V_1 mit der Eigenschaft

$$\{u_n\}_{n\in\mathbb{N}} \subset U_1 \subset U \ , \qquad \{v_n\} \subset V_1 \subset V$$

und

$$S \subset U_1 \ , \qquad T \subset V_1 \ ,$$

wobei S und T wieder die Mengen der jeweiligen Häufungspunkte seien. Analog zum Beweis von (22) folgt zunächst (11) und dann (ii). Aus Satz 1. ergibt sich somit (iii) und die Existenz einer Teilfolge $\{n_k\} \subset \mathbb{N}$, so daß gilt

$$\lim_{k\to\infty} W(G_{n_k}) = W(G_o) \ .$$

Der Konvergenzbeweis zu (iv) folgt aus Satz 2., wenn man dort ersetzt:

$$(U,V,p_n) \quad \text{durch} \quad (U_1,V_1,p_n) =: (G_n^1)$$

und

$$(U_o,V_o,p) \text{ durch } (S,T,p) =: (G_o^1) \ .$$

Wir erhalten dann aus (23):

$$\lim_{n\to\infty} W(G_n^1) = W(G_o^1)$$

Wegen

$$W(G_n) = p_n(u_n,v_n) \quad \text{und} \quad (u_n,v_n) \in U_1 \times V_1 \ ,$$

ist

$$W(G_n) = W(G_n^1)$$

und somit gilt

$$\lim_{n\to\infty} W(G_n) = W(G_o^1) \ .$$

Da nun aber eine Teilfolge von $\{W(G_n)\}_{n\in\mathbb{N}}$ gegen $W(G_o)$ konvergiert, muß $W(G_o^1) = W(G_o)$ sein. (w.z.b.w.)

Eine Funktion $f: U \to \mathbb{R}$ heißt 'gleichmäßig konvex', wenn U konvex ist und es eine stetige monotone Funktion $\delta: [0,\infty) \to [0,\infty)$ gibt mit $\delta(0) = 0$, $\delta(t) > 0$ für $t > 0$, so daß für alle $u_1, u_2 \in U$ gilt

$$f((u_1+u_2)/2) \leq \frac{1}{2} f(u_1) + \frac{1}{2} f(u_2) - \delta(\|u_1-u_2\|) .$$

Entsprechend heißt f 'gleichmäßig konkav', wenn (-f) gleichmäßig konvex ist.

Satz 4.

Seien U_o und V_o nicht leer,
U, V stark abgeschlossen und konvex,
$p_U(u)$, $p_V(v)$ stark halbstetig nach unten und konvex,
p(u,v) gleichmäßig konkav (konvex), stark halbstetig und beschränkt nach oben (unten) in u (v) .

Dann gilt über die Aussagen von Satz 3. hinaus:

(i) (G_o) hat eine eindeutige Lösung (u_o, v_o).

(ii) (G_n) hat eine eindeutige Lösung (u_n, v_n), $n \in \mathbb{N}$.

(iii) u_n (v_n) konvergiert stark gegen u_o (v_o), für $n \to \infty$.

Beweis:

Genügt p den Voraussetzungen, so auch p_n. p und p_n sind koerziv in allen $(\tilde{u},\tilde{v}) \in U \times V$, falls U,V nicht beschränkt sind.
Die Voraussetzungen von Satz 3. sind erfüllt. Da gleichmäßig konvexe Funktionen streng konvex sind, folgt daraus dann (i) und (ii).
Ebenfalls nach Satz 3., bzw. dem Beweis zu (22), hat jede Teilfolge von $\{v_n\}_{n\in\mathbb{N}}$ eine konvergente Teilfolge $\{v_{n_k}\}_{k\in\mathbb{N}}$, so daß gilt

$$u_{n_k} \to u_o , \quad v_{n_k} \to v_o ,$$

und

$$W(G_{n_k}) \to W(G_o) , \quad \text{für } k \to \infty .$$

Aus der Definition der gleichmäßigen Konvexität folgt für feste $u \in U$ die Existenz einer Funktion $\delta_u: [0,\infty) \to [0,\infty)$, $\delta_u(0) = 0$, $\delta(t) > 0$ für $t > 0$, so daß

$$(*) \qquad \delta_u(\|v_{n_k} - v_o\|) \le \frac{1}{2}\, p(u,v_{n_k}) + \frac{1}{2}\, p(u,v_o) - p(u, (v_{n_k}+v_o)/2)$$

Nach (17) ist

$$p(u,v_{n_k}) \le W_{n_k} = W(G_{n_k}) \; , \quad \text{für alle} \quad u \in U_o \; .$$

Nehmen wir nun für u speziell die optimale Strategie $u_o \in U_o$, deren Existenz nach (i) gesichert ist, so erhalten wir aus (*)

$$(**) \qquad \delta_{u_o}(\|v_{n_k}-v_o\|) \le \frac{1}{2}\, W_{n_k} + \frac{1}{2}\, p(u_o,v_o) - p(u_o, (v_{n_k}+v_o)/2) \; .$$

Mit $v_{n_k} \rightharpoonup v_o$ gilt natürlich: $(v_{n_k}+v_o)/2 \rightharpoonup v_o$. Die untere Halbstetigkeit von p in v ergibt

$$\underline{\lim}_{k\to 0}\, p(u_o, (v_{n_k}+v_o)/2) \ge p(u_o,v_o)$$

Unter Beachtung von $p(u_o,v_o) = W(G_o)$ erhalten wir dann aus (**)

$$\begin{aligned} 0 &\le \overline{\lim}_{n\to\infty}\, \delta_{u_o}(\|v_{n_k} - v_o\|) \\ &\le \frac{1}{2}\, \overline{\lim}_{k\to\infty}\, W_{n_k} + \frac{1}{2}\, W(G_o) - \underline{\lim}_{k\to\infty}\, p(u_o, (v_{n_k}+v_o)/2) \\ &\le \frac{1}{2}\, W(G_o) + \frac{1}{2}\, W(G_o) - W(G_o) = 0 \; , \end{aligned}$$

also $\|v_{n_k} - v_o\| \to 0$, für $k \to \infty$.

Da dies für jede schwach konvergente Teilfolge gilt, jede Folge eine schwach konvergente Teilfolge hat und v_o eindeutig ist, konvergiert somit v_n stark gegen v_o.

Entsprechend zeigt man die Konvergenz der u_n. (w.z.b.w.)

Zur Beschränktheitsbedingung der Funktion p machen wir noch folgende

(26) <u>Bemerkung:</u> Ist eine auf einem reflexiven Banachraum gleichmäßig konvexe Funktion stetig, so ist sie beschränkt nach unten. (LEVETIN, POLJAK [15])

Ist eine auf einer konvexen Teilmenge D eines Banachraums konvexe Funktion halbstetig nach unten, so ist sie auf dem rel int(D) stetig, vorausgesetzt rel int(D) $\neq \emptyset$. (BRØNSTED [2])

Anhand zweier Beispiele soll nun die Anwendbarkeit obiger Sätze gezeigt werden.

a) Sei x linear gesteuert,

$$(27) \qquad \dot{x} = A(t)x(t) + B(t)u(t) + C(t)v(t) \;, \quad x(0) = x_o \;, \; 0 \le t \le 1,$$

mit geeigneten Matrizen A, B, C.

Diese Gleichung hat eine eindeutige Lösung auf $[0,1]$:

$$(28) \qquad x(t) = Y(t)\,(x_o + \int_0^t Y^{-1}(s)\,(B(s)u(s) + C(s)v(s))\,ds \;,$$

wobei Y die Fundamentallösung des linearen homogenen Systems ist.

Die auftretenden Matrizen mögen stetig sein.

Mit den konvexen Funktionen

$$H(t,.) : \mathbb{R}^m \to R^I \;, \quad K(t,.) : \mathbb{R}^n \to R^J \;, \quad 0 \le t \le 1 \;,$$

wobei $I \subset \mathbb{R}$, $J \subset \mathbb{R}$ vorgegebene Indexmengen sind, seien die Steuerungen u und v beschränkt:

$$(29) \qquad U_o = \{u \in U \mid \int_0^1 H_i(t, u(t))\,dt \le 0, \text{ für alle } i \in I\}$$

$$V_o = \{v \in V \mid \int_0^1 K_j(t, v(t))\,dt \le 0, \text{ für alle } j \in J\} \;.$$

U_o und V_o seien nicht leer, U und V konvex und stark abgeschlossen.

Als Penalty-Funktionen nehmen wir zum Beispiel

$$(30) \qquad P_U(u) = \max\,(0, \sup_{i \in I} \int_0^1 H_i(t, u(t))\,dt) \;,$$

$$P_V(v) = \max\,(0, \sup_{j \in J} \int_0^1 K_j(t, v(t))\,dt) \;.$$

P_U und P_V sind konvex und halbstetig nach unten. (Wir nehmen an, daß sie überall auf U, V wohl definiert sind.)

Die Auszahlung hänge ab zum einen vom Endpunkt der Trajektorie, den der Spieler I (II) möglichst groß (klein) halten will, zum anderen vom Energieverbrauch, den beide Spieler möglichst gering halten wollen, z.B.

$$\tilde{p}(x,u,v) = \sum_{i=1}^{q} x_i(1) + \int_0^1 \sum_{j=1}^{n} v_j^2(t)\,dt - \int_0^1 \sum_{k=1}^{m} u_k^2(t)\,dt \;.$$

x(1) ist wegen (28) eine schwach stetige Funktion von u bzw. v und linear in u und v.

$p(u,v) = \tilde{p}(x,u,v)$ ist dann glm. konkav - glm. konvex und stetig in beiden Variablen. Mit (26) sind so die Voraussetzungen von Satz 4. erfüllt, und wir erhalten unter anderem: Das Spiel

$$(\tilde{G}) = \{(U_o, V_o, \tilde{p}(x,u,v)), \text{ x genügt (27)}\}$$

hat eine Lösung , und die (immer existierenden) Lösungen der Penalty -Spiele

$$\{(U, V, \tilde{p}_n(x,u,v)), \text{ x genügt (27)}\}$$

konvergieren stark gegen die Lösung von $(\tilde{G})$, selbst wenn U und V die ganzen Räume $L_2^m[0,1]$, bzw. $L_2^n[0,1]$ sind.

b) Sei x wieder durch (27) bzw. (28) gegeben, der Zustandsraum $X = C^q[0,1]$, d.h. $x_i(t)$ ist stetig auf $[0,1]$, $i = 1,\dots,q$, und X mit der Supremumsnorm ausgestattet.

Die Funktion $h(t,.): \mathbb{R}^m \times \mathbb{R}^n \to \mathbb{R}$, $0 \le t \le 1$, sei konkav-konvex, und die Auszahlung sei gegeben durch

$$\tilde{p}(x,u,v) = \int_0^1 \sum_{i=1}^q a_i(t)x_i(t) + h(t, u(t), v(t))\, dt \quad ,$$

für eine feste Ladungsverteilung a(t).

Die Steuerungen mögen in keinem Punkt unbeschränkt groß werden können, d.h. z.B.

$$(31) \qquad U = \{u \in L_2^m[0,1] \mid |u_i(t)| \le 1,\ i = 1,\dots,m\}$$

$$V = \{v \in L_2^n[0,1] \mid |v_j(t)| \le 1,\ j = 1,\dots,n\} .$$

U und V sind konvex, beschränkt und schwach abgeschlossen.

Die Abbildung

$$(u,v) \to x$$

von $U \times V$ in $C^q[0,1]$,definiert durch (28), ist stetig in jeder Variablen in Bezug auf die schwache Topologie des $L_2[0,1]$. Da sie außerdem linear in jeder Variablen ist, ist

$p(u,v) = \tilde{p}(x,u,v)$ konkav-konvex und schwach halbstetig nach oben (unten) in u (v) auf den schwach kompakten Mengen U und V.

Sind die Gesamtkapazitäten der einzelnen Komponenten der Steuerungen beschränkt, etwa

$$(32) \qquad \int_0^1 u_i^2(t)\,dt \le c_i \;, \qquad i = 1,\dots,m \;,$$

$$\int_0^1 v_j^2(t)\,dt \le d_j \;, \qquad j = 1,\dots,n \;,$$

so haben U_o, V_o, gegeben durch (31) und (32), numerisch unangenehmere Struktur als U,V und es bietet sich so, schon für diesen einfachen und häufig auftretenden Fall von Nebenbedingungen, die Penalty-Methode an, etwa mit

$$P_U(u) = \sum_{i=1}^{m} \max\,(0, \int_0^1 u_i^2(t)\,dt - c_i) \;,$$

$$P_V(v) = \sum_{j=1}^{n} \max\,(0, \int_0^1 v_j^2(t)\,dt - d_j) \;.$$

Penalty-Methoden bei Kontrollproblemen

Ist nur ein Spieler in Aktion, sagen wir Spieler II, z.B. wenn der Spieler I nur eine Strategie u_o zur Verfügung hat ($U = \{u_o\}$), so werden die Open-Loop-Differential-Spiele zu gewöhnlichen Kontrollproblemen. Die wesentlichen Aussagen über Penalty-Methoden bei Kontrollproblemen sind somit in obigen Sätzen enthalten. Zu (13), (22) und (23) analoge Aussagen für Kontrollprobleme wurden bereits von BELTRAMI [1] und GLASHOFF [10] angegeben. Jedoch benötigt ersterer im Beweis die schwache Stetigkeit des Zielfunktionals, und letzterer setzt die schwache Stetigkeit der Penalty-Funktionen voraus, was z.B. im allgemeinen nicht erlaubt, Zustandsbeschränkungen in dem Penalty-Term aufzunehmen, wenn der Steuerungsoperator nicht kompakt, bzw. schwach stetig ist.

Der Vollständigkeit halber geben wir hier einen Konvergenzsatz für die Penalty-Methoden bei Kontrollproblemen an, zumal es sich zeigt, daß die Zielfunktion selber in den Minimalstellen der 'unrestringierten' Probleme konvergiert und die Penalty-Funktion dort schneller gegen 0 konvergiert als r_n gegen ∞, obwohl die Folge der Minimalstellen selbst nicht zu konvergieren braucht.

Wir behalten die bisherige Bezeichnung bei, lassen die Variable u weg und schreiben $P(v)$ statt $P_V(v)$.

Satz 5.

Gegeben sei das Kontrollproblem

(K) $\min \{p(v) \mid v \in V,\ P(v) = 0\}, \quad P \geq 0$.

Dabei sei V nicht leer, konvex, abgeschlossen und beschränkt, p und P schwach nach unten halbstetig, und es gebe einen Punkt $\tilde{v} \in V$ mit $P(\tilde{v}) = 0$.
$\{r_n\}_{n\in\mathbb{N}} \subset \mathbb{R}$, $r_n > 0$, sei eine beliebige Folge mit $r_n \to \infty$ für $n \to \infty$.

Dann gilt:

(i) Die 'unrestringierten' Probleme

(K_n) $\min \{p_n(v) = p(v) + r_n P(v) \mid v \in V\}$

haben eine Lösung v_n, $n \in \mathbb{N}$.

(ii) Es existiert eine Lösung v_o des Ausgangsproblems (K) gegen die eine Teilfolge $\{v_{n_k}\}$ von $\{v_n\}$ schwach konvergiert.

(iii) $\lim_{n\to\infty} p_n(v_n) = p(v_o)$

(iv) $\lim_{n\to\infty} p(v_n) = p(v_o)$

(v) $\lim_{n\to\infty} r_n P(v_n) = 0$

(Natürlich folgt (iii) aus (iv) und (v)).

Beweis:

V ist schwach kompakt, p_n nimmt so das Minimum über V an. (i) ist hier äquivalent zur Bedingung (10), so daß (ii) und (iii) aus Satz 2.folgen, wenn man beachtet, daß hier $W(G_n) = p_n(v_n)$ und $W(G_o) = p(v_o)$ ist.

Nach (17) gilt für alle $v \varepsilon V_o$, ($V_o = \{v \varepsilon V \mid P(v) = 0\}$),

$$p(v_n) \leq p_n(v_n) \leq p(v) .$$

Aus (24) folgt

$$\inf_{v \varepsilon T} p(v) \leq \underline{\lim}_{n \to \infty} p(v_n) ,$$

wobei T wieder die Menge der Häufungspunkte von $\{v_n\}$ ist.

Nun wissen wir aber von (22), daß T in V_o enthalten ist.

Dies ergibt, für alle $v \varepsilon V_o$,

$$\inf_{v \varepsilon V_o} p(v) \leq \inf_{v \varepsilon T} p(v) \leq \underline{\lim}_{n \to \infty} p(v_n) \leq \overline{\lim}_{n \to \infty} p(v_n) \leq p(v)$$

Wegen (ii) ist $\inf_{v \varepsilon V_o} p(v) = p(v_o)$, $v_o \varepsilon V_o$, so daß (iv) hiermit bewiesen ist.

Es gilt die Abschätzung

$$p(v_n) \leq p_n(v_n) - r_n P(v_n) \leq p_n(v_n) .$$

Mit (iii) und (iv) folgt daraus dann (v). (w.z.b.w.)

Literatur

[1] Beltrami, E.J., An Algorithmic Approach to Nonlinear Analysis and Optimization, Acad. Press, New York, 1970

[2] Brønsted, A., Conjugate Convex Functions in Topological Vector Spaces, Mat.-Fys.-Medd. Danske. Vid. Selk. 34, 1964

[3] Courant, R., Variational Methods for the Solution of Problems of Equilibration and Vibration, Bull. AMS 49, 1943

[4] Demyanov, V.F., The Solution of some Optimal Control Problems, SIAM J. Control, 6, 1968

[5] --- , - , Rubinov, A.M., Approximate Methods in Optimization Problems, Elsevier, New York, 1970

[6] Dugošia, Đ., Hartung, J., Eine Penalty-Methode für antagonistische Spiele, erscheint in Oper. Res. Verf.

[7] Fiacco, A.V., Mc Cormick, G.P., Nonlinear Programming: Sequential Unconstrained Minimization Techniques, Wiley, New York, 1968

[8] Friedman, A., Lectures in Differential Games, in [13]

[9] Göpfert, A., Mathematische Optimierung in allgemeinen Vektorräumen, Teubner, Leipzig, 1973

[10] Glashoff, K., Regularisierung und Penalty-Methoden, Dissertation, Hamburg, 1972

[11] Holmes, R.B., A Course on Optimization and Best Approximation, Springer Lecture Notes 257, Berlin, 1972

[12] Isaacs, R., Differential Games, Wiley, New York, 1965

[13] Kuhn, H.W., Szegö, G.P., (ed.), Differential Games and Related Topics, North-Holland, Elsevier, Amsterdam, New York, 1971

[14] Ky Fan, On General Minimax Theorems, Proc. Nat. Acad. Sci. 39, 1953

[15] Levetin, E.S., Poljak, B.T., Convergence of Minimizing Sequences in Conditional Extremum Problems, Dokl. Akad. Nauk SSSR 168, 5, 1966

[16] Luenberger, D.G., Optimization by Vector Space Methods, Wiley, New York, 1969

[17] Moreau, J.J., Théorèmes 'inf-sup', Compt. Rend. Acad. Sci. Paris 258, 1964

[18] Rockafellar, R.T., Minimax Theorems and Conjugate Saddle-Functions, Math. Scand. 14, 1964

[19] Tihonov, A.N., Regularization Methods for Optimal Control Problems, Dokl. Akad. Nauk SSSR, 162, 4, 1965

[20] von Neumann, A., Zur Theorie der Gesellschaftsspiele, Math. Ann. 100, 1928

[21] Zangwill, W.I., Nonlinear Programming via Penalty Functions, Man. Sci. 13, 5, 1967

Ein Existenzsatz für Konvexe Optimierungsprobleme

Gerhard Heindl

Unter einer üblichen Regularitätsvoraussetzung (Slaterbedingung) wird gezeigt, daß auf einer durch stetige lineare und konvexe Restriktionen bestimmten beschränkten Teilmenge D eines reellen normierten Vektorraumes X, für jedes stetige konvexe Funktional $f : X \rightarrow R$ $f|D$ genau dann ein Minimum besitzt, wenn X reflexiv ist.

I. Bezeichnungen.

X bezeichne einen reellen durch $\|\cdot\|$ normierten Raum, X^* seinen stetigen Dualraum. Zu $x \in X$ und reellem $r > 0$ sei $K_r(x) := \{y \in X : \|y-x\| < r\}$. Für $C \subset X$ bezeichne $\bar{C}$ die abgeschlossene Hülle, C° den offenen Kern von C. N steht für die Menge der natürlichen, R für die Menge der reellen Zahlen. $N_0 := N \cup \{0\}$. Eine konvexe Funktion $f : X \rightarrow R$ wird (nach oben) <u>beschränkt</u> genannt, wenn es ein $y \in X$ und reelle nicht negative Zahlen r, M gibt, so daß für jedes $x \in X$ $f(x) \leq r + M\|x-y\|$ gilt. Nach [2] S. 92 Prop. 2 sind beschränkte konvexe Funktionen stetig.

II. Problemstellung.

Ziel der vorliegenden Arbeit ist der Beweis des folgenden Satzes:

<u>1. Satz: Voraussetzung: Für eine konvexe abgeschlossene Menge</u> $C \subset X$, <u>lineare Funktionale</u> $l_i \in X^*$, $i=1,\dots,k$, <u>konvexe stetige Funktionen</u> $f_i : X \rightarrow R$, $i=1,\dots,m$ <u>und reelle</u> c_i, $i=1,\dots,k$, <u>gelte mit</u>

$$L := \{x \in X : l_i(x) = c_i\ ,\ i=1,\dots,k\} \text{ und}$$
$$K := \{x \in X : f_i(x) \leq 0\ ,\ i=1,\dots,m\} :$$

a) $(C \cap K)^\circ \cap L \neq \emptyset$, b) $D := C \cap K \cap L$ <u>ist beschränkt</u>.

<u>Behauptung: Genau dann besitzt für jedes stetige konvexe Funktional</u> $f : X \rightarrow R$ $f|D$ <u>ein Minimum, wenn</u> X <u>reflexiv ist</u>.

III. Beweis von Satz 1.

Ein Teil der Behauptung von Satz 1 ist einfach zu beweisen und wohlbekannt: Da C, L und K als abgeschlossene und konvexe Mengen auch schwach abgeschlossen sind, ist im Falle der Reflexivität von X D schwach kompakt. Jede auf D eingeschränkte (schwach unterhalb) stetige konvexe Funktion $f : X \rightarrow R$ besitzt dann ein Minimum.
Die Umkehrung wird sich als Spezialfall eines allgemeineren Satzes

(Korollar 5) erweisen, zu dessen Beweis einige Vorbereitungen nötig sind.

2. Lemma: Gilt für eine Folge F von Kugeln $K_{r_i}(x_i) \subset X$, $i \in N_o$, $\overline{K_{r_{i+1}}(x_{i+1})} \subset K_{r_i}(x_i)$ für jedes $i \in N_o$, und $\lim r_i = 0$, so gibt es eine konvexe Funktion $f : X \to R$ mit den Eigenschaften:

1) $\|x-x_o\| \leq f(x) \leq \text{Max}\{r_o, \|x-x_o\|\}$ für alle $x \in X$,

2) $\{x^* \in X : f(x^*) = \inf\{f(x) : x \in X\}\} = \bigcap_{i \in N_o} \overline{K_{r_i}(x_i)}$.

Beweis: Es wird gezeigt, daß mit $c_o := r_o$, $c_1 := \|x_1-x_o\| + r_1$,

$$c_{i+1} := c_i - \frac{(c_{i-1}-c_i)(r_i-r_{i+1}-\|x_{i+1}-x_i\|)}{r_{i-1}-r_i+\|x_i-x_{i-1}\|}, \quad i \in N,$$

$$f_o : X \ni x \mapsto \|x-x_o\| \in R, \text{ und}$$

$$f_i : X \ni x \mapsto c_i + \frac{(c_{i-1}-c_i)(\|x-x_i\| - r_i)}{r_{i-1}-r_i+\|x_i-x_{i-1}\|} \in R, \quad i \in N,$$

$f := f_F := \sup\{f_i : i \in N_o\}$ 1) und 2) erfüllt. Der Nachweis stützt sich auf folgende, durch einfache Induktionsschlüsse und Abschätzungen zu verifizierende Ungleichungen:

Für jedes $i \in N$ gilt:

(I) $c_{i-1} > c_i > 0$,

(II) $f_{i-1}(x) \leq c_{i-1}$ falls $\|x-x_{i-1}\| \leq r_{i-1}$,

(III) $f_{i-1}(x) > c_{i-1}$ falls $\|x-x_{i-1}\| > r_{i-1}$,

(IV) $f_i(x) \leq c_{i-1}$ falls $\|x-x_{i-1}\| \leq r_{i-1}$,

(V) $f_i(x) \leq f_{i-1}(x)$ falls $\|x-x_{i-1}\| \geq r_{i-1}$,

(VI) $f_i(x) \geq f_{i-1}(x)$ falls $\|x-x_i\| \leq r_i$.

Aus (I) folgt die Konvexität der Funktionen f_i, $i \in N$. f genügt genau dann der Bedingung 1), wenn für jedes $i \in N_o$ und $x \in X$

$$f_i(x) \leq \text{Max}\{r_o, \|x-x_o\|\}$$

gilt. Das Bestehen dieser Ungleichungen folgt durch Induktion aus (IV), (I) und (V).

Zu 2): a) Sei $f(x^*) = \inf\{f(x) : x \in X\}$ und $x^* \notin \bigcap_{i \in N_o} \overline{K_{r_i}(x_i)}$ angenommen. Es gibt dann also ein $i \in N_o$ mit $\|x^*-x_i\| > r_i$. (III) zeigt $c_i < f_i(x^*) \leq f(x^*)$. Andererseits hat man für jedes $x \in \overline{K_{r_i}(x_i)}$ und $j \leq i$: $\|x-x_j\| \leq r_j$; infolge (II) und (VI) also $f_j(x) \leq c_i$ für jedes

$j \leq i$. Aus (IV) folgt $f_{i+1}(x) \leq c_i$. Steht für ein $j > i$ bereits $f_j(x) \leq c_i$ fest, so ergibt sich im Fall $\| x-x_j \| \leq r_j$ aus (IV) und (I) $f_{j+1}(x) \leq c_j < c_i$, im Fall $\| x-x_j \| > r_j$ aus (V) $f_{j+1}(x) \leq f_j(x) \leq c_i$. Für jedes $x \in \overline{K_{r_i}(x_i)}$ ist somit $f(x) \leq c_i$, im Widerspruch zu $f(x^*) > c_i$. Es muß daher $x^* \in \bigcap_{i \in N_o} \overline{K_{r_i}(x_i)}$ sein.

b) Ist $y \in \bigcap_{i \in N_o} \overline{K_{r_i}(x_i)}$, so gibt es wegen $\lim r_i = 0$ für jedes $x \in X \setminus \{y\}$ ein $i \in N_o$ mit $\| x-x_i \| > r_i$. Wegen $y \in \overline{K_{r_i}(x_i)}$ gilt $f(y) \leq c_i < f_i(x) \leq f(x)$ (vergl. a)!), d.h. $f(y) = \inf \{ f(x) : x \in X \}$.

<u>3. Korollar: X ist genau dann vollständig, wenn für jede Folge F von Kugeln mit den im Lemma 2 betrachteten Eigenschaften, die im Beweis des Lemmas konstruierte Funktion f_F ein Minimum besitzt.</u>

<u>Beweis:</u> 1. Ist X vollständig und F eine Folge von Kugeln mit den betrachteten Eigenschaften, so ist $\bigcap_{i \in N_o} \overline{K_{r_i}(x_i)}$ einelementig, also auch $\{ x^* \in X : f_F(x^*) = \inf \{ f_F(x) : x \in X \} \}$.

2. Besitze nun für jede Kugelfolge F mit den im Lemma 2 betrachteten Eigenschaften f_F ein Minimum. Jede Cauchyfolge $\{y_v\}_{v \in N_o}$ in X besitzt eine Teilfolge $\{y_{v_i}\}_{i \in N_o}$ mit der Eigenschaft

$$\overline{K_{2^{-i-1}}(y_{v_{i+1}})} \subset K_{2^{-i}}(y_{v_i}) \quad \text{für jedes } i \in N_o$$

(Induktionsbeweis!). Gilt für $F := \{ K_{2^{-i}}(y_{v_i}) \}_{i \in N_o}$ $f_F(x^*) = \inf \{ f_F(x) : x \in X \}$, so folgt aus Lemma 2 $x^* \in \bigcap_{i \in N_o} \overline{K_{2^{-i}}(y_{v_i})}$ und daraus $x^* = \lim y_{v_i} = \lim y_v$.

<u>4. Satz: Gibt es eine schwach abgeschlossene Teilmenge C von X mit $C^o \neq \emptyset$ und der Eigenschaft, daß für jedes beschränkte konvexe Funktional $f : X \to R$ $f|C$ ein Minimum besitzt, so ist X reflexiv.</u>

<u>Beweis:</u> C habe die im Satz betrachteten Eigenschaften. Es wird zuerst die Vollständigkeit von X bewiesen. Sei $\{y_v\}_{v \in N_o}$ eine Cauchyfolge in X, $\{y_{v_i}\}_{i \in N_o}$ eine Teilfolge mit

$$\overline{K_{2^{-i-1}}(y_{v_{i+1}})} \subset K_{2^{-i}}(y_{v_i}) \quad \text{für jedes } i \in N_o,$$

$\overline{K_{r_o}(x_o)} \subset C$, $x_i := x_o + r_o(y_{v_i} - y_{v_o})$ und $r_i := r_o 2^{-i}$ für jedes $i \in N_o$.

Es gilt dann $\overline{K_{r_{i+1}}(x_{i+1})} \subset K_{r_i}(x_i)$ für alle $i \in N_o$ und $\{y_\nu\}_{\nu \in N_o}$ ist genau dann konvergent, wenn $\{x_i\}_{i \in N_o}$ konvergiert. Setzt man $F := \{K_{r_i}(x_i)\}_{i \in N_o}$, so ist $f := f_F$ ein beschränktes konvexes Funktional auf X. Es gibt daher ein $x^* \in C$ mit $f(x^*) = \inf\{f(x) : x \in C\}$. Aus $\overline{K_{r_o}(x_o)} \subset C$ und Lemma 2 schließt man $x^* \in \bigcap_{i \in N_o} \overline{K_{r_i}(x_i)}$, also $x^* = \lim x_i$, woraus die Vollständigkeit von X folgt.

Da insbesondere für jedes $f \in X^*$ $f|C$ ein Minimum (Maximum) besitzt, ist C nach einem Satz von R. C. James+) ([5]) schwach kompakt. Schwach kompakte Mengen mit inneren Punkten (bzgl. der Normtopologie) gibt es aber nach einem wohlbekannten Satz von Eberlein nur in reflexiven Räumen.

Bemerkung: Mit Hilfe einer von J. Blatter angegebenen Charakterisierung abzählbar-konvex-kompakter konvexer Mengen ([1]++) S. 102, Theorem A.6.), die in normierten Räumen mit schwach kompakten konvexen Mengen identisch sind (vergl. etwa [4] S. 58 Theorem 1), läßt sich zeigen:

Gibt es eine konvexe Teilmenge C von X mit $C^\circ \neq \emptyset$ und der Eigenschaft, daß jedes konvexe unterhalbstetige Funktional $f : C \to R$ ein Minimum besitzt, so ist X reflexiv.

Satz 4 kann daraus (C konvex vorausgesetzt) nicht gefolgert werden. Z.B. gibt es auf dem Einheitskreis des R^2 definierte konvexe unterhalbstetige Funktionen, die sich nicht zu (beschränkten) konvexen Funktionen auf R^2 fortsetzen lassen (vergl. etwa [7] S. 195 Figure 7.9).

5. Korollar: Voraussetzung: U sei ein abgeschlossener Unterraum von X, für den X/U reflexiv ist und für den mit einem $C \subset X$ und einem $x_o \in X$ $D := C \cap (x_o + U)$ schwach abgeschlossen und $C^\circ \cap (x_o + U) \neq \emptyset$ ist.

Behauptung: Besitzt für jedes beschränkte konvexe Funktional $f : X \to R$ $f|D$ ein Minimum, so ist X reflexiv.

Beweis: Es genügt, die Reflexivität von U nachzuweisen (vergl. [3] S.243 Ex. 6e !), was mit Hilfe von Satz 4 geschehen soll. $\tilde{C} := (C - x_o) \cap U$ ist

---

+) Der Satz besagt, daß eine nicht leere schwach abgeschlossene Teilmenge D eines reellen Banachraumes X genau dann schwach kompakt ist, wenn für jedes $f \in X^*$ $f|D$ ein Maximum besitzt. Diese Charakterisierung schwach kompakter Mengen gilt nicht in beliebigen normierten Räumen (vergl. [6] !).

++) Auf diese Arbeit hat mich Herr K.-H. Hoffmann aufmerksam gemacht.

eine rel. U schwach abgeschlossene Teilmenge von U, deren rel. U offener Kern nicht leer ist. Ist $f : U \to R$ ein beschränktes konvexes Funktional, so gibt es (vergl. etwa [8] S. 40 Satz 2.19 !) eine Erweiterung von f zu einem beschränkten konvexen Funktional $g : X \to R$. $\tilde{g} : X \ni x \mapsto g(x-x_o) \in R$ ist ebenfalls beschränkt und konvex. Für jedes $y \in D$ mit $\tilde{g}(y) = \inf\{\tilde{g}(x) : x \in D\}$, ist $y-x_o \in \tilde{C}$ und $f(y-x_o) = \inf\{f(x) : x \in \tilde{C}\}$. Besitzt also für jedes beschränkte konvexe Funktional $h : X \to R$ $h|D$ ein Minimum, so besitzt auch für jedes beschränkte konvexe Funktional $f : U \to R$ $f|\tilde{C}$ ein Minimum und U ist nach Satz 4 reflexiv.

Korollar 5 führt nun unmittelbar zum vollständigen Beweis von Satz1. Man hat dazu nur $U := \{x \in X : l_i(x) = 0 , i=1,\ldots,k\}$ zu setzen und zu berücksichtigen, daß X/U wegen $\dim X/U \leq k < \infty$ reflexiv ist. Mit einem $x_o \in L$ ist $D = (C \cap K) \cap (x_o+U)$ eine schwach abgeschlossene (konvexe) Menge und es gilt $(C \cap K)^{o} \cap (x_o+U) \neq \emptyset$.

Man beachte, daß für den soeben bewiesenen Teil des Satzes die Voraussetzung b) nicht benötigt wird. Die Beschränktheit von D ergibt sich als zusätzliche Folgerung.

Literatur

1. Blatter, J. Grothendieck Spaces in Approximation Theory. Memoirs of the Amer. Math. Soc. 120, (1972).
2. Bourbaki, N. Eléments de Mathématique: Livre V, Espaces vectoriels topologiques, Paris (1953-1955).
3. Brown, A.L., Page, A. Elements of Functional Analysis. London (1970).
4. Day, M.M. Normed linear spaces. 3rd ed. Berlin-Heidelberg-New York (1973).
5. James, R.C. Weakly compact sets. Trans. Amer. Math.Soc. 113,129-140 (1964).
6. James, R.C. A counterexample for a sup theorem in normed spaces. Israel J.Math. 9, 511-512 (1971).
7. Luenberger, D.G. Optimization by vector space methods. New York-London-Sydney-Toronto (1969).
8. Valentine, F.A. Konvexe Mengen, Mannheim (1968).

Gerhard Heindl
Mathematisches Institut
der Technischen Universität München
8 München 2
Arcisstraße 21

Ein Approximationssatz und seine Anwendung in der Kontrolltheorie

K.-H. Hoffmann, E. Jörn, E. Schäfer, H. Weber

I. Einleitung

Zur numerischen Behandlung von Kontrollaufgaben werden in neuerer Zeit verstärkt Diskretisierungsverfahren untersucht. Die praktische Erprobung der entsprechenden Algorithmen scheint uns bisher nur unzureichend durchgeführt. Da man nämlich bei feiner Diskretisierung auf Optimierungsaufgaben geführt wird, die viele Variable und eine große Zahl von Nebenbedingungen besitzen, versagen viele Methoden der endlichdimensionalen Optimierungstheorie zur Lösung der diskreten Kontrollprobleme. Wir werden daher in dieser Arbeit für eine Klasse von Kontrollaufgaben die Konvergenzgeschwindigkeit gewisser Diskretisierungsverfahren untersuchen und ein Verfahren angeben, das sich bei der numerischen Lösung einiger diskreter Probleme bewährt hat.

Diese Note ist Kurzfassung einer Arbeit, die unter dem Titel " Differenzenverfahren zur Behandlung von Kontrollproblemen " bei der Zeitschrift Numerische Mathematik zur Publikation eingereicht wurde. Die ausführlichen Beweise der Sätze und weitere Beispiele sind dort nachzulesen.

II. Problemstellung und ein Beispiel

Wir betrachten Kontrollaufgaben des folgenden Typs:

(OC) Minimiere die Kostenfunktion

$$f(x,u) := h(x(T)) + \int_0^T g(t,x(t),u(t))dt$$

unter allen auf $I := [0,T]$ meßbaren beschränkten Steuerungen u und den Nebenbedingungen

$$u(t) \in U(t) \ , \ t \in I \ ,$$
$$A(x,u,z) = 0 \ .$$

Dabei ist A definiert durch die Differentialgleichung

$$\dot{x} - F(t,x,u) = 0 \quad , \ t \in I \ ,$$

und den Randoperator

$$B(x(0),x(T),u(0),u(T)) - z = 0 \ , \ z \in Z \ .$$

Es sind $h: \mathbb{R}^n \to \mathbb{R}$, $g: I\times \mathbb{R}^n \times \mathbb{R}^m \to \mathbb{R}$, $F: I\times \mathbb{R}^n \times \mathbb{R}^m \to \mathbb{R}^n$, $B: \mathbb{R}^{2n} \times \mathbb{R}^{2m} \to \mathbb{R}^n$ gegebene Abbildungen und $U(t) \subset \mathbb{R}^m$, $Z \subset \mathbb{R}^n$.

Auf das folgende Beispiel beziehen sich später unsere numerischen Rechnungen.

Zweistufige chemische Reaktion:

$$\begin{aligned} f(x,u) &:= - x_2(2) \ , \ x = (x_1,x_2) \ , \\ \dot{x}_1 &= x_1 u \ , \ x_1(0) = 1 \\ \dot{x}_2 &= x_1 u - 2.5 x_2 u^s \ , \ x_2(0) = 0.01 \end{aligned}$$

Dabei werden später $s := 1.5(1 - 1/3^{20})$ und noch die Einschränkungen an die Steuerungen u gewählt.

III. Das Diskretisierungsschema

Kontrollaufgaben vom Typ (OC) wurden bisher von verschiedenen Autoren mit Diskretisierungsmethoden behandelt. So benutzen etwa BOSARGE - JOHNSON [1971] Variationsmethoden, während BUDAK u.a. [1969] , CULLUM [1969] , KRABS [1973] , ESSER [1973] Differenzenverfahren theoretisch untersuchen. Wir knüpfen an die Untersuchungen von KRABS und ESSER an.

1. Diskretisierungsschritt:

Als Steuerungen werden nur Polynome vom Höchstgrad r , $u \in P_r$, zugelassen.

$(OC)^r$ ist dann die Steuerungsaufgabe (OC) mit $u \in P_r$.

2. Diskretisierungsschritt:

Im Problem $(OC)^r$ wird das Intervall I durch ein endliches Gitter $I_h \subset I$ der maximalen Gitterbreite h ersetzt. Es entsteht die endliche Optimierungsaufgabe

$(OC)_h^r$ Minimiere die Kostenfunktion

$$f_h(x_h,u_h) := h(x_h(T)) + Q_h(g(\cdot,x_h,u_h)) \ , \ u_h := u_{|I_h} \text{ für } u \in P_r \ ,$$

unter den Nebenbedingungen

$$u_h(t) \in U(t) \ , \ t \in I_h \ ,$$
$$A_h(x_h, u_h, z) = o \ .$$

Q_h beschreibt ein konvergentes Quadraturverfahren und A_h eine Diskretisierung der Randwertaufgabe. Wir betrachten im weiteren eindimensionale Steuerbereiche der Form

$U(t) := \{ s \in \mathbb{R} \mid g_1(t) \leqq s \leqq g_2(t) \}$ mit $g_1, g_2 \in C(I)$ und $g_1 < g_2$ und führen die Bezeichnungen ein:

$$E := \inf \{ f(x,u) \mid (x,u) \text{ zulässig für (OC) }, u \in C(I) \} ,$$
$$E^r := \inf \{ f(x,u) \mid (x,u) \text{ zulässig für } (OC)^r \} ,$$
$$E_h^r := \inf \{ f_h(x,u) \mid (x_h,u_h) \text{ zulässig für } (OC)_h^r \} .$$

Es gilt dann der folgende auf KRABS und ESSER zurückgehende Konvergenzsatz.

<u>Satz:</u> In (OC) sei h stetig, $f: C(I) \times C(I) \to \mathbb{R}$ stetig und $L^{-1}: C(I) \times \mathbb{R} \to C(I)$ auf kompakten Teilmengen gleichmäßig stetig.
In $(OC)^r$ sei $\lim_{h \to 0} \text{dist}(0, I_h) = \lim_{h \to 0} \text{dist}(T, I_h) = 0$, und für jedes $z \in Z$ sei $L_h^{-1}(u,z) \to L^{-1}(u,z)$ bzw.
$Q_h(g(\cdot,x,u)) \to \int_0^T g(t,x(t),u(t))dt$ gleichmäßig bzgl. u bzw. (x,u) auf kompakten Teilmengen.

Dann gelten:

(i) $\lim_{r \to \infty} E^r = E$,

(ii) $\lim_{h \to 0} E_h^r = E^r$.

Mit L^{-1} haben wir dabei den Operator bezeichnet, der jeder Steuerung u und jedem Zielpunkt z vermittels des Operators A die entsprechende Phase x zuordnet. Wir setzen dabei voraus, daß x durch u und z eindeutig bestimmt wird. Analog ist L_h^{-1} Lösungsoperator der Diskretisierung.
Es liegt nun nahe, durch geeignete Koppelung der Parameter r und h eine Konvergenzaussage

$$\lim_{r \to \infty} E_{h(r)}^r = E$$

zu gewinnen. ESSER [1973] zeigte, daß für äquidistante Gitter

$I_h := \{ ih \mid i=0(1)T/h \}$ und hinreichend glatte Kontrollaufgaben bei Wahl von $k := r^3$, wenn k die Anzahl der Diskretisierungspunkte ist, die Konvergenzgüte

$$| E - E_h^r | = O(r^{-2})$$

folgt.

Um den Vorteil dieser quadratischen Konvergenz zu nutzen, muß man sehr fein diskretisieren, d.h. man hat bei jedem Schritt eine endliche Optimierungsaufgabe mit sehr vielen Nebenbedingungen zu lösen. Wir werden die Konvergenzaussage von ESSER verbessern.

IV. Der Approximationssatz zur Abschätzung der Konvergenzgeschwindigkeit

Wesentliches Hilfsmittel unserer Überlegungen ist der

Approximationssatz:

Sei $J := [-1,+1]$ und

(i) $g_1, g_2 \in C(J)$ mit $g_1 < g_2$,

(ii) $J_k := \{ t_i := \cos((2i-1)\pi/2k) \mid i=1(1)k \}$, $k \in \mathbb{N}$
(Nullstellen der Tschebyscheff-Polynome 1.Art)

Dann gilt:

Es gibt ein $c>0$ und ein $r_o \in \mathbb{N}$ mit der Eigenschaft:

$\wedge (r>r_o) \wedge (k\in \mathbb{N} : r/k \leqq q < 1) \wedge (p \in P_r :$

(iii) $\wedge\, t \in J_k \quad g_1(t) \leqq p(t) \leqq g_2(t))$

$(\wedge (t\in J) \quad g_1(t) - c((r/k)^2 + \rho_r) \leqq p(t) \leqq g_2(t) + c((r/k)^2 + \rho_r) \quad).$

Es wurden $\rho_r := \max (\rho_r((g_1+g_2)/2), \rho_r(2/(g_2-g_1)))$ und $\rho_r(w) := \min_{p \in P_r} \| w-p \|_J$ gesetzt.

Beweisskizze:

Definiere $A := (g_1+g_2)/2$, $D := (g_2-g_1)/2$ und betrachte die folgenden Beweisschritte:

1. $\wedge\, r>r_o \quad \vee (a\in P_r : \| A-a \| \leqq \rho_r) \quad \vee (d \in P_r : \| 1/D-d \| \leqq \rho_r)$.

Dies folgt unmittelbar aus der Definition von ρ_r

2. $\| (p-A)/D-(p-a)d \| \leqq c_1 \rho_r$

Zum Beweis benutzt man eine Abschätzung von EHLICH-ZELLER [1964] über Schwankungen von Polynomen zwischen Gitterpunkten.

3. Sei $t_o \in J$ mit $|(p(t_o)-a(t_o))d(t_o)| = \|(p-a)d\|$ und $\eta_o \in [0,\pi]$ mit $t_o = \cos\eta_o$.
Dann gilt:
$\forall\, t' \in J_k \quad (\|(p-a)d\| \leqq |(p(t')-a(t'))d(t')| / \cos 2r(\eta'-\eta_o)$ und $|\eta'-\eta_o| \leqq \pi/2k$), wobei $t' =: \cos\eta'$.

Diese Ungleichung beweist man mit Hilfe einer Ungleichung von RIESZ [1914].

Unter Verwendung der Hilfsüberlegungen 1.-3. leitet man die behaupteten Ungleichungen her.

Die Sätze von JACKSON machen Aussagen über die Konvergenzgeschwindigkeit von $\rho_r \to 0$. Für $g_1, g_2 \in C^2$ folgt etwa $\rho_r = O(r^{-2})$. Damit kann man unter Verwendung des Approximationssatzes für gewisse Klassen glatter Kontrollaufgaben (vgl. die Arbeit der Autoren [1974]) die gegenüber ESSER verschärfte Konvergenzaussage

$$| E - E_h^r | = O(r^{-2})$$

herleiten für eine (nicht äquidistante) Diskretisierung mit nur $k=r^2$ Punkten.

V. Numerische Realisierbarkeit von Differenzenverfahren

Für das Beispiel der chemischen Reaktion wählen wir

$$g_1(t) = 0.4 \;,\; g_2(t) = 0.2$$

und die Diskretisierung

$(OC)_k^r$ Zielfunktional: $f(a) = -x_2(t_k) = -x_2(2)$.

Mittelpunktsformel:

$$x_{1,i+1} = x_{1,i}[1+\tfrac{1}{2}h_i^2\, u_{i+1/2}\,(u_i-2/h_i)]\;,$$

$$x_{2,i+1} = x_{2,i}[1+\tfrac{1}{2}h_i^2\, 2.5u_{i+1/2}^s\,(2.5u_i^s - 2/h_i)] -$$
$$- x_{1,i}(h_i^2/2)\,[u_{i+1/2}(u_i - 2/h_i) + 2.5u_{i+1/2}^s\, u_i]\;.$$

Nebenbedingungen:

$$G_i(a) := (u_i-g_1(t_i))(u_i-g_2(t_i)) \leqq 0\;,\; i=0(1)k\;.$$

Dabei ist $u(t) = \sum_{j=o}^{r} a_j t^j$, $a := (a_o,\dots,a_r)^T \in \mathbb{R}^{r+1}$, $0=t_o<\dots<t_k=2$ mit a) $t_i := 1+\cos((k-i)\pi/k)$ und b) $t_i:=2i/k$ für $i=0(1)k$, $h_i:=t_{i+1}-t_i$, $x_{j,i}:=x_j(t_i)$, $u_i:=u(t_i)$, $u_{i+1/2}:= \,:=u(t_i+h_i/2)$ für $j=1,2$ und $i=0(1)k$.

Zur Lösung des Problems $(OC)_k^r$ benutzen wir ein Verfahren von ROCKAFELLAR [1973] , das wir geeignet modifizieren, um seinen Anwendungsbereich zu erweitern.

Modifizierter Algorithmus:

1. Wähle $\varkappa \in \mathbb{R}_+$, $a^o \in \mathbb{R}^{r+1}$, $\varepsilon > 0$, $l_o \in \mathbb{N}$.
   Setze $\mu^o := 0.5(1+\text{sign}G(a^o))$, $n := 1$.
2. Setze $\mu(\beta) := [\mu^{n-1} + 2\beta G(a^{n-1})]_+$ und bestimme μ^n so, daß

$$R(a^{n-1}, \mu^n) = \min_{\beta \geqq 0} R(a^{n-1}, \mu(\beta)) ,$$

$$R(a,\mu) := \| \nabla f(a) - \nabla G(a)\mu \|_2^2 + (\mu^T G(a))^2 .$$

3. Setze $a^n := M^{l_o}(a^{n-1})$ (= l_o- Schritte eines Minimierungsverfahrens M für die Funktion $W(a,\mu^n,\varkappa)$ [1], beginnend bei a^{n-1}).
4. Wenn $R(a^n,\mu^n) \leqq \varepsilon$ Ende, sonst $n := n+1$ und weiter bei 2.

In dem folgenden numerischen Beispiel wurde für M ein konjugiertes Gradientenverfahren und $l_o = r+2$ benutzt. Das Beispiel wurde mit Tschebyscheff- und äquidistanten Knoten gerechnet. Der exakte Minimalwert beträgt

$$E = -0.3061076 .$$

Es wurde $\Delta := E_k^r - E$ gesetzt.

r	k	a) Tschebyscheff-Knoten E_k^r	$\Delta 10^2$	$r^2\Delta 10^2$	$r^3\Delta 10^2$	$r^4\Delta 10^2$
2	4	-0.293771	1.234	4.93	9.9	20
3	9	-0.303144	0.296	2.67	8.0	24
4	16	-0.305099	0.107	1.61	6.4	26
5	25	-0.305676	0.043	1.07	5.4	27
6	36	-0.305874	0.023	0.84	5.1	30
7	49	-0.305949	0.015	0.78	5.4	38
8	64	-0.305985	0.012	0.79	6.3	50

1) Dabei ist $W(a,\mu,\varkappa) := f(a) + 1/(4\varkappa)\sum_{i=0}^{k+1}([\mu_i + 2\varkappa G_i(a)]_+^2 - \mu_i^2)$.

r	k	b) Äquidistante Knoten E_k^r	$\Delta 10^2$	$r^2\Delta 10^2$	$r^3\Delta 10^2$	$r^4\Delta 10^2$
2	4	-0.297429	0.868	3.47	6.9	14
3	9	-0.304512	0.159	1.44	4.3	13
4	16	-0.305508	0.060	0.96	3.8	15
5	25	-0.305834	0.027	0.68	3.4	17
6	36	-0.305943	0.016	0.59	3.6	21
7	49	-0.305984	0.012	0.61	4.2	30
8	64	-	-	-	-	-

Die Rechnungen wurden an einer TR 440 - Anlage des Leibnitz-Rechenzentrums der Bayerischen Akademie der Wissenschaften in München durchgeführt.

Literatur:

BOSARGE,Jr.,W.E., O.G.JOHNSON [1971] :
Error bounds of high order accuracy for the state regulator problem via piecewise polynomial approximations.
SIAM J. Control 9(1971) 15-28 .

BUDAK,B.M., E.M. BERKOVICH, E.N. SOLOV'EVA [1969] :
Difference Approximations in optimal control problems.
SIAM J. Control 7(1969) 18-31 .

CULLUM,J. [1969] :
Discrete approximations to continuous optimal control problems.
SIAM J. Control 7(1969) 32-49 .

EHLICH,H., K. ZELLER [1964] :
Schwankung von Polynomen zwischen Gitterpunkten.
Math. Z. 86(1964) 41-44 .

ESSER,H. [1973] :
Zur Diskretisierung von Extremalproblemen. In R.Ansorge,W.Törnig:
Numerische, insbesondere approximationstheoretische Behandlung von Funktionalgleichungen.
Berlin,u.a., 1973 .

HOFFMANN,K.-H., E. JÖRN, E. SCHÄFER, H. WEBER [1974] :
Differenzenverfahren zur Behandlung von Kontrollproblemen.
Eingereicht: Numerische Mathematik .

KRABS,W. [1973] :
Stabilität und Stetigkeit bei nichtlinearer Optimierung.
In: Methoden des Operation Research XVII
Herausgeber R.Henn, 1973 .

RIESZ,M. [1914] :
Eine trigonometrische Interpolationsformel und einige
Ungleichungen für Polynome.
J.-ber. Deutsch. Math.-Verein 23(1914) 354-368 .

ROCKAFELLAR,R.T. [1973] :
The multiplier method of Hestenes and Powell applied to
convex programming.
J. Optimization Theory Appl. 12(1973) 555-562 .

Mathematisches Institut
der Universität München

8 München 2
Theresienstraße 39

Zur Störungstheorie Nichtlinearer Variationsungleichungen

Hansgeorg Jeggle

Behandelt wird eine allgemeine Störungstheorie für die folgenden Variationsungleichungen: Gesucht ist $u \varepsilon K$, so daß

$$(U) \quad (Lv + A(u),\ v - u) \geqslant (f, v - u) \ \forall\, v \varepsilon K \cap D(L),$$

wobei K eine konvexe Teilmenge eines Banach-Raums E ist und L eine lineare, A eine i.a. nichtlineare Abbildung von E in seinen dualen Raum E^* bedeutet. Es geht also um die numerische Approximation bzw. um die stetige Abhängigkeit von Operatoren und Daten der im Sinne von Lions [6] schwachen Lösungen von parabolischen Evolutionsungleichungen. Als approximierende Probleme verwenden wir die folgenden Aufgaben in den Räumen E_ι und E_ι^* : Gesucht ist $u_\iota \varepsilon K_\iota$, so daß

$$(U_\iota) \quad (L_\iota v_\iota + A_\iota(u_\iota),\ v_\iota - u_\iota) \geqslant (f_\iota, v_\iota - u_\iota) \ \forall\, v_\iota \varepsilon K_\iota \cap D(L_\iota),\ \iota \varepsilon \Lambda_o,$$

bzw. $u_\iota \varepsilon D(L_\iota)$, so daß

$$(G_\iota) \quad L_\iota u_\iota + A_\iota(u_\iota) + \alpha_\iota P_\iota(u_\iota) = f_\iota,\ \iota \varepsilon \Lambda_o,$$

wobei P_ι eine Folge von Penalty-Operatoren ist. Als Resultat erzielen wir Bedingungen für die Operatoren L, L_ι und A, A_ι sowie die rechten Seiten f, f_ι, unter denen die Existenz einer Lösung von (U) und die Konvergenz der Lösungen von (U_ι) bzw. (G_ι) erschlossen werden kann. Als angemessenen Rahmen für eine so allgemeine Störungstheorie legen wir dabei die diskrete Approximation normierter Räume von Stummel [12], [13] mit den von Mirgel [7] für den vorliegenden Problemkreis angepaßten Ergänzungen zugrunde.

Störungen bzw. diskrete Approximationen der Lösungen von nichtlinearen Variationsungleichungen wurden von Mosco [8] und Aubin [1] für stationäre Aufgaben mit monotonen Operatoren behandelt, in diese Richtung gehen auch die Arbeiten von Brézis und Sibony [3], [10]. Eine wesentlich allgemeinere Problemklasse studiert Mirgel [7] mit Hilfe des auf ihn zurückgehenden Begriffs der diskreten Pseudomonotonie von Operatoren A, A_ι. Dieser wird von uns aufgegriffen und bei den instationären Problemen (U) herangezogen. Unter dem Aspekt der numerischen Approximation von Lösungen von (U) scheint uns das Studium der Folge von Gleichungen (G_ι) besonders interessant zu sein, steht dafür doch die ganze Theorie der Näherungsverfahren für nichtlineare Gleichungen zur Verfügung [5].

Zwischen den Lösungen von Variationsungleichungen (U) und denen von Extremal- bzw. Minimaxproblemen bestehen enge Zusammenhänge, wie z.B.

in den Arbeiten [9],[10],[2] ausgeführt wird. Insbesondere für Kontrollprobleme bei partiellen Differentialgleichungen scheint der Weg über äquivalente Variationsungleichungen der passende numerische Zugang zu sein [10],[11]. Auf die Bedeutung von Untersuchungen wie dieser im Zusammenhang mit der numerischen Behandlung, bzw. mit simultanen Störungen von Operatoren und Daten von Variationsungleichungen für nichthomogene Randwertaufgaben braucht man nicht näher einzugehen [4],[6], [7].

Das Anliegen dieser Arbeit ist, auf dem Hintergrund der skizzierten Motivationen den Rahmen einer allgemeinen Störungstheorie für Aufgaben (U) bereitzustellen. Dies ist mehr im Sinne von eher abstrakten "Arbeitssätzen" zu verstehen, als daß hier schon feine Details und gewichtige konkrete Anwendungen vorgeführt werden. Das ist im Anschluß an diese Arbeit vorgesehen.

1. Vorbereitendes.

Im folgenden stellen wir die wichtigsten Definitionen und Eigenschaften der diskreten Approximation dualer Paare normierter Räume zusammen. Dabei halten wir uns eng an die Arbeiten [12] und [7] von Stummel bzw. Mirgel.

Es bedeuten E, E_ι stets reelle Banach-Räume und E^*, E_ι^* ihre topologischen Duale. Für die Normen von E bzw. E_ι und E^* bzw. E_ι^* verwenden wir stets dieselben Symbole $\|.\|$ und $\|.\|_*$, ebenso bedeutet $(\cdot,\cdot)$ ohne weitere Unterscheidung die Dualität auf $E^* x E$ bzw. $E_\iota^* x E_\iota$. Λ_o ist eine abzählbare, linear geordnete Indexfolge, $\Lambda, \Lambda', \ldots$ bedeuten Teilfolgen von Λ_o, jedoch nicht notwendig stets dieselben. K, K_ι sind konvexe Mengen.

Ausgangspunkt ist die folgende Definition.

(A) Das Paar E, E_ι heißt diskrete Approximation, wenn folgende Bedingungen erfüllt sind.

(i) Es gibt eine Abbildung $r: E \to \Pi E_\iota$ mit
$\forall\, u \varepsilon E \;\exists\, (r_\iota(u)) \varepsilon \Pi E_\iota : r(u) := (r_\iota(u))$.

(ii) $\forall\, u \varepsilon E : \|r_\iota(u)\| \to \|u\| (\iota \varepsilon \Lambda_o)$

(iii) $\forall\, u, v \varepsilon E\; \forall\, \alpha, \beta \varepsilon \mathbb{R} : \|\alpha r_\iota(u) + \beta r_\iota(v) - r_\iota(\alpha u + \beta v)\| \to 0 (\iota \varepsilon \Lambda)$.

In einer diskreten Approximation wird folgender Konvergenzbegriff erklärt.

(1) Die Folge $u_\iota \varepsilon E_\iota$ heißt (diskret) stark konvergent gegen $u \varepsilon E$, $u_\iota \to u (\iota \varepsilon \Lambda)$, wenn $\|r_\iota(u) - u_\iota\| \to 0 (\iota \varepsilon \Lambda)$.

Die Definitionen sind so getroffen, daß man eine permanente Erweiterung der üblichen starken Konvergenz in normierten Räumen hat, sie also auch alle bekannten Eigenschaften besitzt. Wegen der Einzelheiten verweisen wir auf [12]. Dort findet man auch zahlreiche Beispiele.

Wir benötigen im folgenden vor allem aber auch einen schwachen Konvergenzbegriff, im allgemeinen Fall kann man ihn jedoch in einer diskreten Approximation nicht definieren [12]. Nimmt man hingegen eine Einschränkung auf den Spezialfall reflexiver Räume vor, so kann man von diskreten Approximationen E, E_ι und E^*, E^*_ι ausgehen und mit Hilfe einer Verträglichkeitsbedingung zwischen beiden eine diskrete Approximation für das duale Paar (E^*, E) erklären [7].

(D) Das Paar (E^*, E), (E^*_ι, E_ι) heißt diskrete Approximation des dualen Paares (E^*, E), wenn folgende Bedingungen erfüllt sind.

(i) E, E_ι und E^*, E^*_ι sind jeweils diskrete Approximationen im Sinne von (A).

(ii) Für alle $u \varepsilon E$, alle $f \varepsilon E^*$ und alle Folgen $u_\iota \varepsilon E_\iota$ sowie $f_\iota \varepsilon E^*_\iota$ gilt $u_\iota \to u (\iota \varepsilon \Lambda)$, $f_\iota \to f (\iota \varepsilon \Lambda) \Longrightarrow (f_\iota, u_\iota) \to (f, u) (\iota \varepsilon \Lambda)$.

Dabei ist die Konvergenz jeweils im Sinne der diskreten Approximationen E, E_ι und E^*, E^*_ι bzw. von $\mathbb{R}$ verstanden. Entsprechend bleibt in der diskreten Approximation eines dualen Paares der starke Konvergenzbegriff von Elementen bzw. von Funktionalen derjenige von E, E_ι bzw. E^*, E^*_ι. Zunächst können wir den schwachen Konvergenzbegriff für Funktionale aus [12] hier einordnen.

(2) Die Folge $f_\iota \varepsilon E^*_\iota$ heißt (diskret) schwach konvergent gegen $f \varepsilon E^*$, $f_\iota \rightharpoonup f (\iota \varepsilon \Lambda)$, wenn für alle $u \varepsilon E$ und alle Folgen $u_\iota \varepsilon E_\iota$ gilt

$$u_\iota \to u (\iota \varepsilon \Lambda) \Longrightarrow (f_\iota, u_\iota) \to (f, u) (\iota \varepsilon \Lambda).$$

Also sind stark konvergente Folgen von Funktionalen wegen (D,ii) auch schwach konvergent. Hinter der Definition der schwachen Konvergenz von Elementen steht die von Funktionalen, wobei die Reflexivität der beteiligten Räume die wesentliche Rolle spielt.

(3) Die Folge $u_\iota \varepsilon E_\iota$ heißt (diskret) schwach konvergent gegen $u \varepsilon E$, $u_\iota \rightharpoonup u (\iota \varepsilon \Lambda)$, wenn für alle $f \varepsilon E^*$ und alle Folgen $f_\iota \varepsilon E^*_\iota$ gilt

$$f_\iota \to f (\iota \varepsilon \Lambda) \Longrightarrow (f_\iota, u_\iota) \to (f, u) (\iota \varepsilon \Lambda).$$

Dieser schwache Konvergenzbegriff für Elemente in diskreten Approximationen dualer Paare besitzt alle bei einer solchen Begriffsbildung erwarteten Eigenschaften. Wir führen einige wenige zur Verdeutlichung vor, die Beweise finden sich in [7], unter Rückgriff auf [12].

(4) $$u_\iota \to u (\iota \varepsilon \Lambda) \Longrightarrow u_\iota \rightharpoonup u (\iota \varepsilon \Lambda)$$

(5) $u_\iota \rightharpoonup u(\iota\in\Lambda) \Longrightarrow \exists\ \alpha>0\ \forall\ \iota\in\Lambda : \|u_\iota\| \leq \alpha.$

(6) $u_\iota \rightharpoonup u^1(\iota\in\Lambda),\ u_\iota \rightharpoonup u^2(\iota\in\Lambda) \Longrightarrow u^1 = u^2.$

Setzt man zusätzlich noch voraus, daß E und E_ι separabel sind, dann hat man auch die bekannten schwachen Kompaktheitseigenschaften [7].

(7) <u>Falls</u> E <u>und</u> $E_\iota, \iota\in\Lambda_o$, <u>reflexiv und separabel sind und</u> $(E^*,E),(E_\iota^*,E_\iota)$ <u>eine diskrete Approximation im Sinne von</u> (D) <u>ist, dann gilt</u>

$$u_\iota\in E_\iota,\ \iota\in\Lambda,\ \underline{\text{beschränkt}} \Longrightarrow \exists\Lambda'\subset\Lambda\ \exists u\in E : u_\iota \rightharpoonup u(\iota\in\Lambda').$$

Aus [13] übernehmen wir schließlich noch für Teilmengen $G_\iota\subset E_\iota,\ \iota\in\Lambda_o$, die Begriffe des starken (diskreten) Limes inferior,

$$s - \text{Lim inf } G_\iota : = \{u\in E / \forall\iota\in\Lambda_o\ \exists u_\iota\in G_\iota : u_\iota \to u(\iota\in\Lambda_o)\}$$

und des schwachen (diskreten) Limes superior,

$$w - \text{Lim sup } G_\iota : = \{u\in E / \exists u_\iota\in G_\iota,\ \iota\in\Lambda\subset\Lambda_o : u_\iota \rightharpoonup u(\iota\in\Lambda)\}.$$

In diesem Zusammenhang kann man auch auf [7] und [8] verweisen.

Für Folgen von Operatoren werden wir die folgenden Begriffe heranziehen.

(8) <u>Die Folge</u> $A_\iota : K_\iota\subset E_\iota \to E_\iota^*$ <u>heißt beschränkt, wenn für jede Folge</u> $u_\iota\in K_\iota$ <u>aus</u>

$$u_\iota,\ \iota\in\Lambda,\ \underline{\text{beschränkt}} \Longrightarrow A_\iota(u_\iota),\ \iota\in\Lambda,\ \underline{\text{beschränkt}}.$$

(9) <u>Die Folge</u> $A_\iota : K_\iota\subset E_\iota \to E_\iota^*$ <u>heißt konvergent gegen</u> $A : K\subset E \to E^*$, <u>wenn für jedes</u> $u\in K$ <u>und jede Folge</u> $u_\iota\in K_\iota$ <u>aus</u>

$$u_\iota \to u(\iota\in\Lambda) \Longrightarrow A_\iota(u_\iota) \to A(u)(\iota\in\Lambda).$$

Im folgenden setzen wir stets E und $E_\iota,\ \iota\in\Lambda_o$, reflexiv und separabel voraus, ferner nehmen wir an, daß $(E^*,E),\ (E_\iota^*,E_\iota)$ eine diskrete Approximation des dualen Paares (E^*,E) im Sinne von (D) ist.

2. <u>Diskret pseudomonotone Operatoren.</u>

In diesem Abschnitt definieren und beschreiben wir die im Mittelpunkt der vorgelegten Theorie stehende Klasse von Operatoren und approximierender Folgen von Operatoren. Dabei geht die folgende Definition auf die Arbeit [7] von Mirgel zurück.

(1) <u>Das Paar</u> A, A_ι <u>heißt approximationspseudomonoton (a-pseudomonoton) auf</u> K, K_ι, <u>wenn folgende Bedingungen erfüllt sind.</u>

(i) <u>Die Folge</u> A_ι <u>ist beschränkt,</u> (ii) <u>für jede Folge</u> $u_\iota\in K_\iota$ <u>mit</u>

$$u_\iota \rightharpoonup u\in K(\iota\in\Lambda),\quad \limsup_\Lambda (A_\iota(u_\iota),\ u_\iota - w_\iota) \leq 0$$

für eine Folge $w_\iota \varepsilon K_\iota$ mit $w_\iota \to u(\iota\varepsilon\Lambda)$ gilt

$$(A(u),\ u-v) \leqslant \lim\inf_\Lambda\ (A_\iota(u_\iota),\ u_\iota - v_\iota)$$

für alle $v\varepsilon K$ und alle Folgen $v_\iota \varepsilon K_\iota$ mit $v_\iota \to v(\iota\varepsilon\Lambda)$.

In engem Zusammenhang damit steht die folgende Konvergenzbedingung, auf die wir gelegentlich zurückgreifen werden.

(2) Für jede Folge $u_\iota \varepsilon K_\iota$ mit $u_\iota \rightharpoonup u\varepsilon K(\iota\varepsilon\Lambda)$ gilt

$$(A_\iota(v_\iota),\ v_\iota - u_\iota) \to (A(v), v-u)(\iota\varepsilon\Lambda)$$

für alle $v\varepsilon K$ und alle Folgen $v_\iota \varepsilon K_\iota$ mit $v_\iota \to v(\iota\varepsilon\Lambda)$.

Die Konvergenzbedingung ist erfüllt, wenn die Folge A_ι im Sinne von 1.(9) gegen A konvergiert. Mit Hilfe dieser Konvergenz wurde in [5] von Grigorieff die Stetigkeit von A bewiesen. Hier gilt für den in diesem Zusammenhang wesentlichen Stetigkeitsbegriff eine entsprechende Aussage.

(3) Die Folge A_ι sei beschränkt und die Konvergenzbedingung (2) mit $K_\iota := E_\iota$, $K := E$ erfüllt. Dann ist A hemistetig in E.

Beweis. Wir nehmen an, die Behauptung sei falsch, d.h., es existieren Zahlen $t_o, t_j \varepsilon [0,1]$, $j\varepsilon\mathbb{N}$, und Elemente r,s und $\hat{u}\varepsilon E$, so daß

$$t_j \to t_o(j\varepsilon\mathbb{N}),\quad \forall j\varepsilon\mathbb{N}: |(A(t_o r+(1-t_o)s)-A(t_j r+(1-t_j)s, \hat{u})| > 2\varepsilon.$$

Nun nimmt man Folgen $r_\iota, s_\iota, \hat{u}_\iota \varepsilon E_\iota$ mit $r_\iota \to r$, $s_\iota \to s$, $\hat{u}_\iota \to \hat{u}$ $(\iota\varepsilon\Lambda_o)$ und setzt für jedes $\iota\varepsilon\Lambda_o$ und jedes $j\varepsilon\mathbb{N}$

$$v := t_o r + (1-t_o)s,\quad v^j := t_j r + (1-t_j)s,\quad u := v - \hat{u},\quad u^j := v^j - \hat{u},$$

$$v_\iota := t_o r_\iota + (1-t_o)s_\iota,\quad v_\iota^j := t_j r_\iota + (1-t_j)s_\iota,\quad u_\iota := v_\iota - \hat{u}_\iota,\quad u_\iota^j := v_\iota^j - \hat{u}_\iota.$$

Dabei haben wir

$$v^j \to v(j\varepsilon\mathbb{N}),\quad u^j \to u(j\varepsilon\mathbb{N}),$$

$$v_\iota \to v(\iota\varepsilon\Lambda_o), v_\iota^j \to v^j(\iota\varepsilon\Lambda_o),\ j\varepsilon\mathbb{N},\ u_\iota \to u(\iota\varepsilon\Lambda_o),\ u_\iota^j \to u^j(\iota\varepsilon\Lambda_o),\ j\varepsilon\mathbb{N}.$$

Deshalb können wir für jedes $j\varepsilon\mathbb{N}$ ein Endstück $\Lambda_j \subset \Lambda_o$ finden mit

$$\|v_\iota^j - v_\iota\| \leqslant \|v^j - v\| + \frac{1}{j},\quad \iota\varepsilon\Lambda_j,\ j\varepsilon\mathbb{N},$$

und

$$|(A_\iota(v_\iota),\ v_\iota - u_\iota) - (A_\iota(v_\iota^j),\ v_\iota^j - u_\iota^j)| > \varepsilon,\ \iota\varepsilon\Lambda_j,\ j\varepsilon\mathbb{N},$$

wobei wir die Annahme und die Konvergenzbedingung (2) herangezogen haben. Von den Endstücken Λ_j dürfen wir annehmen, daß $\Lambda_{j+1} \subset \Lambda_j$, $j\varepsilon\mathbb{N}$. Setzt man nun

$$\tilde{v}_\iota := \begin{cases} v_\iota & \forall\ \iota\varepsilon\Lambda_o \setminus \Lambda_1, \\ v_\iota^j & \forall\ \iota\varepsilon\Lambda_j \setminus \Lambda_{j+1},\ j\varepsilon\mathbb{N}, \end{cases} \qquad \tilde{u}_\iota := \begin{cases} u_\iota & \forall\ \iota\varepsilon\Lambda_o \setminus \Lambda_1, \\ u_\iota^j & \forall\ \iota\varepsilon\Lambda_j \setminus \Lambda_{j+1},\ j\varepsilon\mathbb{N}, \end{cases}$$

dann haben wir

$$v_\iota \to v,\ u_\iota \to u,\ \hat{v}_\iota \to v,\ \hat{u}_\iota \to u\ (\iota\in\Lambda_o),$$

jedoch

$$\forall\iota\in\Lambda_o : |(A_\iota(v_\iota),\ v_\iota - u_\iota) - (A_\iota(\hat{v}_\iota),\ \hat{v}_\iota - \hat{u}_\iota)| > \varepsilon.$$

Dieser Widerspruch zur Konvergenzbedingung zeigt, daß die Annahme falsch ist. ✱

Mit Hilfe des eben bewiesenen Lemmas können wir ein Kriterium für die a-Pseudomonotonie eines Paares A, A_ι bewiesen.

(4) Es sei (i) $K \subset$ s-Lim inf K_ι, (ii) die Folge A_ι beschränkt, (iii) die Bedingung (2) durch das Paar A, A_ι erfüllt und (iv) für alle Folgen $u_\iota \in K_\iota$ mit $u_\iota \rightharpoonup u \in K(\iota\in\Lambda)$

$$\lim\inf_\Lambda (A_\iota(u_\iota) - A_\iota(v_\iota),\ u_\iota - v_\iota) \geqslant 0$$

für alle $v\in K$ und alle Folgen $v_\iota \in K_\iota$ mit $v_\iota \to v(\iota\in\Lambda)$. Dann ist das Paar A, A_ι a-pseudomonoton auf K, K_ι.

Beweis. Wegen (ii) müssen wir nur (1,ii) nachweisen. Dazu nehmen wir an, daß $u_\iota \in K_\iota$ eine Folge ist mit

(5) $\quad u_\iota \rightharpoonup u\in K(\iota\in\Lambda),\ \lim\sup_\Lambda\ (A_\iota(u_\iota),\ u_\iota - w_\iota) \leqslant 0$

für eine Folge $w_\iota \in K_\iota$ mit $w_\iota \to u(\iota\in\Lambda_o)$. Mit $v_\iota := w_\iota$, $v := u$ und der Bedingung (2) erhalten wir aus (iv)

(6) $\quad \lim\inf_\Lambda\ (A_\iota(u_\iota),\ u_\iota - w_\iota) \geqslant 0.$

Zusammen mit (5) hat man dann

$$(A_\iota(u_\iota),\ u_\iota - w_\iota) \to 0(\iota\in\Lambda).$$

Wählt man jetzt $v\in K$, so existiert wegen (i) eine Folge $v_\iota \in K_\iota$ mit $v_\iota \to v(\iota\in\Lambda_o)$ und wegen der Konvexität von K_ι ist

$$\forall t\in[0,1]\ \forall\iota\in\Lambda_o : tv_\iota + (1-t)w_\iota \in K_\iota.$$

Weiter entnimmt man (iv) die Existenz einer Teilfolge $\Lambda \subset \Lambda_o$ mit

$$(A_\iota(u_\iota) - A_\iota(tv_\iota + (1-t)w_\iota),\ u_\iota - tv_\iota - (1-t)w_\iota) \geqslant 0,$$

daher ist für alle $\iota\in\Lambda$

$$(A_\iota(u_\iota), u_\iota - w_\iota) - (A_\iota(tv_\iota + (1-t)w_\iota),\ u_\iota - tv_\iota - (1-t)w_\iota)$$
$$\geqslant -t(A_\iota(u_\iota),\ w_\iota - v_\iota).$$

Mit Hilfe der Konvergenzbedingung (2) können wir daraus erschließen

$$(A(tv+(1-t)u),\ u-v) \leqslant \lim\inf_\Lambda\ (A_\iota(u_\iota),\ w_\iota - v_\iota).$$

Unter Anwendung von (6) gelangen wir zu

$$(A(tv+(1-t)u),\ u-v) \leqslant \lim\inf_\Lambda\ (A_\iota(u_\iota),\ u_\iota - v_\iota).$$

Aus (3) kann man nun noch entnehmen, daß die linke Seite eine stetige Funktion von t ist. Mit $t \to 0$ erschließen wir dann die Behauptung. *

Wir bemerken, daß die Monotonie aller Abbildungen A_ι hinreichend für die Bedingung (4,iv) ist.

Der folgende Satz gibt ein Beispiel für Störungen gegenüber denen die Eigenschaft eines Paares A, A_ι a-pseudomonoton zu sein, erhalten bleibt.

(7) <u>Es sei</u> (i) $K \subset$ s-Lim inf K_ι, (ii) <u>das Paar</u> A, A_ι <u>a-pseudomonoton auf</u> K, K_ι, (iii) <u>die Folge</u> B_ι <u>beschränkt,</u> (iv) <u>die Bedingung</u> (2) <u>durch das Paar</u> B, B_ι <u>erfüllt und</u> (v) <u>für alle Folgen</u> $u_\iota \varepsilon K_\iota$ <u>mit</u> $u_\iota \rightharpoonup u \varepsilon K (\iota \varepsilon \Lambda)$

$$\liminf_\Lambda (B_\iota(u_\iota) - B_\iota(v_\iota), u_\iota - v_\iota) \geq 0$$

<u>für alle</u> $v \varepsilon K$ <u>und alle Folgen</u> $v_\iota \varepsilon K_\iota$ <u>mit</u> $v_\iota \to v (\iota \varepsilon \Lambda)$. <u>Dann ist das Paar</u> $A+B$, $A_\iota + B_\iota$ <u>a-pseudomonoton auf</u> K, K_ι.

Beweis. Wir gehen wieder aus von einer Folge $u_\iota \varepsilon K_\iota$ mit den Eigenschaften

(8) $$u_\iota \rightharpoonup u \varepsilon K (\iota \varepsilon \Lambda), \quad \limsup_\Lambda (A_\iota(u_\iota) + B_\iota(u_\iota), u_\iota - w_\iota) \leq 0$$

für eine Folge $w_\iota \varepsilon K_\iota$ mit $w_\iota \to u (\iota \varepsilon \Lambda)$. Man hat für jedes $\iota \varepsilon \Lambda_0$

$$(A_\iota(u_\iota), u_\iota - w_\iota) = A_\iota(u_\iota) + B_\iota(u_\iota), u_\iota - w_\iota)$$

$$- (B_\iota(u_\iota) - B_\iota(w_\iota), u_\iota - w_\iota) - (B_\iota(w_\iota), u_\iota - w_\iota).$$

Daraus folgern wir unter Anwendung von (8), (v) und (iv)

$$\limsup_\Lambda (A_\iota(u_\iota), u_\iota - w_\iota) \leq 0.$$

Die a-Pseudomonotonie von A, A_ι führt jetzt zu

$$(A(u), u-v) \leq \liminf_\Lambda (A_\iota(u_\iota), u_\iota - v_\iota)$$

für alle $v \varepsilon K$ und alle Folgen $v_\iota \varepsilon K_\iota$ mit $v_\iota \to v (\iota \varepsilon \Lambda)$. Setzen wir speziell $v := u$ und $v_\iota := w_\iota$, dann können wir den beiden vorstehenden Ungleichungen entnehmen, daß

$$(A_\iota(u_\iota), u_\iota - w_\iota) \to 0 \ (\iota \varepsilon \Lambda).$$

Zieht man dies neben (8) heran, so hat man

$$\limsup_\Lambda (B_\iota(u_\iota), u_\iota - w_\iota) \leq 0.$$

Wegen (i), (iii), (iv) und (v) ist B, B_ι a-pseudomonoton auf K, K_ι, aus der vorstehenden Relation folgt also

$$(B(u), u-v) \leq \liminf_\Lambda (B_\iota(u_\iota), u_\iota - v_\iota)$$

für alle $v \varepsilon K$ und alle Folgen $v_\iota \varepsilon K_\iota$ mit $v_\iota \to v (\iota \varepsilon \Lambda)$. Zusammen haben wir somit aus (8) erschlossen, daß

$$(A(u) + B(u), u-v) \leq \liminf_\Lambda (A_\iota(u_\iota) + B_\iota(u_\iota), u_\iota - v_\iota)$$

für alle $v \in K$ und $v_\iota \in K_\iota$ mit $v_\iota \to v (\iota \in \Lambda)$. Damit ist (1,ii) nachgewiesen.*

Der letzte Hilfssatz gibt Bedingungen an, unter denen aus der schwachen Konvergenz einer Folge approximierender Gleichungen auf die Existenz einer Lösung der zugehörigen Gleichung geschlossen werden kann.

(9) Das Paar A, A_ι sei a-pseudomonoton auf E, E_ι und für eine Folge $u_\iota \in E_\iota$ und $u \in E$ sowie $f \in E^*$ gelte

$$u_\iota \rightharpoonup u (\iota \in \Lambda), \; A_\iota(u_\iota) \rightharpoonup f (\iota \in \Lambda), \; \lim \sup_\Lambda (A_\iota(u_\iota), u_\iota) \leqslant (f, u).$$

Dann ist $A(u) = f$.

Beweis. Zu $u \in E$ existiert eine Folge $w_\iota \in E_\iota$ mit $w_\iota \to u (\iota \in \Lambda_o)$. In

$$(A_\iota(u_\iota), u_\iota) = (A_\iota(u_\iota), u_\iota - w_\iota) + (A_\iota(u_\iota), w_\iota)$$

führen wir unter Verwendung der Voraussetzung den Grenzübergang durch mit dem Resultat

$$\lim \sup_\Lambda (A_\iota(u_\iota), u_\iota - w_\iota) \leqslant \lim \sup_\Lambda (A_\iota(u_\iota), u_\iota) - (f, u) \leqslant 0.$$

Damit haben wir aufgrund der a-Pseudomonotonie von A, A_ι

$$(A(u), u - v) \leqslant \lim \inf_\Lambda (A_\iota(u_\iota), u_\iota - v_\iota)$$

für alle $v \in E$ und alle Folgen $v_\iota \in E_\iota$ mit $v_\iota \to v (\iota \in \Lambda)$. Nehmen wir beliebige v, v_ι und argumentieren so wie oben für u, w_ι, so führt das zu

$$\lim \sup (A_\iota(u_\iota), u_\iota - v_\iota) \leqslant (f, u - v).$$

Dies bedeutet

$$\forall v \in E \; (A(u), u - v) \leqslant (f, u - v)$$

und damit haben wir die Behauptung.*

3. Konvergenzsätze.

Ziel dieses Abschnittes ist es, aus der Existenz von Lösungen einer Folge gestörter Ungleichungen bzw. approximierender Gleichungen auf die Existenz einer Lösung des ungestörten Problems zu schließen. Diese Existenzaussagen sind konstruktiv in dem Sinne, daß man zumindest schwach konvergente Teilfolgen von approximierenden Lösungen erhält.

Die folgenden zwei Eigenschaften müssen wir im folgenden von der Folge L_ι durchweg voraussetzen.

(L_i) Für jedes $\iota \in \Lambda_o$ sei L_ι eine lineare, abgeschlossene Abbildung mit der Definitionsmenge $D(L_\iota)$ von E_ι in E_ι^* mit

$$\forall u \in D(L_\iota) \; (L_\iota u, u) \geqslant 0, \quad \forall u \in D(L_\iota^*) \; (L_\iota^* u, u) \geqslant 0.$$

Dabei bedeutet L_ι^* die zu L_ι adjungierte Abbildung. Unter den gegebenen

Voraussetzungen ist jedes L_ι eine beschränkt-lineare Abbildung von $D(L_\iota)$ in $(D(L_\iota^*))^*$, wenn diese Räume mit der Graphennorm $|||.|||$ bzw. der adjungierten Graphennorm $|||.|||_*$ versehen werden.

(Lii) <u>Die Folge</u> $L_\iota: (D(L_\iota), |||.|||) \to ((D(L_\iota^*))^*, |||.|||_*)$, $\iota\varepsilon\Lambda_o$, <u>sei beschränkt.</u>

Zunächst beschäftigen wir uns nun mit Ungleichungen (U_ι). Dafür besteht der folgende Existenzsatz [6].

(1) <u>Es seien</u> E_ι, E_ι^* <u>strikt konvex,</u> $K_\iota \subset E_\iota$ <u>abgeschlossen und konvex.</u> L_ι <u>besitze die Eigenschaft, daß es zu jedem</u> $v_\iota \varepsilon K_\iota$ <u>eine Folge</u> $v_\iota^j \varepsilon K_\iota \cap D(L_\iota)$ <u>gibt mit</u>

$$v_\iota^j \to v_\iota (j\varepsilon\mathbb{N}), \quad \lim\sup_{\mathbb{N}} (L_\iota v^j, v_\iota^j - v_\iota) \leq 0.$$

<u>Ferner sei</u> A_ι <u>eine pseudomonotone Abbildung mit Definitionsmenge</u> K_ι <u>von</u> E_ι <u>in</u> E_ι^*, <u>für die mit einem</u> $v_\iota^o \varepsilon K_\iota \cap D(L_\iota)$ <u>gilt</u>

$$\frac{1}{\|v_\iota\|} (A_\iota(v_\iota), v_\iota - v_\iota^o) \to \infty \quad (\|v_\iota\| \to \infty).$$

<u>Dann besitzt</u> (U_ι) <u>für jedes</u> $f_\iota \varepsilon E_\iota$ <u>eine Lösung.</u>

Im folgenden nehmen wir stets an, daß es eine Teilfolge $\Lambda \subset \Lambda_o$ gibt, für die die Lösungsmengen $S(U_\iota)$ von U_ι nichtleer sind. Es gilt dann der folgende Konvergenzsatz.

(2) <u>Es sei</u> (i) $w\text{-}\mathrm{Lim}\sup_\Lambda K_\iota \subset K$, (ii) $K \subset s\text{-}\mathrm{Lim}\inf K_\iota \cap D(L_\iota)$,

(iii) <u>die Folge</u> f_ι <u>konvergent gegen</u> $f\varepsilon E^*$, $f_\iota \to f(\iota\varepsilon\Lambda_o)$, <u>und</u>

(iv) <u>das Paar</u> A, A_ι <u>a-pseudomonoton auf</u> K, K_ι.

(v) <u>Die Folge</u> A_ι <u>erfülle mit einer beschränkten Folge</u> $v_\iota^o \varepsilon K_\iota \cap D(L_\iota)$ <u>für jede Folge</u> $w_\iota \varepsilon K_\iota \cap D(L_\iota)$ <u>mit</u> $\|w_\iota\| \to \infty$ $(\iota\varepsilon\Lambda)$ <u>die Bedingung</u>

$$\frac{1}{\|w_\iota\|} (A_\iota(w_\iota), w_\iota - v_\iota^o) \to \infty \quad (\iota\varepsilon\Lambda)$$

<u>und</u> (vi) <u>das Paar</u> L, L_ι <u>für alle Folgen</u> $u_\iota \varepsilon K_\iota$ <u>mit</u> $u_\iota \rightharpoonup u\varepsilon K(\iota\varepsilon\Lambda)$ <u>die Bedingung</u>

$$\lim\sup_\Lambda (L_\iota v_\iota, v_\iota - u_\iota) \leq (Lv, v-u)$$

<u>für alle</u> $v\varepsilon K \cap D(L)$ <u>und alle Folgen</u> $v_\iota \varepsilon K_\iota \cap D(L_\iota)$ <u>mit</u> $v_\iota \to v(\iota\varepsilon\Lambda)$. <u>Dann besitzt</u> (U) <u>mindestens eine Lösung</u> $u\varepsilon K$, <u>jede Folge von Lösungen</u> u_ι <u>von</u> (U_ι) <u>ist beschränkt und besitzt eine Teilfolge</u> u_ι, $\iota\varepsilon\Lambda' \subset \Lambda$, <u>die gegen</u> $u\varepsilon S(U)$ <u>schwach konvergiert,</u> $u_\iota \rightharpoonup u(\iota\varepsilon\Lambda')$.

Beweis. Zunächst beweisen wir die Beschränktheit beliebiger Folgen $u_\iota \varepsilon S(U_\iota)$, $\iota\varepsilon\Lambda$. Dazu nehmen wir an, es gebe eine Folge $u_\iota \varepsilon S(U_\iota)$ mit $\|u_\iota\| \to \infty(\iota\varepsilon\Lambda)$. Dann haben wir

$$\lim\inf_\Lambda \frac{1}{\|u_\iota\|} (L_\iota v_\iota^o + A_\iota(u_\iota), u_\iota - v_\iota^o)$$

$$\geq \liminf_{\Lambda} \frac{1}{\|u_\iota\|} (A_\iota(u_\iota), u_\iota - v_\iota^o) - \limsup_{\Lambda} \frac{1}{\|u_\iota\|} (L_\iota v_\iota^o, v_\iota^o - u_\iota)$$

$$\geq \liminf_{\Lambda} \frac{1}{\|u_\iota\|} (A_\iota(u_\iota), u_\iota - v^o) - \text{const.} = \infty,$$

dabei wurden (v) und (Lii) herangezogen. Auf der anderen Seite können wir in (U_ι) speziell $v_\iota := v_\iota^o$ setzen,

$$\frac{1}{\|u_\iota\|} (L_\iota v_\iota^o + A_\iota(u_\iota),\ u_\iota - v_\iota^o) \leq (f_\iota,\ \frac{1}{\|u_\iota\|} (v_\iota^o - u_\iota)).$$

Daraus ist zu entnehmen

$$\liminf_{\Lambda} \frac{1}{\|u_\iota\|} (L_\iota v_\iota^o + A_\iota(u_\iota),\ u_\iota - v_\iota^o) \leq \text{const.}$$

Die beiden Folgerungen aus der Annahme schließen sich gegenseitig aus. Also ist jede Folge $u_\iota \varepsilon S(U_\iota)$, $\iota\varepsilon\Lambda$, beschränkt.

Nach 1.(7) besitzt jede beschränkte Folge $u_\iota \varepsilon S(U_\iota)$ eine schwach konvergente Teilfolge mit dem Grenzwert $u\varepsilon E$. Wir haben nun nachzuweisen, daß dieses u Lösung von (U) ist.
Wegen (i) ist zunächst $u\varepsilon K$. Aus (ii) und der Konvergenzbedingung (vi) folgt die Existenz einer Folge $w_\iota \varepsilon K_\iota \cap D(L_\iota)$ mit $w_\iota \to u(\iota\varepsilon\Lambda_o)$ und

$$\limsup_{\Lambda'} (L_\iota w_\iota,\ w_\iota - u_\iota) \leq 0.$$

Aus (U_ι) für $v:=u, v_\iota := w_\iota$, können wir damit folgern

$$\limsup_{\Lambda'} (A_\iota(u_\iota),\ u_\iota - w_\iota) \leq 0$$

und mit (iv) erhalten wir

(3) $\quad (A(u),\ u-v) \leq \liminf_{\Lambda'} (A_\iota(u_\iota),\ u_\iota - v_\iota)$

für alle $v\varepsilon K \cap D(L)$ und alle Folgen $v_\iota \varepsilon K_\iota \cap D(L_\iota)$ mit $v_\iota \to v(\iota\varepsilon\Lambda)$. Auf der anderen Seite folgt aus (U_ι) unter Verwendung von (iii) und (vi)

$$\limsup_{\Lambda'} (A_\iota(u_\iota),\ u_\iota - v_\iota) \leq (f, u-v) + (Lv, v-u).$$

Dies und (3) erbringen die Behauptung. ✱

Unter stärkeren Voraussetzungen läßt sich das Konvergenzresultat verbessern. Wir zeigen zwei Beispiele.

(4) Es seien die Voraussetzungen von (2) erfüllt, außerdem sei die Konvergenzbedingung 2.(2) gegeben und zu der Folge A_ι existiere eine Zahl $\gamma > 0$, so daß für alle Folgen $u_\iota \varepsilon K_\iota$ mit $u_\iota \rightharpoonup u\varepsilon K$ $(\iota\varepsilon\Lambda)$ gilt

(5) $\quad (A_\iota(u_\iota) - A_\iota(w_\iota),\ u_\iota - w_\iota) \geq \gamma \|u_\iota - w_\iota\|$

für alle Folgen $w_\iota \varepsilon K_\iota \cap D(L_\iota)$ mit $w_\iota \to u(\iota\varepsilon\Lambda)$ und $\limsup_{\Lambda} (A_\iota(u_\iota),\ u_\iota - w_\iota) \leq 0$. Dann besitzt jede Folge $u_\iota \varepsilon S(U_\iota)$ eine Teilfolge u_ι, $\iota\varepsilon\Lambda' \subset \Lambda$, die gegen $u\varepsilon S(U)$ stark konvergiert, $u_\iota \to u(\iota\varepsilon\Lambda')$.

Beweis. In Satz (2) haben wir die Existenz einer Lösung u von (U) und die schwache Konvergenz einer Teilfolge von Lösungen u_ι von (U_ι) bewiesen. Wir nehmen an, die Behauptung, daß sogar $u_\iota \to u(\iota\varepsilon\Lambda')$ gilt, sei falsch. Weiter wählen wir eine Folge $w_\iota \varepsilon K_\iota \cap D(L_\iota)$ mit $w_\iota \to u(\iota\varepsilon\Lambda_0)$. Dann haben wir aufgrund der Annahme die Existenz einer Zahl $\delta > 0$ und eines Endstücks $\Lambda_1 \subset \Lambda_0$ mit

$$0 < \delta \leq \gamma \|u_\iota - w_\iota\| \leq (A_\iota(u_\iota) - A_\iota(w_\iota),\ u_\iota - w_\iota)$$

$$\leq (A_\iota(w_\iota),\ w_\iota - u_\iota) + (f_\iota,\ u_\iota - w_\iota) - (L_\iota w_\iota,\ u_\iota - w_\iota)$$

für alle $\iota\varepsilon\Lambda_1$, dabei wurde (5) herangezogen. Mit Hilfe der Konvergenzbedingungen 2.(2) und (2,vi) erschließt man daraus

$$0 < \delta \leq \lim\sup_{\Lambda_1}(A_\iota(w_\iota),\ w_\iota - u_\iota) + \lim\sup_{\Lambda_1}(L_\iota w_\iota, w_\iota - u_\iota) \leq 0.$$

Die Annahme ist also falsch. ✱

Wir bemerken, daß die starke Monotonie aller A_ι mit derselben Funktion hinreichend für (5) ist. Aus 2.(2), (5) und der Beschränktheit der Folge A_ι folgt übrigens die a-Pseudomonotonie des Paares A, A_ι.

Die weitere Folgerung aus Satz (2) müssen wir erst vorbereiten.

(6) <u>Die Folge</u> E_ι <u>heißt gleichgradig uniform konvex, wenn es zu jedem</u> $\varepsilon\varepsilon[0,2]$ <u>ein</u> $\delta(\varepsilon) > 0$ <u>gibt, so daß für alle Folgen</u> $u_\iota, v_\iota \varepsilon E_\iota$, $\iota\varepsilon\Lambda$, <u>aus</u>

$$\forall \iota\varepsilon\Lambda: \|u_\iota\|, \|v_\iota\| \leq 1,\ \|u_\iota - v_\iota\| \geq \varepsilon \Rightarrow \exists\, \Lambda' \subset \Lambda: \tfrac{1}{2}\|u_\iota + v_\iota\| \leq 1 - \delta(\varepsilon).$$

Nun besteht folgendes Resultat von Mirgel [7].

(7) <u>Wenn die Folge</u> E_ι <u>gleichgradig uniform konvex ist, dann gilt für jede Folge</u> $u_\iota \varepsilon E_\iota$

$$u_\iota \rightharpoonup u(\iota\varepsilon\Lambda),\ \|u_\iota\| \to \|u\|(\iota\varepsilon\Lambda) \Rightarrow u_\iota \to u(\iota\varepsilon\Lambda).$$

In diesem Zusammenhang erhalten wir folgendes Korollar zu (2).

(8) <u>Es seien die Voraussetzungen von</u> (2) <u>erfüllt, außerdem liege</u> 2.(2) <u>und folgende Bedingung vor: Die Folge</u> E_ι <u>sei gleichgradig uniform konvex und für alle Folgen</u> $u_\iota \varepsilon K_\iota$ <u>mit</u> $u_\iota \rightharpoonup u\varepsilon K(\iota\varepsilon\Lambda)$ <u>gelte</u>

(9) $$\lim\inf_\Lambda (A_\iota(u_\iota) - A_\iota(w_\iota),\ u_\iota - w_\iota) \geq 0$$

<u>und</u>

(10) $$(A_\iota(u_\iota) - A_\iota(w_\iota),\ u_\iota - w_\iota) \to 0(\iota\varepsilon\Lambda) \Rightarrow \|u_\iota\| \to \|u\|\ (\iota\varepsilon\Lambda)$$

<u>für alle Folgen</u> $w_\iota \varepsilon K_\iota \cap D(L_\iota)$ <u>mit</u> $w_\iota \to u(\iota\varepsilon\Lambda)$ <u>und</u> $\lim\sup_\Lambda (A_\iota(u_\iota),\ u_\iota - w_\iota) \leq 0$. <u>Dann besitzt jede Folge</u> $u_\iota \varepsilon S(U_\iota)$ <u>eine Teilfolge</u> u_ι, $\iota\varepsilon\Lambda' \subset \Lambda$, <u>die gegen</u> $u\varepsilon S(U)$ <u>stark konvergiert,</u> $u_\iota \to u(\iota\varepsilon\Lambda')$.

Beweis. Wie im Beweis von (4) beschränken wir uns darauf, für $u_\iota \in S(U_\iota)$ zu zeigen, daß $u_\iota \to u \in S(U)(\iota\in\Lambda')$. Aus dem Beweis von (2) wissen wir, daß bei der Folge u_ι gilt $\lim\sup_\Lambda (A_\iota(u_\iota), u_\iota - w_\iota) \leq 0$ für eine Folge $w_\iota \in K_\iota \cap D(L_\iota)$ mit $w_\iota \to u(\iota\in\Lambda_o)$. Dafür ergibt sich dann

$$\lim\sup_\Lambda (A_\iota(u_\iota) - A_\iota(w_\iota), u_\iota - w_\iota)$$
$$\leq \lim\sup_\Lambda (A_\iota(u_\iota), u_\iota - w_\iota) + \lim\sup_\Lambda (A_\iota(w_\iota), u_\iota - w_\iota) \leq 0$$

wegen der Bedingung 2.(2). Ziehen wir nun (9) und anschließend (10) heran, so ist mit (7) die Behauptung bewiesen.✱

Wir bemerken, daß man (9) und (10) aus der folgenden Bedingung erschließen kann: Für alle Folgen $u_\iota \in K_\iota$ mit $u_\iota \rightharpoonup u(\iota\in\Lambda)$ gilt

(11) $$(A_\iota(u_\iota) - A_\iota(w_\iota), u_\iota - w_\iota) \geq (\varphi(\|u_\iota\|) - \varphi(\|w_\iota\|))(\|u_\iota\| - \|w_\iota\|)$$

mit einer strikt monotonen Funktion $\varphi : \mathbb{R}_+ \to \mathbb{R}_+$ mit $\varphi(r) \to \infty$ $(r \to \infty)$ und für alle Folgen $w_\iota \in K_\iota \cap D(L_\iota)$ mit $w_\iota \to u(\iota\in\Lambda)$. Die Eigenschaft (11) ihrerseits liegt vor, wenn alle A_ι φ-monoton sind mit derselben Funktion φ. Aus (11), 2.(2) und der Beschränktheit der Folge A_ι folgt übrigens auch wieder die a-Pseudomonotonie des Paares A, A_ι.

Es ist klar, daß die bisher erzielten Resultate den Fall stationärer Probleme umfassen. Die entsprechende Aussage entspricht im wesentlichen zentralen Resultaten der Arbeit [7] von Mirgel und lautet folgendermaßen.

(12) Es sei (i) w-Lim sup $K_\iota \subset K$, (ii) $K \subset$ s-Lim inf K_ι, (iii) die Folge $f_\iota \in E_\iota^*$ konvergent gegen $f \in E^*$, $f_\iota \to f(\iota\in\Lambda)$, (iv) das Paar A, A_ι a-pseudomonoton auf K, K_ι. (v) Mit einer beschränkten Folge $v_\iota^o \in K_\iota$ gelte

$$\frac{1}{\|w_\iota\|}(A_\iota(w_\iota), w_\iota - v_\iota^o) \to \infty \quad (\|w_\iota\| \to \infty,\ w_\iota \in K_\iota,\ \iota\in\Lambda).$$

Dann besitzt die Aufgabe

$$\forall v \in K: (A(u), v-u) \geq (f, v-u)$$

eine Lösung, jede Folge von Lösungen der Aufgaben

$$\forall v_\iota \in K_\iota: (A_\iota(u_\iota), v_\iota - u_\iota) \geq (f_\iota, v_\iota - u_\iota),\ \iota\in\Lambda_o,$$

ist beschränkt und besitzt eine Teilfolge u_ι, $\iota\in\Lambda' \subset \Lambda$, die gegen $u \in S(U)$ schwach konvergiert, $u_\iota \rightharpoonup u(\iota\in\Lambda')$.

Die Folgerungen (4) und (8) bleiben offensichtlich auch für den Fall stationärer Probleme bestehen.

Anschließend beschäftigen wir uns mit Penalty-Methoden. Um anstelle von approximierenden Ungleichungen auf K_ι nun eine Folge von Gleichungen auf E_ι heranziehen zu können, bedienen wir uns der Penalty-Operatoren aus der folgenden Definition.

(13) Das Paar P,P_ι heißt Paar konvergenter Penalty-Operatoren für K,K_ι, wenn folgende Bedingungen erfüllt sind. (i) Die Folge P_ι sei beschränkt, (ii) das Paar P,P_ι erfülle die Bedingung 2.(2), (iii) für alle Folgen $u_\iota, v_\iota \varepsilon E_\iota$ gelte

$$\lim\inf_\Lambda\ (P_\iota(u_\iota)-P_\iota(v_\iota),\ u_\iota-v_\iota) \geq 0,$$

und (iv) sei $K \subset$ s-Lim inf K_ι, K = Kern (P) und K_ι = Kern (P_ι) für alle $\iota\varepsilon\Lambda_o$.

Wegen 2.(3) ist P in einem Paar konvergenter Penalty-Operatoren P,P_ι hemistetig in E, und wegen 2.(4) bzw. 2.(5) haben wir a-pseudomonotone Paare P,P_ι bzw. $A+P$, $A_\iota+\alpha_\iota P_\iota$ mit einem a-pseudomonotonen Paar A,A_ι und einer Zahlenfolge $a_\iota \varepsilon \mathbb{R}_+$.

Wir beschäftigen uns jetzt neben (U) mit Gleichungen (G_ι). Dafür steht etwa der folgende Existenzsatz [6] zur Verfügung (Auch hier wird (L,i) benötigt).

(14) Es sei E_ι und E_ι^* strikt konvex, L_ι eine lineare, abgeschlossene Abbildung mit Definitionsmenge $D(L_\iota)$ von E_ι in E_ι^* und $A_\iota+\alpha_\iota P_\iota$ eine pseudomonotone Abbildung von E_ι in E_ι^* mit

$$\frac{1}{\|u_\iota\|}\ (A_\iota(u_\iota)+\alpha_\iota P_\iota(u_\iota),\ u_\iota-v_\iota^o) \to \infty \quad (\|u_\iota\| \to \infty,\ u_\iota \varepsilon D(L_\iota))$$

für ein Element $v_\iota^o \varepsilon D(L_\iota)$. Dann besitzt (G_ι) mindestens eine Lösung.

Wir nehmen für die anschließenden Ausführungen wieder an, daß zumindest eine Folge von Aufgaben (G_ι), $\iota\varepsilon\Lambda$, eine Lösung besitzt. Es besteht dann der folgende Konvergenzsatz.

(15) Es sei (i) $\alpha_\iota \varepsilon \mathbb{R}_+$ eine Zahlenfolge mit $\alpha_\iota \to \infty$ $(\iota\varepsilon\Lambda_o)$,
(ii) die Folge $f_\iota \varepsilon E_\iota^*$ konvergent gegen $f\varepsilon E^*$, $f_\iota \to f(\iota\varepsilon\Lambda)$,
(iii) P,P_ι ein konvergentes Paar von Penalty-Operatoren für K,K_ι,
(iv) für die Folge A_ι für jede Folge $w_\iota \varepsilon D(L_\iota)$ mit $\|w_\iota\| \to \infty$ $(\iota\varepsilon\Lambda)$

$$\frac{1}{\|w_\iota\|}\ (A_\iota(w_\iota),\ w_\iota-v_\iota^o) \to \infty \quad (\iota\varepsilon\Lambda)$$

mit einer beschränkten Folge $v_\iota^o \varepsilon K_\iota \cap D(L_\iota)$,
(v) das Paar A,A_ι a-pseudomonoton auf E,E_ι,
(vi) $K \subset$ s-Lim inf $K_\iota \cap D(L_\iota)$ und
(vii) für alle Folgen $u_\iota \varepsilon K_\iota$ mit $u_\iota \rightharpoonup u(\iota\varepsilon\Lambda)$

$$\lim\sup_\Lambda\ (L_\iota v_\iota,\ v_\iota-u_\iota) \leq (Lv,\ v-u)$$

für alle $v\varepsilon K \cap D(L)$ und alle Folgen $v_\iota \varepsilon K_\iota \cap D(L_\iota)$ mit $v_\iota \to v(\iota\varepsilon\Lambda)$.
Dann besitzt die Aufgabe (U) eine Lösung $u\varepsilon K$, jede Folge von Lösungen u_ι von (G_ι) ist beschränkt und besitzt eine Teilfolge u_ι, $\iota\varepsilon\Lambda' \subset \Lambda$, die gegen $u\varepsilon S(U)$ schwach konvergiert, $u_\iota \rightharpoonup u(\iota\varepsilon\Lambda)$.

Beweis. Zunächst zeigen wir, daß jede Folge von Lösungen von (G_ι) beschränkt ist. Dazu nehmen wir an, es gebe eine Folge von Lösungen $u_\iota \varepsilon S(G_\iota)$ mit $\|u_\iota\| \to \infty$ $(\iota\varepsilon\Lambda)$. Dann haben wir mit der Folge v_ι^o aus (iv)

$$\liminf_\Lambda \frac{1}{\|u_\iota\|} (L_\iota u_\iota + A_\iota(u_\iota) + \alpha_\iota P_\iota(u_\iota),\ u_\iota - v_\iota^o)$$

$$\geq \liminf_\Lambda \frac{1}{\|u_\iota\|} (A_\iota(u_\iota),\ u_\iota - v_\iota^o)$$

$$+ \liminf_\Lambda \frac{\alpha_\iota}{\|u_\iota\|} (P_\iota(u_\iota) - P_\iota(v_\iota^o),\ u_\iota - v_\iota^o)$$

$$+ \liminf_\Lambda \frac{1}{\|u_\iota\|} (L_\iota(u_\iota - v_\iota^o),\ u_\iota - v_\iota^o)$$

$$- \limsup_\Lambda \frac{1}{\|u_\iota\|} (L_\iota v_\iota^o,\ v_\iota^o - u_\iota)$$

$$\geq \infty .$$

Dabei muß (iv), $v_\iota^o \varepsilon K_\iota$, (13,iv) und (13,iii), (L,i) und (L,ii) herangezogen werden. Verwenden wir auf der anderen Seite, daß $u_\iota \varepsilon S(G_\iota)$, so haben wir

$$\liminf_\Lambda \frac{1}{\|u_\iota\|} (L_\iota u_\iota + A_\iota(u_\iota) + \alpha_\iota P_\iota(u_\iota),\ u_\iota - v_\iota^o)$$

$$= \liminf_\Lambda \frac{1}{\|u_\iota\|} (f_\iota, u_\iota - v_\iota^o)$$

$$\leq \text{const.}$$

Diese beiden Relationen stehen im Widerspruch zueinander, damit ist die Annahme falsch. Jetzt haben wir also, daß jede Folge $u_\iota \varepsilon S(G_\iota)$ beschränkt ist. Dies gilt wegen (13,i) und (v) dann auch von den Folgen $P_\iota(u_\iota)$ und $A_\iota(u_\iota)$. Aufgrund von 1.(7) existiert also eine Teilfolge $\Lambda' \subset \Lambda$ mit

$$u_\iota \rightharpoonup u\varepsilon E(\iota\varepsilon\Lambda'),\quad P_\iota(u_\iota) \rightharpoonup r\varepsilon E^*(\iota\varepsilon\Lambda'),\quad A_\iota(u_\iota) \rightharpoonup s\varepsilon E^*(\iota\varepsilon\Lambda').$$

Wir haben nun noch zu beweisen, daß $u\varepsilon K$ und eine Lösung von (U) ist. Dazu ziehen wir zunächst (G_ι) heran und erschließen aus (13,i), (v), (L,i), (L,ii) und (i) zunächst, daß $\|P_\iota(u_\iota)\|_* \to 0(\iota\varepsilon\Lambda)$ konvergiert. Daraus entnehmen wir, daß $r = 0$ ist. Ferner haben wir

$$\limsup_\Lambda (P_\iota(u_\iota),\ u_\iota)$$

$$\leq \limsup_\Lambda \frac{1}{\alpha_\iota} ((f_\iota, u_\iota) - (A_\iota(u_\iota), u_\iota)) - \liminf \frac{1}{\alpha_\iota} (L_\iota u_\iota, u_\iota) \leq 0.$$

Damit haben wir alle Voraussetzungen von Hilfssatz 2.(9) zur Anwendung auf P, P_ι, u_ι, u und $f = 0$ verifiziert und erhalten daher $P(u) = 0$, wegen (13,iv) also $u\varepsilon K$. Aufgrund von (vi) haben wir eine Folge $w_\iota \varepsilon K_\iota \cap D(L_\iota)$ mit $w_\iota \to u(\iota\varepsilon\Lambda_o)$. Mit ihr erhalten wir für $\iota\varepsilon\Lambda$

$$(16)\qquad \begin{aligned}&(L_\iota w_\iota + A_\iota(u_\iota) - f_\iota,\ w_\iota - u_\iota)\\ &= (L_\iota u_\iota + A_\iota(u_\iota) - f_\iota,\ w_\iota - u_\iota) + (L_\iota(w_\iota - u_\iota),\ w_\iota - u_\iota) \geq 0,\end{aligned}$$

denn $u_\iota \varepsilon S(G_\iota)$, ferner wurde (L,i) verwendet. Aufgrund der Konvergenzbedingung (vii) und (ii) entnehmen wir daraus weiter

$$\lim\sup_\Lambda\ (A_\iota(u_\iota),\ u_\iota - w_\iota) \leq 0.$$

Nun ziehen wir (v) heran mit dem Ergebnis

$$(17)\qquad (A(u),\ u-v) \leq \lim\inf_\Lambda\ (A_\iota(u_\iota),\ u_\iota - v_\iota)$$

für jedes $v\varepsilon K \cap D(L)$ und alle Folgen $v_\iota \varepsilon K_\iota \cap D(L_\iota)$ mit $v_\iota \to v(\iota\varepsilon\Lambda)$.

Auf der anderen Seite folgt aus (vi), daß es zu jedem $v\varepsilon K \cap D(L)$ mindestens eine Folge $v_\iota \varepsilon K_\iota \cap K(L_\iota)$ mit $v_\iota \to v(\iota\varepsilon\Lambda_o)$ gibt. Wie bei (16) erhalten wir dann für $\iota\varepsilon\Lambda$

$$(L_\iota v_\iota + A_\iota(u_\iota) - f_\iota,\ v_\iota - u_\iota) \geq 0,$$

was dann letzten Endes zu

$$\begin{aligned}\lim\sup_\Lambda(A_\iota(u_\iota),\ u_\iota - v_\iota) &\leq \lim\sup_\Lambda(f_\iota, u_\iota - v_\iota) + \lim\sup_\Lambda(L_\iota v_\iota, v_\iota - u_\iota)\\ &\leq (f,\ u-v) + (Lv,\ v-u)\end{aligned}$$

führt. Dies ergibt mit (17) zusammen gerade die Behauptung. ✱

Auch für die approximierenden Folgen von Lösungen von (G_ι) kann man unter verschärften Bedingungen die starke Konvergenz behaupten.

(18) <u>Die Voraussetzungen von</u> (15) <u>seien erfüllt, außerdem liege</u> 2.(2) <u>und</u> (5) <u>vor. Dann besitzt jede Folge</u> $u_\iota \varepsilon S(G_\iota)$ <u>eine Teilfolge, die gegen</u> $u\varepsilon S(U)$ <u>stark konvergiert,</u> $u_\iota \to u(\iota\varepsilon\Lambda')$.

Beweis. Wir nehmen an, die in (15) nachgewiesene schwach konvergente Folge $u_\iota \varepsilon S(G_\iota)$ sei nicht stark konvergent, d.h. mit einer Folge $w_\iota \varepsilon K_\iota \cap D(L_\iota)$ mit $w_\iota \to u(\iota\varepsilon\Lambda_o)$ nehmen wir die Existenz einer Zahl $\delta > 0$ und eines Endstücks $\Lambda_1 \subset \Lambda_o$ an, so daß für alle $\iota\varepsilon\Lambda_1$ gilt

$$0 < \delta \leq \gamma\, \|u_\iota - w_\iota\| \leq (A_\iota(u_\iota) - A_\iota(w_\iota),\ u_\iota - w_\iota).$$

Hieraus ist mit Hilfe von 2.(2), (15,ii), (L,i), (15,vii) und $w_\iota \varepsilon K_\iota$, (13,iv) und (13,iii) zu entnehmen, daß

$$\begin{aligned}0 < \delta \leq\ &\lim\sup_{\Lambda'}\ (A_\iota(w_\iota),\ w_\iota - u_\iota) + \lim\sup_{\Lambda'}\ (f_\iota,\ u_\iota - w_\iota)\\ &- \lim\inf_{\Lambda'}\ (L_\iota(u_\iota - w_\iota),\ u_\iota - w_\iota) + \lim\sup_{\Lambda'}\ (L_\iota w_\iota,\ w_\iota - u_\iota)\\ &- \lim\inf_{\Lambda'}\ (P_\iota(u_\iota) - P_\iota(w_\iota),\ u_\iota - w_\iota) \leq 0.\end{aligned}$$

Damit ist die Annahme als falsch nachgewiesen. ✱

Eine analoge Aussage zur Folgerung (9) kann man auf entsprechende Weise erzielen: Die starke Konvergenz der Lösungen liegt vor, wenn die Folge E_ι gleichgradig uniform konvex ist und neben den Voraussetzungen von (15) die Bedingungen 2.(2), (9) und (10) bestehen. Abschließend geben wir noch die Formulierung von (15) für den speziellen Fall stationärer Probleme. Die starke Konvergenz der Lösungen erhält man wieder entsprechend zu (18) bzw. der vorstehenden Bemerkung.

(19) Es sei (i) $\alpha_\iota \in \mathbb{R}_+$ eine Zahlenfolge mit $\alpha_\iota \to \infty (\iota \in \Lambda_0)$, (ii) die Folge $f_\iota \in E_\iota^*$ konvergent gegen $f \in E^*$, $f_\iota \to f(\iota \in \Lambda)$, (iii) P, P_ι ein konvergentes Paar von Penalty-Operatoren für K, K_ι, (iv) für die Folge A_ι für jede Folge $w_\iota \in K_\iota$ mit $\|w_\iota\| \to \infty \ (\iota \in \Lambda)$

$$\frac{1}{\|w_\iota\|} (A_\iota(w_\iota), w_\iota - v_\iota^o) \to \infty \ (\iota \in \Lambda)$$

mit einer beschränkten Folge $v_\iota^o \in K_\iota$, und (v) das Paar A, A_ι a-pseudomonoton auf E, E_ι. Dann besitzt die Aufgabe

$$\forall v \in K \cap D(L) \quad (A(u), v-u) \geq (f, v-u)$$

eine Lösung, jede Folge von Lösungen der Aufgaben

$$A_\iota(u_\iota) + P_\iota(u_\iota) = f_\iota \ , \quad \iota \in \Lambda_0,$$

ist beschränkt und besitzt eine Teilfolge u_ι, $\iota \in \Lambda' \subset \Lambda$, die gegen $u \in S(U)$ schwach konvergiert, $u_\iota \rightharpoonup u(\iota \in \Lambda')$.

4. Das Galerkin-Verfahren.

Im letzten Abschnitt führen wir vor, wie die erzielten Resultate bei einer speziellen Auswahl von Räumen E_ι und Operatoren L_ι, A_ι aussehen, obwohl wir dabei den diesbezüglichen Rahmen unserer allgemeinen Störungstheorie in keiner Weise ausschöpfen.
Beim Galerkin-Verfahren liegt folgende Situation vor: In einem reflexiven und separablen Banach-Raum E ist eine Folge beschränkt-linearer Projektionsoperatoren $p_\iota : E \to E$, $\iota \in \mathbb{N}$, gegeben, deren Bilder $E_\iota := p_\iota E$, $\iota \in \mathbb{N}$, dann ihrerseits wieder reflexive und separable Banach-Räume sind, von denen wir fordern

$$E_\iota \subset E_{\iota+1}, \ \iota \in \mathbb{N}, \quad cl(\bigcup_{\iota \in \mathbb{N}} E_\iota) = E.$$

Daraus folgt folgende Eigenschaft der Folge p_ι:

(1) $\qquad \forall u \in E \ \|u - p_\iota u\| \to 0(\iota \in \mathbb{N})$.

Für die zu E, E_ι dualen Räume E^*, E_ι^* gilt $E^* \subset E_\iota^*$, $\iota \in \mathbb{N}$, für die zu p_ι dualen Operatoren $p_\iota^* : E_\iota^* \to E^*$, $\iota \in \mathbb{N}$.

Von Mirgel [7] wurde bewiesen, daß unter den hier vorliegenden Bedingungen (E^*,E), (E_ι^*,E_ι) eine diskrete Approximation im Sinne von (D) ist, falls

(2) $\forall \iota\in\mathbb{N}\ \forall f\in E^*\ \forall v_\iota\in E_\iota\ (f,v_\iota)_{E_\iota^*,E_\iota} = (f,v_\iota)_{E^*,E}$.

In dieser diskreten Approximation sind ferner die diskrete starke bzw. schwache Konvergenz von Folgen $u_\iota\in E_\iota$ äquivalent mit der üblichen starken bzw. schwachen Konvergenz in E.

Die Bedingungen (1) und (2) werden bei den folgenden Überlegungen durchweg vorausgesetzt.

Als Beispiel für die geschilderten Verhältnisse nehmen wir an, in E sei eine Schauder-Basis $\varphi_j, j\in\mathbb{N}$, gegeben. Dann gibt es zu jedem $u\in E$ eindeutig bestimmte Zahlen $a_j^\iota(u)\in\mathbb{R}$, $j = 1,\ldots,\iota, \iota\in\mathbb{N}$, mit denen gilt

$$\left\| u - \sum_{j=1}^{\iota} a_j^\iota(u)\,\varphi_j \right\| \to 0 \quad (\iota\in\mathbb{N}).$$

Daran anschließend können beschränkt-lineare Projektionsoperatoren p_ι definiert werden durch

$$\forall u\in E\ \forall \iota\in\mathbb{N}\quad p_\iota(u) := \sum_{j=1}^{\iota} a_j^\iota(u)\,\varphi_j .$$

Es sind also $E_\iota = \mathrm{span}\,\{\varphi_1,\ldots,\varphi_\iota\}$.

Nun sei $K\subset E$ eine konvexe, abgeschlossene Teilmenge mit $\mathrm{int}(K) \neq \emptyset$. Setzen wir

(3) $\forall \iota\in\mathbb{N}\ K_\iota := K\cap E_\iota$,

dann ist aufgrund unserer Voraussetzungen offenbar

$$\text{w-Lim sup } K_\iota \subset K, \quad K\subset \text{s-Lim inf } K_\iota .$$

Falls man für das folgende die im Hinblick auf die Existenzsätze 3.(1) und 3.(14) angebrachte weitere Voraussetzung trifft, daß alle Räume E,E^*,E_ι,E_ι^* strikt konvex sind, so existiert zu jeder konvexen, abgeschlossenen Teilmenge $K\subset E$ stets ein beschränkter, monotoner und hemistetiger Operator $P: E\to E$ mit $K = \mathrm{Kern}\ (P)$, der mit Hilfe eines Dualitätsoperators $\mathfrak{J}$ definert werden kann [6],

$$\forall u\in E\quad P(u) := \mathfrak{J}(u-\pi_K(u)),$$

wobei mit $\pi_K(u)$ das eindeutig bestimmte Element aus K mit

$$\forall k\in K\quad \|u-\pi_K(u)\| \leq \|u-k\|$$

bezeichnet wird. Wir erhalten dann ein konvergentes Paar von Penalty-Operatoren P,P_ι für K,K_ι, indem wir setzen

(4) $\forall \iota\in\mathbb{N}\quad P_\iota := Pp_\iota,\quad K_\iota := \mathrm{Kern}\ (P_\iota)$.

Bei den Operatoren setzt man beim Galerkin-Verfahren

$$\forall \iota\in\mathbb{N} \quad A_\iota := A|E_\iota, \quad L_\iota := L|K_\iota \cap D(L).$$

Diese Abbildungen operieren von E_ι in E_ι. Unter Berücksichtigung von Sätzen von Lions [6] lassen sich die Bedingungen (L,i), (L,ii) und 3.(2,vi) bzw. 3.(15,vii) bereits aus der Voraussetzung von (L,i) für L selbst erschließen. Ferner folgt offenbar aus der Forderung der Beschränktheit und Pseudomonotonie von A in K, daß auch alle A_ι in K_ι pseudomonoton sind und das Paar A,A a-pseudomonoton auf K,K_ι ist. Unter Verwendung des Begriffs der Koerzivität von A,

$$\frac{1}{\|w\|}(A(w),\ w-v^o) \to \infty \ (\|w\| \to \infty),\ v\in K\cap D(L),$$

erhalten wir dann die beiden folgenden Konvergenzsätze für das Galerkin-Verfahren.

(5) Es sei A eine in K definierte beschränkte, pseudomonotone und koerzitive sowie L eine in D(L) definierte lineare und abgeschlossene Abbildung in E^* mit

$$\forall u\in D(L) \quad (Lu,u) \geqslant 0, \qquad \forall u\in D(L^*) \quad (L^*u,u) \geqslant 0$$

sowie

$$\forall v\in K \quad \exists\, v^j\in K \cap D(L) \quad v^j \to v(j\in\mathbb{N}),\ \limsup_{\mathbb{N}}\ (Lv^j, v^j-v) \leqslant 0.$$

Ferner sei $f_\iota := p_\iota f$, $\iota\in\mathbb{N}$, für $f\in E^*$. Dann besitzen die Aufgaben (U) und (U_ι) mit K_ι aus (3) jeweils eine Lösung, jede Folge von Lösungen $u_\iota\in S(U_\iota)$ ist beschränkt und besitzt eine Teilfolge, die gegen $u\in S(U)$ schwach konvergiert, $u_\iota \rightharpoonup u(\iota\in\mathbb{N}')$.

(6) Die Voraussetzungen von (5) seien erfüllt. Dann besitzen die Aufgaben (U) und (G_ι) mit einer Zahlenfolge $\alpha_\iota\in\mathbb{R}_+$ mit $\alpha_\iota \to \infty$ $(\iota\in\mathbb{N})$ und den Penalty-Operatoren P_ι aus (4) jeweils eine Lösung, jede Folge von Lösungen $u_\iota\in S(G_\iota)$ ist beschränkt und besitzt eine Teilfolge, die gegen $u\in S(U)$ schwach konvergiert, $u_\iota \rightharpoonup u(\iota\in\mathbb{N}')$.

Die Bedingungen, die man stellen muß, um beim Galerkin-Verfahren die starke Konvergenz von Teilfolgen von Lösungen zu erhalten, sind ebenfalls klar: A kann stetig sein mit

$$\forall u,v\in K \quad (A(u)-A(v),\ u-v) \geqslant \gamma\|u-v\|,\ \gamma > 0,$$

bzw. es kann E uniform konvex und A monoton und stetig sein mit der Bedingung: Wenn für eine Folge $u^j\in K$ gilt

$$(A(u^j)-A(u), u^j-u) \to 0(j\in\mathbb{N}) \Longrightarrow \|u^j\| \to \|u\|(j\in\mathbb{N}).$$

Hierfür ist wieder die φ-Monotonie von A hinreichend.

Literatur.

[1] Aubin, J.P.: Approximation des éspaces de distributions et des opérateurs différentials. Bull.Soc.Math. France, Mémoire 12 (1967).

[2] Auslender, A.: Problémes de minimax via l'analyse convexe et les inégalités variationelles: Theorie et algorithmes. Lecture Notes in Economics and Mathematical Systems, Vol.77. Springer, Berlin 1972.

[3] Brézis, H.-Sibony, M.: Méthodes d'approximation et d'itération pour les opérateurs monotones. Arch.Rat.Mech.Anal. 28, 59-82 (1968).

[4] Duvant, G. - Lions, J.L.: Les inéquations en mécanique et en physique. Dunod, Paris 1972.

[5] Grigorieff, R.D.: Über diskrete Approximationen nichtlinearer Gleichungen 1. Art. Mathematische Nachrichten.

[6] Lions, J.L.: Quelques méthodes de résolution des problémes aux limites non linéaires. Dunod-Gauthier-Villars, Paris 1969.

[7] Mirgel, W.: Eine allgemeine Störungstheorie für Variationsungleichungen. Dissertation Frankfurt a.M. 1971.

[8] Mosco, U: Convergence of convex sets and variational inequalities. Adv.Math. 3, 510-585 (1969).

[9] Sibony, M.: Contrôle des systémes gouvernés par des équations aux derivées partielles. Rend.Sem.Mat.Univ. Padova 43, 277-339 (1970).

[10] Sibony, M.: Sur l'approximation d'équations et d'inéquations aux derivées partielles non linéaires de type monotone. J.Math.Anal.Appl. 34, 502-564 (1971).

[11] Sibony, M.: Some numerical techniques for optimal control governed by partial differential equations. Numer.Meth. bei Optimierungsaufgaben, ed. L. Collatz, W. Wetterling, ISNM Vol. 17. Birkhäuser, Basel 1973.

[12] Stummel, F.: Diskrete Konvergenz linearer Operatoren I. Math. Ann. 190, 45-92 (1970).

[13] Stummel, F.: Discrete convergence of mappings. Topics in numerical analysis, ed. J.J.H. Miller. Academic Press, New York 1973.

Prof. Dr. H. Jeggle
Fachbereich Mathematik der Technischen Universität
1000 B e r l i n 12
Straße des 17. Juni 135

On Optimal Control Problems with State Space Constraints

H.W. Knobloch

1. Introduction.

The purpose of this report is to give a summary on a recent contribution to optimal control theory. All details and related literature can be found in a forthcoming paper [2]. The type of problems treated and the basic principles are not new, nevertheless the method we are going to present is more than an exploitation of existing techniques and seems especially suited for attacking questions of qualitative nature.

Let us first explain the type of control problems we are going to consider. We will assume that the following data are given

(i) A differential equation $\dot{x} = f(t,x;u)$, where f is a vector-valued function of t, of the state variable $x = (x^1,\dots,x^n)$ and of the control variable $u = (u^1,\dots,u^m)$. f is supposed to be continuous and have continuous partial derivatives with respect to x and u on the whole (t,x,u)-space.

(ii) The control set U, that is an arbitrary non-empty set in u-space. An admissible control-function is a piecewise continuous function u(t) which assumes values in U.

(iii) Two manifolds of the x-space M_0, M_1 (initial and terminal manifold).

(iv) A scalar function $\gamma(t,x)$ which is defined and of class C^2 on the whole (t,x)-space.

We say that a control problem is defined by these data. A pair (x(t), u(t)) is called a solution of the control problem, if u(t) is an admissible control function and x(t) a solution of the initial value problem

(1.1) $\dot{x} = f(t,x;u(t))$, $x(t_0) \in M_0$,

and if there exists a $\tilde{t} > 0$ such that x(t) satisfies the conditions

$$x(\tilde{t}) \in M_1,\ \gamma(t,x(t)) \le 0 \text{ for all } t \in [t_0,\tilde{t}].$$

For shortness we will refer to x(t) as an admissible trajectory, if there exists an admissible control function u(t) such that (u(t),x(t)) is a solution in the sense just explained.

We further assume that there is given a cost functional in the usual form (that is, the values of the functional can be identified with the terminal values of one component of the augmented state variable).
An optimal solution of the control problem is then a solution which minimizes this functional.

In section 4 of this paper we introduce a set of first order necessary conditions for optimal solutions, which differ essentially from the type of conditions appearing elsewhere in the literature. The main difference concerns the form and the use of the Hamiltonian function H. This function appears in our version in the same form as for control problems without space contraints, however the maximum principle then assumes a somewhat modified form. The usual approach is different and involves keeping the familiar form of the maximum condition at the expense of a modification of H, which then involves an additional Lagrange multiplier λ. From the viewpoint of applications the absence of λ is the main advantage of our method as will be shown by an example in section 5. The elimination of the additional multiplier is achieved by the usage of a new kind of cone of attainability. This is a convex cone (in x-space) which can be associated with any (not necessarily optimal) solution, provided the trajectory x(t) satisfies three conditions. That we cannot work without these conditions is a certain disadvantage of our method. The third condition will be discussed in detail in the next section, the two first will be stated now.

We consider the scalar function $\gamma_o(t) = \gamma(t,x(t))$ and assume that the following statements are true.

(1.2) The set $\{t\in[t_o,\tilde{t}]\,|\,\gamma_o(t) = 0\}$ is the union of finitely many disjoint closed intervals (some may reduce to isolated points).

The left (right) endpoint of such an interval is called a left (right) entry point. Then the second condition can be phrased this way

(1.3) If $t^*\in(t_o,\tilde{t}]$ is a left entry point, then $\dot{\gamma}(t^* - 0) > 0$. If $t^*\in[t_o,\tilde{t})$ is a right entry point then $\dot{\gamma}(t^*+0) < 0$.

It should be mentioned that condition (1.3) can be weakened, however our main theorem (section 4) then will take a somewhat complicated form Even in its present form (1.3) is not as restrictive as it may look. In many applications (1.3) turns out to be a consequence of the fact that the bang-bang-principle holds along those parts of an optimal trajectory which are in the interior of the region $\{t,x\,|\,\gamma(t,x)\le 0\}$. Before we begin with the discussion of the third condition, we wish to make a few remarks concerning our notation. At points of discontinuity (with respect to t) the value of a function is always equal to the left-hand limit. Matrix notation is used exclusively, in particular matrices of one column will be called vectors and denoted by x,y etc., with $\|x\|$ signifying the maximum norm. The transpose of a matrix A is denoted by A^T.

2. The Condition on the Constraint Function.

We introduce the following notation

$$g(t,x) = (\gamma_{x^1}(t,x),\ldots,\gamma_{x^n}(t,x))^T$$

(2.1)

$$\Gamma(t,x,u) = \gamma_t(t,x)+g(t,x)^T f(t,x;u).$$

g is the gradient of γ with respect to x and Γ the derivative of γ with respect to the differential equation (1.1). We are now going to state a further condition, which concerns the behaviour of the function Γ in a neighbourhood of a given solution $(u(t),x(t))$ of our control problem. The construction of the cone of attainability associated with that solution depends essentially upon this condition.

(2.2) For every t^* which is such that $\gamma_o(t) = \gamma(t,x(t))$ vanishes on some interval $[t^*,t^*+\varepsilon]\subseteq[t_o,\tilde{t}]$ one can find a neighbourhood N of $(t^*,x(t^*))$ and a function $u(t,x)$ with these properties:
(i) u is continuous and has a continuous partial derivative with respect to x, for all $(t,x)\in N$,
(ii) $u(t,x)\in U$ if $(t,x)\in N$,
(iii) $\Gamma(t,x,u(t,x)) \leq 0$ if $(t,x)\in N$ and $\gamma(t,x) \leq 0$,
(iv) $u(t,x(t)) = u(t)$ if $(t,x(t))\in N$ and $t>t^*$

Similarly, if $\gamma_o(t)$ vanishes on some interval $(t^*-\varepsilon, t^*]$ there exists a function $u(t,x)$ with the properties (i) - (iii) and
(iv') $u(t,x(t)) = u(t)$ if $(t,x(t))\in N$ and $t<t^*$.

In passing we note, that the existence of a function $u(t,x)$ with all properties (2.2) can often be inferred from the implicit function theorem. Indeed, if there exists a m-dimensional vector v such that the directional derivative of $\Gamma(t^*,x(t^*),u)$ (regarded as a function of u) in the direction of v does not vanish at $u = u(t^*)$, then for an arbitrary non-negative function $\rho(t,x)$ the equation

(2.3) $\Gamma(t,x,u(t) + \lambda v) = \gamma(t,x)\rho(t,x)$

has a solution $\lambda = \lambda(t,x)$, which vanishes if $x = x(t)$ and $t \geq t^*$. The last statement follows simply from the fact, that $\gamma_o(t) = 0$ and hence $\dot{\gamma}_o(t) = \Gamma(t,x(t),u(t)) = 0$ for all $t\in(t^*,t^*+\varepsilon)$. Hence the relations (iii) and (iv) will certainly hold if one takes as $u(t,x)$ the function $u(t) + \lambda(t,x)v$.

The hypothesis (2.2) allows the "embedding" of the given control function $u(t)$ into a suitable family of admissible control functions. This is achieved by means of a function $\tilde{u}(t,x)$ which depends upon t a n d the state variable and which has the following properties

(2.4) (i) $\tilde{u}$ is defined on a neighbourhood of the set $\{t,x \mid x=x(t),\ t\in[t_0,\tilde{t}]\}$ and is continuous except for finitely many jumps which occur at hyperplanes t = const. Except for the points of discontinuity $\tilde{u}$ has continuous partial derivatives with respect to x.

(ii) $\tilde{u}(t,x(t)) = u(t)$ for all $t\in[t_0,\tilde{t}]$.

(iii) $\tilde{u}(t,x) = u(t)$ whenever $\gamma_0(t) < 0$.

(iv) If t^* is such that $\gamma_0(t)$ vanishes on a whole right neighbourhood of t^*, then $\tilde{u}(t,x)$ coincides on a set of the form $\{(t,x)\in N \mid t>t^*\}$ with a function $u(t,x)$ satisfying all conditions (2.2). Similarly if γ_0 vanishes on a left neighbourhood of t^* then $\tilde{u}$ coincides with u on $\{(t,x)\in N \mid t<t^*\}$.

The existence of such a function follows immediately from the hypothesis (2.2). It should be noted however that $u(t,x)$ and hence $\tilde{u}(t,x)$ is not determined uniquely by the given control function. This arbitrariness will enter into the necessary conditions.

3. Restrictions on the Control Variable.

The variational technique on which the proof of our main theorem (see section 4) is based requires the "embedding" property not only for the given control function u(t) but also for those values u of U, which enter into the variations. We therefore introduce certain subsets U_t' of the control set U.

Definition. For every $t^*\in[t_0,\tilde{t}]$ the set U_{t^*}' consist of all $u^*\in U$ having the following property.

One can find a neighbourhood N of $(t^*,x(t^*))$ and a function $v(t,x)$ which is defined on N and satisfies these conditions:

(i) v is continuous and has a continuous partial derivative with respect to x,

(ii) $v(t,x)\in U$ for all $(t,x)\in N$,

(iii) $\Gamma(t,x,v(t,x)) \leq \Gamma(t^*,x(t^*),u^*)$ if $\gamma(t,x) \leq 0$ and $(t,x)\in N$,

(iv) $v(t^*,x(t^*)) = u^*$.

Note that $u(t^*)\in U_{t^*}'$ if γ_0 vanishes on a left neighbourhood of t^* and $u(t^*+0)\in U_{t^*}'$ if γ_0 vanishes on a right neighbourhood of U_{t^*}'. This follows immediately from the hypothesis (2.2). Furthermore it can be shown by an argument similar to the one used in the previous section that the following conditon is sufficient in order that $u^*\in U_{t^*}'$.

(3.1) There exists a v such that $u^*+\lambda v\in U$ for $|\lambda|$ small enough. The directional derivative of $\Gamma(t^*,x(t^*),u)$ in the direction of v does not vanish for $u = u^*$.

Before concluding this section we note, that the condition (3.1) implies an additional property of u^*:

(3.2) For every $\epsilon>0$ there exists a $t \neq t^*$ and a $u\in U'_t$ such that $|t-t^*|\leq\epsilon$, $\|u-u^*\|\leq\epsilon$ and $\Gamma(t,x(t),u)\leq\Gamma(t^*,x(t^*),u^*)$.

4. Necessary Conditions for Optimal Solutions.

We now assume that there is given, in addition to the data mentioned in section 1, a scalar function $f^o(t,x;u)$ which has the same differentiability properties as f. We wish to characterize those solutions which minimize the integral $\int_{t_o}^{\tilde{t}} f^o(t,x;u)dt$. To be more specific, we consider a solution $(u(t),x(t))$, which exists on some interval $[t_o,\tilde{t}]$, satifies the terminal and state constraint conditions and which is such that the value of the integral cannot be diminished by choosing some other solution or some other interval. Furthermore, we assume that this solution satisfies the conditions (1.2), (1.3) and (2.2). This implies, among other things, that u(t) can be embedded in a state-dependent control function $\tilde{u}(t,x)$ (cf. (2.4)). We also will assume that the statement (3.2) holds for all $u^*\in U'_{t^*}$ whenever γ_o vanishes in a full neighbourhood of t^*. Using these hypotheses one can now construct the cone of attainability. In order to establish its separation property one needs certain conditions on the initial and terminal manifold. So we assume that the M_i are locally defined by equations and inequalities and that $x(t_o),x(\tilde{t})$ respectively are regular points of M_o, M_1 respectively. There exists then a tangent cone T_o at M_o in $x(t_o)$ and a tangent cone T_1 at M_1 in $x(\tilde{t})$ (we use the notion of tangent cone and regular point in the sense of [1], Chapter VII, p. 320). Finally the restrictions for the initial and terminal point have to be compatible with the state space restriction. This means that M_o, M_1 have to satisfy some further condition.

The following one is convenient for our purposes

$$(6.1)\quad \begin{aligned} M_o &\subseteq \{x \mid \gamma(t_o,x) \le 0\}, \\ M_1 &\subseteq \{x \mid \gamma(t,x) \le 0 \text{ for all } t\in[\tilde{t}-\delta,\ \tilde{t}+\delta]\}, \end{aligned}$$

where δ is some positive number.

Before we are in a position to state the main theorem we have to introduce the Hamiltonian functions which play the essential role in the sequel. We choose independent variables $y^o, y=(y^1,\ldots,y^n)$ and put

$$H(t,x,y,y^o;u) = \sum_{j=0}^{n} y^j f^j(t,x;u) = y^o f^o(t,x;u)+y^T f(t,x;u),$$

$$\tilde{H}(t,x,y,y^o)=H(t,x,y,y^o;\tilde{u}(t,x)), \quad \tilde{H}_t(t,x,y,y^o) = \partial\tilde{H}(t,x,y,y^o)/\partial t.$$

<u>Theorem.</u> Let all assumptions just mentioned be satisfied. Then there exists a number y^o and a function $y(t) = (y^1(t),\ldots y^n(t))$ such that the statements (i) - (viii) hold.

(i) $y^o \le 0$, $(y^o, y(\tilde{t})) \neq (0,0)$. $y(\cdot)$ is continuous except possibly for those $t\in(t_o,\tilde{t})$ which are left entry points and no right entry points at the same time. At each point of continuity of $u(\cdot)$ the pair $(x(t),y(t))$ is a solution of the Hamiltonian system

$$\dot{x}^i = \frac{\partial\tilde{H}(t,x,y,y^o)}{\partial y^i}, \quad \dot{y}^i = -\frac{\partial\tilde{H}(t,x,y,y^o)}{\partial x^i}, \quad i = 1,\ldots,n.$$

(ii) $H(t,x(t),y(t),y^o;u(t)) \ge H(t,x(t),y(t);u)$ for all $u\in U$, provided t is a limit point of non-zeros of the function $\gamma_o(\cdot)$.

(iii) Given numbers t_i, ζ_i and elements $u_i\in U'_{t_i}$ which satisfy the inequalities

$$\zeta_i\ge 0,\ i=1,\ldots,r,\quad t_o < t_1 \le t_2 \ldots \le t_r < \tilde{t},$$

$$\sum_{i=1}^{p} \zeta_i \Gamma(t_i,x(t_i),u_i) \le 0,\ p = 1,\ldots,r.$$

Furthermore let $\gamma_o(t)$ vanish on a neighbourhood of $[t_1,t_r]$. Then

$$\sum_{i=1}^{r} \zeta_i[H(t_i,x(t_i),y(t_i);u(t_i))-H(t_i,x(t_i),y(t_i);u_i)] \ge 0.$$

(iv) (Special case of (iii), $r = 1$). If $\gamma_o(\cdot)$ vanishes in some neighbourhood of t, then

$H(t,x(t),y(t),y^o;u(t)) \geq H(t,x(t),y(t),y^o;u)$

if $u \in U'_t$ and $\Gamma(t,x(t),u) \leq 0$.

(v) The function $h(t) = H(t,x(t),y(t),y^o;u(t))$ is continuous at all points where $y(t)$ is continuous (cf. (i)). At points of discontinuity we have the jump conditions

$y(t+0) - y(t) = \mu g(t,x(t))$, $h(t+0)-h(t) = -\mu \gamma_t(t,x(t))$

with some $\mu \geq 0$ (depending upon t).

(vi) $h(\tilde{t}) = 0$

(vii) $\dot{h}(t)$ exists and is equal to $H_t(t,x(t),y(t),y^o;u(t))$ whenever $u(t)$, f_t are continuous and $\gamma_o(t) < 0$. $\dot{h}(t)$ exists and is equal to $\tilde{H}_t(t,x(t),y(t),y^o)$ whenever $f(t,x;\tilde{u}(t,x))$ and $f^o(t,x;\tilde{u}(t,x))$ have continuous partial derivatives with respect to all variables (in a neighbourhood of $(t,x(t))$).

(viii) We have the transversality conditions

$y(\tilde{t})^T c \geq 0$ for all $c \in T_1$, $y(t_o)^T c \leq 0$ for all $c \in T_o$.

5. Example.

We consider an autonomous 2-dimensional system which is controlled by a scalar control variable u. The control set U is the interval [-1,1]. We will write $x = (\xi,\eta)$ for the state variable and $y = (\varphi,\psi)$ for the adjoint variable. We assume that the state of the system has to move in accordance with the relations

(5.1) $\dot{\xi} = \sigma(x) = \sigma(\xi,\eta)$, $\dot{\eta} = u$

(5.2) $\alpha \leq \eta \leq \beta$.

Here σ is supposed to be a sufficiently smooth function whose partial derivative σ_η with respect to η vanishes nowwhere, α and β are real numbers with $\alpha < \beta$. The differential equation (5.1) may be viewed upon as a simplified model for the operation of an engine, whose state ξ depends upon the amount η of fuel which is injected per time unit. So the system is actually controlled through the variable η, which is not only subject to (5.2) but also to the restriction $|\dot{\eta}| \leq 1$. That is, the rate of change of η is also limited. Hence from the beginning we have not a standard control problem, but we can transform it into one by regarding $\dot{\eta}$ instead of η as control variable.

Let us now consider the problem of getting from an initial state x_o to a terminal state x_1 in shortest time. Since (5.2) can be replaced by the single inequality $\gamma(x) = (\eta-\alpha)(\eta-\beta) \leq 0$ we have a special optimal control problem of the type discussed in the previous sections. It is easily verified that all hypotheses of our theorem are satisfied in this case and that u assumes the values ± 1 whenever x is in the interior of the admissible region, and that $u = 0$ whenever x rests on the boundary $\eta = \alpha$ or $\eta = \beta$. That the bang-bang principle holds along any interior segment follows immediately from $\sigma_\eta \neq 0$. The compatibility (6.1) of terminal and constraint restrictions amounts simply to the inequalities $\gamma(x_o) \leq 0$, $\gamma(x_1) \leq 0$. Furthermore we have

(5.3) $g(t,x) = (0, 2\eta - \alpha - \beta)^T$, $\Gamma(t,x,u) = u(2\eta - \alpha - \beta)$,

and the condition (2.2) can be satisfied by choosing $u(t,x) = 0$. The "embedding" control $\tilde{u}(t,x)$ therefore reduces to $u(t)$ and the Hamiltonian function $\tilde{H}(t,x,y,y^o)$ coincides with $H(t,x,y,y^o;u(t))$, where

(5.4) $H(t,x,y,y^o;u) = y^o + \varphi\sigma(x) + \psi u.$

The differential equation for the adjoint variables assumes the same form as for the corresponding problem without state space constraint, namely

(5.5) $\dot{\varphi} = -\varphi\sigma_\xi(x(t))$, $\dot{\psi} = -\varphi\sigma_\eta(x(t))$.

Furthermore the Hamiltonian function vanishes along the optimal trajectory, in view of parts (v) - (vii) of the theorem (section 4). Hence we have the following relation between the values of the state variables, the adjoint variables and the control variable

(5.6) $y^o + \varphi(t)\sigma(x(t)) + \psi(t)u(t) = 0$ for all $t \in [t_o, \tilde{t}]$.

That $\varphi(t)$ is continuous on the whole interval $[t_o, \tilde{t}]$ can be seen immediately from the jump condition (v) and from the special form of g(see (5.3)). Since $\varphi(t)$ is a solution of the first of the differential equations (5.5) this function is either zero for all t or it vanishes nowhere.

We now claim: If $\gamma_o(\cdot)$ does not vanish on a left neighbourhood of $\tilde{t}$ (= terminal time), then $(y^o, \varphi(\tilde{t})) \neq (0,0)$. If in addition $u(t)$ is not constant +1 or -1 on the whole interval $[t_o, \tilde{t}]$, then $\varphi(\tilde{t}) \neq 0, y^o < 0$. Indeed, if $\gamma_o(t) \neq 0$ for all t which are sufficiently close to $\tilde{t}$, then $u(\tilde{t}) = \pm 1$ and it follows from (5.6) and from part (i) of our theorem that $(y^o, \varphi(\tilde{t})) \neq (0,0)$. Furthermore, if $u(t)$ is not constant ± 1 on $[t_o, \tilde{t}]$, then $\psi(t)u(t)$ has a zero on this interval, say at $t = t^*$. Using (5.6) once more one sees that either $y^o, \varphi(t^*)$ and hence also $y^o, \varphi(\tilde{t})$ are both zero or none of them is zero.

Before we proceed let us remark that $y^o < 0$ implies

(5.7) $\varphi(t)\sigma(x(t)) > 0$ whenever $\psi(t)u(t) = 0$.

This is also a consequence of (5.6).

We now turn to the discussion of the second component of the adjoint variable.

Theorem. ψ is continuous except possibly for those t which are left entry points (but not right entry points at the same time). ψ vanishes at all points of discontinuity of u(t), especially at all entry points (right or left). If η is constant and equal to α (equal to β) on an open interval, then ψ is non-positive and increasing (non-negative and decreasing) on this interval.

Proof. The first statement follows from part (v) of the theorem in section 4 and from the special form of g (cf. (5.3)). The second statement follows from the fact, that $\psi(t)u(t)$ is continuous on $[t_o,\tilde{t}]$, in view of the relation (5.6). Note that if t is a point of discontinuity for $\psi(\cdot)$, then t is a left entry point and cannot be a right entry point at the same time and this implies $u(t) = \pm 1$, $u(t+0) = 0$. The last statement can be phrased this way. If $\eta(t)=\alpha(\eta(t)=\beta)$ for all t in a neighbourhood of $[t_1,t_2]$, $t_1<t_2$, then $\psi(t_1)\leq\psi(t_2)\leq 0$ $(\psi(t_1)\geq\psi(t_2)\geq 0)$.
It will turn out that this is a consequence of the modified maximum-principle (part (iii) and (iv) of the general theorem). We restrict ourselves to the case $\eta(t) = \alpha$ and first note that the relations

(5.8) $\quad \mathrm{int}\,(U)\subseteq U_t' \quad , \quad \Gamma(t,x(t),u) = u(\alpha-\beta)$

hold for all $t\in[t_1,t_2]$. This can be seen immediately from (5.3) and from what has been said in connection with the definition of U_t' (see section 3). Furthermore $u(t) = 0$ for all $t\in[t_1,t_2]$, and it follows from (5.6) that the function $H(t,x(t),y(t),y^o;u)$ reduces to $\psi(t)u$. The modified maximum principle for r=2 therefore implies that the inequality

(5.9) $\quad \zeta_1\psi(t_1)u_1 + \zeta_2\psi(t_2)u_2 \leq 0$

holds whenever ζ_i,u_i satisfy the conditions

$$\zeta_i \geq 0,\ |u_i| < 1 \text{ for } i = 1,2,\quad \zeta_1u_1 \geq 0,\quad \zeta_1u_1+\zeta_2u_2 \geq 0.$$

There are two possible choices of the ζ_i,u_i which are in accordance with these conditions and which lead immediately to the desired result, if inserted into (5.9):

$$\zeta_1u_1 = 0,\quad \zeta_2u_2 = 1 \quad\text{and}\quad \zeta_1u_1 = 1,\quad \zeta_2u_2 = -1.$$

Corollary. Let us assume that φ does not vanish identically. If $t^*\in(t_o,\tilde{t})$ is a right (left) entry point or if $\psi(t^*) = 0$, $\gamma(x(t^*))<0$, then $u(t) = \text{const} = \pm 1$ on $(t^*,\tilde{t}]$ $([t_o,t^*))$.

Proof. If $\varphi \not\equiv 0$, then ψ is strictly monotone as can be seen from (5.5). Let us consider the point of discontinuity $\hat{t}$ of u(t) which is closest to t* and which is to the right (left) of t*. ψ is then continuous and strictly monotone in the interval between $\hat{t}$ and t^* but vanishes in the endpoints, which is of course not possible. Hence there can be no $\hat{t}$.

Using the results we have obtained so far it is not difficult to find out how an optimal trajectory may look like provided the terminal point x_1 does not belong to a boundary segment (that means, $\gamma_0(\cdot)$ does not vanish on a left neighbourhood of $\tilde{t}$). Reversing the time one obtains the analogous result in case x_0 does not belong to a boundary segment. Finally, if x_0 and x_1 both belong to boundary segments one can apply the foregoing arguments to each portion of the optimal trajectory. One then arrives at four types of solutions to which all possible candidates for time-optimal solutions belong. The schedule below contains all information about the values of $u(t), \eta(t)$ and sign $[\sigma(x(t))\sigma_\eta(x(t))]$ which can be inferred from the first order necessary conditions for each candidate. The symbols I_1, I_2, I_3 refer to three intervals of the form $[t_0,t_1]$, $(t_1, t_2]$, $(t_2, \tilde{t}]$ respectively. We have left aside the case that $u(t)$ is constant and equal to +1, -1 or 0 on the whole interval $[t_0,\tilde{t}]$. This case can occur for special boundary conditions only. In order to determine the sign of $\sigma\sigma_\eta$ along boundary segments one has to combine (5.7) and the statement of the last theorem with the fact, that ψ satisfies the differential equation (5.5).

type	1			2			3			4		
interval	I_1	I_2	I_3	I_1	I_2	I_3	I_1	I_2	I_3	I_1	I_2	I_3
u	+1	0	-1	-1	0	+1	0	+1	0	0	-1	0
η		β			α		α		β	β		α
sign $(\sigma\sigma_\eta)$		+1			-1		-1		+1	+1		-1

Addendum to Section 5.

The mere existence of a zero t^* of $\psi(t)u(t)$ does not guarantee that $y^o < 0$. However if $u(t)$ vanishes on an open interval I which contains t^*, then we have $\sigma(x(t^*)) \neq 0$ and all conclusions can be drawn as before. That σ cannot vanish in a point of a boundary segment can be seen as follows. Assume $\eta(t) = \alpha$ for all $t \in I$ and let $x(t^*) = (\xi^*, \alpha)$ with $\sigma(\xi^*, \alpha) = 0$. ξ^* is then a stationary point of the differential equation $\dot{\xi} = \sigma(\xi, \alpha)$ and we would therefore have $x(t) = \text{const} = (\xi^*, \alpha)$ for all $t \in I$. But this is impossible in view of the time optimal character of the solution.

Bibliography.

[1] Knobloch, H.W.; Kappel, F: Gewöhnliche Differentialgleichungen. B.G.Teubner, Stuttgart 1974.

[2] Knobloch, H.W.: Das Pontryaginsche Maximumprinzip für Probleme mit Zustandsbeschränkung I, II. To appear in Z. Angew. Math. Mech.

Author's address: Mathematisches Institut der Universität,
D-87 Würzburg, Am Hubland.

Explicit Approximation of Optimal Control Processes

Michael Köhler

1. Introduction and Problem Statement

We consider an optimal control problem of the following type:

$$P_Q : \quad \text{Min!}\ F(x,u)$$

$$T(x,u)=0 \ , \qquad (1.1)$$

$$(x,u) \in Q \ . \qquad (1.2)$$

The equation (1.1) given by the operator $T:X\times U\to P$ defines the process, $F:X\times U\to R$ is a real-valued cost function, and the constraint (1.2) satisfies $Q = Q_X\times Q_U \subset X\times U$, $Q \neq \emptyset$. X is the Banach space of the state variables, U is the Banach space of the control variables, and P is a further Banach space.

The symbols $\|\cdot\cdot\|_X$, $\|\cdot\cdot\|_U$, $\|\cdot\cdot\|_P$ denote, respectively, the norms in the appropriate spaces.

For arbitrary fixed $\delta>0$, define

$$\{Q_X,\delta\}:=\{x \in X \mid \inf\{\|x-x^*\|_X \mid x^* \in Q_X\} \leqq \delta\} \ . \qquad (1.3)$$

Throughout this paper, let the following assumptions hold (cf. [6], section 1):

<u>A1.</u> For each $u \in Q_U$, there exists one and only one $x \in X$ such that $T(x,u)=0$ follows.

<u>A2.</u> $Q_X = \overline{Q}_X \neq X$, and $0 \in T(Q_X\times Q_U)$

<u>A3.</u> There exists a number $\gamma_0>0$ such that F is continuous and bounded from below on $\overline{T^{-1}(0)\cap(\{Q_X,\gamma_0\}\times Q_U)}$.

<u>A4.</u> There exists an infinite subset N' of the natural numbers N with the following property:

For each $l \in N'$, there exists a set $Q_U(l) \subset Q_U$ satisfying

$$\overline{\bigcup_{l\in N'} (T^{-1}(0)\cap(\{Q_X,\delta\}\times Q_U(l)))} = \overline{T^{-1}(0)\cap(\{Q_X,\delta\}\times Q_U)}$$

for all $\delta>0$.

(If A is a set of a Banach space, then $\overline{A}$ denotes the closure of A, and $\overset{\circ}{A}$ denotes the interior of A.)

Assumption A1 implies the existence of a map $\Gamma:Q_U\to X$ defined by

$$\Gamma u := x \text{ with } T(x,u)=0. \tag{1.7}$$

Applying a general discretization method to P_Q, we define finite-dimensional approximate control problems such that their extreme values are converging to the extreme value of the original problem. The approximate control problems are explicitly given, even if $Q_X \neq X$. (Problems with $Q_X = X$ are considered in [6], sections 4 and 6.) The extreme value of P_Q is given by

$$e(P_Q) := \begin{cases} \inf\{F(x,u) \mid (x,u) \in T^{-1}(0)\cap Q\} & \text{if } T^{-1}(0)\cap Q \neq \emptyset, \\ +\infty & \text{if } T^{-1}(0)\cap Q = \emptyset. \end{cases} \tag{1.8}$$

It follows from assumptions A2 and A3 that

$$-\infty < e(P_Q) < +\infty . \tag{1.9}$$

2. Application of a General Discretization Method to the Original Problem

The definitions in this section are similar to those given in [7].

Definition 2.1. A discretization method V applicable to P_Q consists of an infinite sequence $\{X_n\times U_n, \xi_n, \phi_n, \tau_n, P_n, \pi_n\}_{n\in N'}$, where

(i) N' has the property required in assumption A4;

(ii) X_n, U_n, P_n are finite-dimensional Banach spaces with $\dim P_n = \dim X_n$;

(iii) $\xi_n := \chi_n \times \partial_n$ with linear $\chi_n : X\to X_n$ and linear $\partial_n : U\to U_n$ (e.g. $\xi_n : X\times U \to X_n\times U_n$ is a mapping defined by $\xi_n(x,u) := (\chi_n(x), \partial_n(u))$) and $\pi_n : P\to P_n$ are linear mappings with $\lim_{n\to\infty} \|\xi_n(x,u)\|_{X_n\times U_n} := \lim_{n\to\infty}(\|\chi_n(x)\|_{X_n} + \|\partial_n(u)\|_{U_n} = \|(x,u)\|_{X\times U} := \|x\|_X + \|u\|_U$ for each fixed $(x,u)\in X\times(\bigcup_{l\in N'} Q_U(l))$, and $\lim_{n\to\infty}\|\pi_n(p)\|_{P_n} = \|p\|_P$ for each fixed $p\in P$;

(iv) $\tau_n : \{X\times U\to P\} \to \{X_n\times U_n \to P_n\}$ are mappings with T in the domain of all τ_n, $\phi_n : \{X\times U \to R\} \to \{X_n\times U_n \to R\}$ are mappings with F in the domain of all ϕ_n.

Definition 2.2. A discretization method $V = \{X_n\times U_n, \xi_n, \phi_n, \tau_n, P_n, \pi_n\}_{n\in N'}$ applicable to P_Q is called consistent at $(x,u)\in X\times U$ if

$$\lim_{n\to\infty} \|\tau_n(T)(\xi_n(x,u)) - \pi_n(T(x,u))\|_{P_n} = 0, \text{ and}$$

$$\lim_{n\to\infty} |\phi_n(F)(\xi_n(x,u)) - F(x,u)| = 0.$$

Definition 2.3. A discretization method $V = \{X_n\times U_n, \xi_n, \phi_n, \tau_n, P_n, \pi_n\}_{n\in N'}$ applicable to P_Q is called stable at $\{(x_n,u_n)\}_{n\in N'} \subset X_n\times U_n$ if the following condition holds: There exist constants S and $r > 0$ such that, uniformly for all $n\in N'$,

$$\|(x_n^1,u_n)-(x_n^2,u_n)\|_{X_n\times U_n} \leq S\|\tau_n(T)(x_n^1,u_n)-\tau_n(T)(x_n^2,u_n)\|_{P_n}$$

for all (x_n^1,u_n), $(x_n^2,u_n) \in X_n\times U_n$ satisfying

$$\|\tau_n(T)(x_n^i,u_n)-\tau_n(T)(x_n,u_n)\|_{P_n} < r \ , \ i=1,2.$$

Let $V=\{X_n\times U_n,\xi_n,\phi_n,\tau_n,P_n,\pi_n\}_{n\in N'}$ be a discretization method applicable to P_Q. Define T_n and F_n by $T_n:=\tau_n(T)$, $n \in N'$, and $F_n:=\phi_n(F)$, $n \in N'$. Further, let Q_{U_n} denote the set

$$Q_{U_n}:=\partial_n(Q_U(n)),\ n \in N'. \tag{2.1}$$

The following assumptions are required (with γ_0 from section 1):

A5. For each $\delta \in [0,\gamma_0]$, $\Gamma^{-1}(\{Q_X,\delta\})\cap Q_U$ is weakly sequentially closed and weakly sequentially compact.

A6. For each $\delta \in [0,\gamma_0]$, the weak convergence of a sequence $\{u^k\}_{k\in N} \subset \Gamma^{-1}(\{Q_X,\delta\})\cap Q_U$ to $u \in \Gamma^{-1}(\{Q_X,\delta\})\cap Q_U$ implies

$$\liminf_{k\to\infty} F(\Gamma u^k,u^k) \geq F(\Gamma u,u).$$

A7. $Q_U(l) \subset Q_U(l')$ for all l, $l'\in N'$ with $l\leq l'$.

A8. For each $(x,u) \in T^{-1}(0)\cap(X\times Q_U)$, V is stable at $\{(\chi_n(x),\partial_n(u))\}_{n\in N'}$.

A9. For every $\varepsilon>0$, there is an $n_0\in N'$ such that $n\geq n_0, n\in N'$, implies

$$|F(x,u)-F_n(\xi_n(x,u))|<\varepsilon \text{ for all } (x,u) \in T^{-1}(0)\cap(X\times Q_U(n)),$$

and

$$\|T_n(\xi_n(x,u))\|_{P_n}<\varepsilon \text{ for all } (x,u)\in T^{-1}(0)\cap(X\times Q_U(n)).$$

A10. For each $n\in N'$, F_n is bounded from below on $(\chi_n(\{Q_X,\gamma_0\})\times\partial_n(Q_U(n)))\cap T_n^{-1}(0)$. There is a constant $K\in R$ such that, uniformly for all $n\in N'$,

$$|F_n(x_n^1,u_n)-F_n(x_n^2,u_n)|\leq K\|x_n^1-x_n^2\|_{X_n}$$

for all $(x_n^i,u_n) \in \chi_n(\{Q_X,\gamma_0\})\times\partial_n(Q_U(n))$, $i=1, 2$.

A11. For each $n\in N'$, the following condition holds:
For each $u_n\in Q_{U_n}$, there is one and only one $x_n\in X_n$ with $T_n(x_n,u_n)=0$.

A12. $\{x_n\in X_n \mid \inf\{\|x_n-x_n^*\|_{X_n} \mid x_n^*\in\chi_n(\{Q_X,\delta\})\}\leq\varepsilon\} \subset \chi_n(\{\{Q_X,\delta\},\varepsilon\})$ for all $\varepsilon>0$ and all $n\in N'$ whenever $\delta \geq 0$.

A13. For each $\delta \geq 0$, the following condition holds:
For every $\varepsilon>0$, there is an $n_0\in N'$ such that $n\geq n_0$, $n\in N'$, implies

$\Gamma u \in \{\{Q_X,\delta\},\varepsilon\}$ for all $u \in Q_U(n)$ with

$\partial_n(u) \in \{u_n \mid \exists\, x_n \in X_n$ such that $(x_n,u_n) \in T_n^{-1}(0) \cap (\chi_n(\{Q_X,\delta\}) \times Q_{U_n})\}$.

From the assumptions above one obtains immediately the following existence theorem.

Theorem 2.1. Suppose that P_Q satifies the conditions A5 and A6. Then there exists an $u^* \in \Gamma^{-1}Q_X \cap Q_U$ such that

$e(P_Q) = F(\Gamma u^*,u^*)$.

Proof. See Theorem 1.4.1. in [3].

Assertion 2.1. Suppose that the following conditions hold:

(i) for every bounded set $B \subset X \times Q_U$, there is a constant $K_B \in R$ such that

$|F(x,u)-F(x^*,u)| \leq K_B \|x-x^*\|_X$ for all $(x^*,u),(x,u) \in B$;

(ii) for fixed $x \in \{Q_X,\delta\}$, $\delta \in [0,\gamma_0]$, weak convergence of a sequence $\{u^k\}_{k\in N} \subset \Gamma^{-1}(\{Q_X,\delta\} \cap Q_U$ to u implies $\liminf_{k\to\infty} F(x,u^k) \geq F(x,u)$;

(iii) weak convergence of a sequence $\{u^k\}_{k\in N} \subset \Gamma^{-1}(\{Q_X,\delta\}) \cap Q_U$ $(\delta \in [0,\gamma_0])$ to u implies $\lim_{k\to\infty} \|\Gamma u^k - \Gamma u\|_X = 0$.

Then weak convergence of a sequence $\{u^k\}_{k\in N} \subset \Gamma^{-1}(\{Q_X,\delta\}) \cap Q_U$ to u implies

$$\liminf_{k\to\infty} F(\Gamma u^k,u^k) \geq F(\Gamma u,u).$$

The assertion follows immediately from the assumptions.

The following known result is useful for applications.

Theorem 2.2. Let Q_U be closed and convex. Assume that, for fixed $x \in \{Q_X,\gamma_0\}$, $F(x,.)$ as a function of u is convex on Q_U. Then weak convergence of a sequence $\{u^k\}_{k\in N} \subset Q_U$ to $u \in Q_U$ implies

$$\liminf_{k\to\infty} F(x,u^k) \geq F(x,u).$$

Proof. Satz 4(3.X1) in [5] implies that Q_U is weakly closed. Following the Lemma in [2], page 121, one obtains the assertion.

Lemma 2.1. Suppose that the discretization method V applicable to P_Q satisfies assumptions A8 and A9, and that, in addition, A11 holds. For each $n \in N'$, let $\varepsilon_n(u)$ be defined for each $(x,u) \in \bigcup_{n\in N'} (T^{-1}(0) \cap (X \times Q_U(n)))$ by $\varepsilon_n(u) := (x_n,u) - (\chi_n(x),\partial_n(u))$ with $u_n := \partial_n(u)$ and $T_n(x_n,u_n) = 0$. Then for every $\varepsilon > 0$, there is an $n_0 \in N'$ such that $n \geq n_0$, $n \in N'$, implies

$$\|\varepsilon_n(u)\|_{X_n \times U_n} < \varepsilon \text{ for all } (x,u) \in T^{-1}(0) \cap (X \times Q_U(n)).$$

Proof. The assertion follows from assumption A8 and, applying A9, from the proof of Satz 2.1 in [6].

Generally, the lemma below holds.

Lemma 2.2. Let $Q_k \subset Q_U$ be weakly closed and weakly sequentially compact for all $k \in N$. Furthermore, suppose that $Q_k \supset Q_{k+1}$ for all $k \in N$, and that

$$\tilde{Q} := \bigcap_{k \in N} Q_k \subset Q_U, \quad \tilde{Q} \neq \emptyset .$$

Assume, in addition, that $\{u^k\}_{k \in N} \subset Q_U$ is a sequence satifying $u^k \in Q_k$ for all $k \in N$. Then there exists a subsequence of $\{u^k\}_{k \in N}$ converging weakly to an element of $\tilde{Q}$.

Proof. Since Q_1 is weakly closed and weakly sequentially compact, there exists a subsequence $\{u^{k_l}\}_{l \in N} \subset \{u^k\}_{k \in N}$ converging weakly to an $u_0 \in Q_1$. We argue by contradiction, and thus suppose that there is an $l_0 \in N$ such that

$$u_0 \in Q_{k_{l_0}}, \text{ and} \tag{2.2}$$

$$u_0 \notin Q_{k_l} \quad \text{for every } l > l_0,\ l \in N, \tag{2.3}$$

holds. Now,

$$\{u^{k_l}\}_{\substack{l \in N \\ l \geqq l_0+1}} \quad \text{converges weakly to } u_0. \text{ Since } Q_{k_{l_0+1}} \text{ is}$$

weakly closed, we have

$$u_0 \in Q_{k_{l_0+1}}$$

contradicting (2.2) and (2.3).

Let $\mathcal{V}=\{X_n \times U_n, \xi_n, \phi_n, \tau_n, P_n, \pi_n\}_{n \in N}$, be a discretization method applicable to $\mathcal{P}_Q$. Then for every $\delta \in [0, \gamma_0]$, there exists a sequence of problems $\{\mathcal{P}_{Q(\delta)_n}\}_{n \in N}$, generated by $\mathcal{V}$, where $\mathcal{P}_{Q(\delta)_n}$ denotes a finite-dimensional control problem of the following type:

$$\mathcal{P}_{Q(\delta)_n}: \quad \text{Min! } F_n(x_n, u_n)$$

$$T_n(x_n, u_n) = 0, \tag{2.4}$$

$$(x_n, u_n) \in Q(\delta)_n := \{Q_X, \delta\}_n \times Q_{U_n}, \tag{2.5}$$

where

$$\{Q_X, \delta\}_n := \chi_n(\{Q_X, \delta\}). \tag{2.6}$$

Here $Q(\delta)_n$ is explicitly given. Of course, the definition of $Q(\delta)_n$ does not refer to $\{u \in Q_U \mid \exists\, x \in Q_X \text{ with } (x,u) \in T^{-1}(0) \cap Q\}$, the set of admissible controls of $\mathcal{P}_Q$ (cf. section 2 in [6]).

3. Explicit Approximation of P_Q

Define for every $\delta\in(0,\gamma_0]$ the problem $P_{Q(\delta)}$ by

$$P_{Q(\delta)}: \quad \text{Min! } F(x,u)$$
$$T(x,u)=0,$$
$$(x,u)\in Q(\delta):=\{Q_X,\delta\}\times Q_U.$$

Then A2 and A3 imply that $-\infty < e(P_{Q(\delta)}) < +\infty$ for every $\delta\in(0,\gamma_0]$.

We obtain the following result:

Theorem 3.1. Let $\{\delta_k\}_{k\in N}\subset R$ be a monotonic decreasing sequence with $\delta_1\leqq\gamma_0$ converging to zero. Suppose, in addition, that assumptions A5 and A6 are satisfied. Then

$$\lim_{k\to\infty} e(P_{Q(\delta_k)}) = e(P_Q).$$

Proof. Since $\{\delta_k\}_{k\in N}$ is monotonic decreasing and converging to zero, we have

$$\Gamma^{-1}\{Q_X,\delta_k\}\cap Q_U \supset \Gamma^{-1}\{Q_X,\delta_{k+1}\}\cap Q_U \supset \ldots\ldots \supset \Gamma^{-1}Q_X\cap Q_U, \tag{3.10}$$

and

$$\bigcap_{k\in N}(\Gamma^{-1}\{Q_X,\delta_k\}\cap Q_U)=\Gamma^{-1}Q_X\cap Q_U. \tag{3.11}$$

(3.10) implies

$$e(P_{Q(\delta_k)})\leqq e(P_{Q(\delta_{k+1})})\leqq\ldots\ldots\leqq e(P_Q) \text{ for all } k\in N. \tag{3.12}$$

Hence, there is an $\alpha\in R$ such that

$$\lim_{k\to\infty} e(P_{Q(\delta_k)}) = \alpha. \tag{3.13}$$

Now, let $\{\varepsilon_k\}_{k\in N}\subset(0,\infty)$ be a monotonic decreasing sequence converging to zero. Then for every $k\in N$, there is an $u^k\in\Gamma^{-1}\{Q_X,\delta_k\}\cap Q_U$ satisfying

$$e(P_{Q(\delta_k)})+\varepsilon_k > F(\Gamma u^k,u^k). \tag{3.14}$$

By Lemma 2.2., there is a weakly convergent subsequence $\{u^{k_l}\}_{l\in N}\subset\{u^k\}_{k\in N}$ with weak limit $u^*\in\Gamma^{-1}Q_X\cap Q_U$. Then A6 implies

$$F(\Gamma u^*,u^*) \leqq \liminf_{l\to\infty} F(\Gamma u^{k_l},u^{k_l}). \tag{3.15}$$

Hence, for every $j\in N$, there is an $l_j\in N$, $l_j<l_{j+1}$ for all $j\in N$, such that

$$F(\Gamma u^*,u^*) < F(\Gamma u^{k_{l_j}},u^{k_{l_j}})+\varepsilon_j. \tag{3.16}$$

It follows from (3.14) and (3.16) that

$$e(P_{Q(\delta_{k_{l_j}})})+\varepsilon_{k_{l_j}} > F(\Gamma u^{k_{l_j}},u^{k_{l_j}}) > F(\Gamma u^*,u^*)-\varepsilon_j \geqq e(P_Q)-\varepsilon_j. \tag{3.17}$$

Therefore ,we have

$$0 \geqq e(P_{Q(\delta_{k_{l_j}})})-e(P_Q) > -\varepsilon_{k_{l_j}} - \varepsilon_j \text{ for all } j\in N.$$

Thus,

$$\lim_{j\to\infty} e(P_{Q(\delta_{k_{1_j}})}) = e(P_Q).$$

Together with (3,13), the assertion follows.

Finally, we obtain the following approximation theorem:

Theorem 3.2. Let P_Q satisfy the conditions A5-A7, suppose that the discretization method V applicable to P_Q satisfies assumptions A8 and A9, and, in addition, assume that assumptions A10-A13 hold. Further, let $\{\delta_k\}_{k\in N'}$ be a monotonic decreasing sequence with $0<\delta_k \leq \frac{\gamma}{2}0$ converging to zero. Then there exists a sequence $\{n_k\}_{k\in N'} \subset N'$ such that

$$\lim_{k\to\infty} e(P_{Q(\delta_k)n_k}) = e(P_Q)$$

follows.

Proof. Let us fix $\delta\in(0,\frac{\gamma}{2}0]$. It follows from A2 -A4 and A7 that there is an $n^*\in N'$ such that $T^{-1}(0)\cap(\{Q_X,\delta\}\times Q_U(n))=\emptyset$ for all $n\geq n^*$, $n\in N'$. Further, let us arbitrarily fix $\varepsilon\in(0,\varepsilon^*]$, where $\varepsilon^*:=\min(2\delta,2\delta K)$ (K from A10). Hence, A4 and A7 imply that, for every $n\geq n^*$, $n\in N'$, there exists an $(x^{(n)},u^{(n)}) \in T^{-1}(0)\cap(\{Q_X,\delta\}\times Q_U(n))$ satifying

$$\frac{\varepsilon}{4}+e(P_{Q(\delta)}) \geq F(x^{(n)},u^{(n)}). \tag{3.18}$$

By A9, there is an $n_1\in N'$ (n_1 independent of δ), $n_1=n_1(\varepsilon)\geq n^*$, such that

$$-\frac{\varepsilon}{4} < F(x^{(n)},u^{(n)})-F_n(\xi_n(x^{(n)},u^{(n)})) < \frac{\varepsilon}{4} \quad \text{for all } n\geq n_1,\ n\in N'. \tag{3.19}$$

(3.18) and (3.19) imply

$$e(P_{Q(\delta)}) \geq F_n(\xi_n(x^{(n)},u^{(n)}))-\frac{\varepsilon}{2} \quad \text{for all } n\geq n_1,\ n\in N'. \tag{3.20}$$

By A11, there exists one and only one $x_n\in X_n$ satisfying

$$T_n(x_n,\partial_n(u^{(n)}))=0. \tag{3.21}$$

Applying Lemma 2.1., there is an $n_2=n_2(\varepsilon)\in N'$ such that

$$\|(x_n,\partial_n(u^{(n)}))-(\chi_n(x^{(n)}),\partial_n(u^{(n)}))\|_{X_n\times U_n}=\|x_n-\chi_n(x^{(n)})\|_{X_n} < \frac{\varepsilon}{2K} \quad \text{for all } n\geq n_2,\ n\in N' \tag{3.22}$$

Since A12 holds, it follows from (3.22) that

$$x_n \in \chi_n(\{\{Q_X,\delta\},\frac{\varepsilon}{2K}\}) \subset \chi_n(\{Q_X,2\delta\}) \quad \text{for all } n\geq n_2,\ n\in N'. \tag{3.23}$$

Further, A10 and (3.22) imply

$$|F_n(x_n,\partial_n(u^{(n)}))-F_n(\chi_n(x^{(n)}),\partial_n(u^{(n)}))|<K\|x_n-\chi_n(x^{(n)})\|_{X_n}<\frac{K}{2K}\varepsilon=\frac{\varepsilon}{2} \quad \text{for all } n\geq n_2,\ n\in N'. \tag{3.24}$$

Defining $n_3(\varepsilon):=\max(n_1(\varepsilon),n_2(\varepsilon))$, we have from (3.20), (3.23), and

(3.24) that

$$e(\mathcal{P}_{Q(\delta)}) \geqq F_n(\xi_n(x^{(n)},u^{(n)})) - \frac{\varepsilon}{2} > F_n(x_n, \partial_n(u^{(n)})) - \varepsilon$$
$$\geqq e(\mathcal{P}_{Q(2\delta)_n}) - \varepsilon \text{ for all } n \geqq n_3,\ n \in N'. \tag{3.25}$$

Furthermore, for every $n \in N'$, $n \geqq n_3$, there exists an

$(x_n(2\delta), u_n(2\delta)) \in Q(2\delta)_n \cap T_n^{-1}(0)$ satisfying

$$\frac{\varepsilon}{2} + e(\mathcal{P}_{Q(2\delta)_n}) > F_n(x_n(2\delta), u_n(2\delta)). \tag{3.26}$$

Now, (2.9) and A2 imply the existence of

$$(x^{(n)}(2\delta), u^{(n)}(2\delta)) \in T^{-1}(0) \cap X \times Q_U(n), \text{ where } u_n(2\delta) = \partial_n(u^{(n)}(2\delta)).$$

By A8, there are r, S according to Definition 2.3. Because of A9, there exists an $n_4 = n_4(\varepsilon) \in N'$ implying

$$-\frac{\varepsilon}{4} < F(x^{(n)}(2\delta), u^{(n)}(2\delta)) - F_n(\xi_n(x^{(n)}(2\delta), u^{(n)}(2\delta))) < \frac{\varepsilon}{4}, \tag{3.27}$$

$$\|T_n(\xi_n(x^{(n)}(2\delta), u^{(n)}(2\delta)))\|_{P_n} < \frac{1}{SK}\gamma \tag{3.28}$$

for all $n \geq n_4$, $n \in N'$, where $\gamma := \min(\frac{\varepsilon}{4}, rSK)$.

A10, A8, and (3.28) imply

$$|F_n(x_n(2\delta), u_n(2\delta)) - F_n(\chi_n(x^{(n)}(2\delta)), \partial_n(u^{(n)}(2\delta)))|$$
$$\leqq K\|x_n(2\delta) - \chi_n(x^{(n)}(2\delta))\|_{X_n} \leqq KS\|T_n(\xi_n(x^{(n)}(2\delta), u^{(n)}(2\delta)))\|_{P_n}$$
$$< KS\frac{1}{KS}\gamma \leqq \frac{\varepsilon}{4} \text{ for all } n \geqq n_4,\ n \in N'. \tag{3.29}$$

By A13, there exists an $n_5 = n_5(\varepsilon) \in N'$ such that

$$x^{(n)}(2\delta) \in \{Q_X, 2\delta + \frac{\varepsilon}{2}\} \subset \{Q_X, 3\delta\} \text{ for all } n \geqq n_5,\ n \in N'. \tag{3.30}$$

Hence, it follows from (3.26), (3.29), (3.27), and (3.30) that

$$\frac{\varepsilon}{2} + e(\mathcal{P}_{Q(2\delta)_n}) > F(x^{(n)}(2\delta), u^{(n)}(2\delta)) - \frac{\varepsilon}{2}, \text{ and}$$

$$x^{(n)}(2\delta) \in \{Q_X, 3\delta\} \text{ for all } n \geqq n_6 = n_6(\varepsilon) := \max(n_4(\varepsilon), n_5(\varepsilon)),\ n \in N',$$

hold. Consequently, we have

$$\varepsilon > e(\mathcal{P}_{Q(3\delta)}) - e(\mathcal{P}_{Q(2\delta)_n}) \text{ for all } n \geqq n_6,\ n \in N', \tag{3.31}$$

and one obtains from (3.31) and (3.25) that

$$\varepsilon > e(\mathcal{P}_{Q(3\delta)}) - e(\mathcal{P}_{Q(2\delta)_n}), \tag{3.32}$$

$$e(\mathcal{P}_Q) - e(\mathcal{P}_{Q(2\delta)_n}) \geqq e(\mathcal{P}_{Q(\delta)}) - e(\mathcal{P}_{Q(2\delta)_n}) > -\varepsilon \tag{3.33}$$

for all $n \geqq n_0 = n_0(\varepsilon) := \max(n_3(\varepsilon), n_6(\varepsilon))$, $n \in N'$.

Now, define $\varepsilon_k > 0$ by

$\varepsilon_k := \min(K\delta_k, \delta_k)$ for $k \in N'$. (3.34)

Defining $n_k := n_0(\varepsilon_k)$ for $k \in N'$, one obtains from (3.32) and (3.33) that

$$\varepsilon_k > e(P_{Q(\frac{3}{2}\delta_k)}) - e(P_{Q(\delta_k)_{n_k}}) \text{ , and} \tag{3.35}$$

$$e(P_Q) - e(P_{Q(\delta_k)_{n_k}}) > -\varepsilon_k \quad \text{for all } k \in N'. \tag{3.36}$$

From Theorem 3.1. we deduce that there is a sequence $\{\gamma_k\}_{k \in N'} \subset R$ converging to zero satisfying

$$\gamma_k \leqq e(P_{Q(\frac{3}{2}\delta_k)}) - e(P_Q) \leqq 0 \quad \text{for all } k \in N'.$$

Thus, we obtain from (3.35) that

$$|e(P_Q) - e(P_{Q(\delta_k)_{n_k}})| < \varepsilon_k + \gamma_k \quad \text{for all } k \in N'.$$

Since (3.34) implies $\lim_{k\to\infty} \varepsilon_k = 0$, we are done.

4. Applications

Let the following optimal control problem be given:

$$\text{minimize} \quad \int_0^1 g(x(t),u(t))dt \tag{4.1}$$

subject to

$$\frac{d}{dt}x(t) = f(x(t),u(t)) \quad \text{a.e. on } [0,1], \tag{4.2}$$

$$x(0) = x_0, \tag{4.3}$$

$$x(t) \in Y(t) \subset R^m \quad \text{for all } t \in [0,1], \tag{4.4}$$

$$u(t) \in S(t) \subset R^r \quad \text{for all } t \in [0,1], \tag{4.5}$$

$$x(1) \in Z \subset R^m, \tag{4.6}$$

$$u \in \{u:[0,1]\to R^r \mid u \text{ Lebesgue-integrable}\}. \tag{4.7}$$

Assume that the problem above satisfies the following conditions:

<u>C1.</u> $Y(t)$ and $S(t)$ are closed and convex for all $t \in [0,1]$, and Hausdorff-continuous for all $t \in [0,1]$; there are bounded sets $Y \subset R^m$, $S \subset R^r$ satisfying $Y(t) \subset Y$ and $S(t) \subset S$ for all $t \in [0,1]$.

<u>C2.</u> (α) $f: R^m \times R^r \to R^m$ is continuous on $D := \{R^m \times S(t) \mid t \in [0,1]\}$;

(β) there exists a constant L_f such that

$\|f(y^*,s) - f(y^{**},s)\|_{R^m} \leqq L_f \|y^* - y^{**}\|_{R^m}$ for all $(y^*,s),(y^{**},s) \in D$;

(γ) there are constants K_1 and K_2 and a function $\lambda:[0,\infty)\to R$, continuous on $[0,\infty)$ but not integrable over $[0,\infty)$, such that

$$\|f(y,s)\|_{R^m} \leqq \frac{K_1}{(\|y\|_{R^m})} + K_2 \quad \text{for all } (y,s) \in D.$$

<u>C3.</u> (α) $g: R^m \times R^r \to R$ is continuous on D;

(β) $g(y,s)$ is bounded from below for all

$(y,s)\in\{\{Y(t),\gamma_0\}\times S(t)\,|\,t\in[0,1],\ \gamma_0>0\}$ (where, for every $\varepsilon>0$,

$\{Y(t),\varepsilon\}:=\{y\in R^m\,|\,\inf\{\|y-y^*\|_{R^m}\,|\,y^*\in Y(t)\}\leq\varepsilon\}$);

(γ) there exists a constant L_g such that

$$|g(y^*,s)-g(y^{**},s)|\leq L_g\|y^*-y^{**}\|_{R^m}$$

for all $(y^*,s),(y^{**},s)\in D$.

C4. The target set Z is closed.

Define:

$$X:=C^m[0,1]:=\{x:[0,1]\to R^m\,|\,x \text{ continuous}\} \text{ with } \|x\|_X:=\sup_{t\in[0,1]}\|x(t)\|_{R^m}; \quad (4.8)$$

$$U:=L_1^r[0,1]:=\{u:[0,1]\to R^r\,|\,u \text{ Lebesgue-integrable}\} \text{ with } \|u\|_U:=\int_0^1\|u(t)\|_{R^r}\,dt; \quad (4.9)$$

$$U(l):=\{u:[0,1]\to R^r\,|\,u(t)=\sum_{i=0}^{l-1}v^i\chi_{J_i^l}(t),\ v^i\in R^r,\|v^i\|_{R^r}<\infty \text{ for all } i=0,1,\ldots,l-1\},\ l\in N', \quad (4.10)$$

where

$J_i^l:=[\frac{i}{l},\frac{i+1}{l})$, $i=0,1,\ldots\ldots,l-2$,

$J_{l-1}^l:=[\frac{l-1}{l},1]$, and

$\chi_{J_i^l}(t)$ denotes the characteristic function of J_i^l;

$$P:=C^m[0,1]\times R^m \quad (\hat{p}\in P \text{ implies that } \hat{p}=(p,p_0) \text{ with } p\in C^m[0,1], \text{and } p_0\in R^m)$$
$$\text{with } \|\hat{p}\|_P:=\sup_{t\in[0,1]}\|p(t)\|_{R^m}+\|p_0\|_{R^m}; \quad (4.11)$$

$T:X\times U\to P$ by

$$T(x,u):=\begin{pmatrix} x(0)-x_0 \\ x(t)-\int_0^t f(x(\tau),u(\tau))d\tau - x_0\end{pmatrix}, \text{ and} \quad (4.12)$$

$F:X\times U\to R$ by

$$F(x,u):=\int_0^1 g(x(t),u(t))dt. \quad (4.13)$$

Further, define

$$Q_U:=\{u\in U\,|\,u(t)\in S(t) \text{ for all } t\in[0,1]\}, \quad (4.14)$$

$$Q_U(l):=Q_U\cap U(l) \text{ for } l\in N, \text{ and} \quad (4.15)$$

$$Q_X:=\{x\in X\,|\,x(t)\in Y(t) \text{ for all } t\in[0,1],\ x(1)\in Z\}. \quad (4.16)$$

Then by (4.8), (4.9), (4.11), (4.12), (4.13), (4.14) and (4,16) the control problem (4.1)-(4.7) is equivalent to the following problem:

CP_Q: Min! $F(x,u)$

$T(x,u)=0$,

$(x,u) \in Q_X \times Q_U$.

The conditions C1, C2, C3 and C4 imply:
CP_Q satisfies assumptions A1 and A3 (see [1], Satz 3.1, Satz 3.3). In [1], pp. 56-59, there are conditions for problem (4.1)-(4.7) implying that CP_Q satisfies A4 with

$$N' := \{k_0 2^{n-1} \mid n \in N,\ k_0 \in N \text{ fixed}\}.$$

Hence, we have by (4.15) that A7 is satisfied.

We assume, in addition, that there exists an $u \in L_1^r[0,1]$ satisfying (4.5) such that the corresponding solution of (4.2), (4.3) is an element of Q_X. Therefore A2 holds.

Finally, we obtain for problem CP_Q the following result:

<u>Lemma 4.1.</u> Let $g: R^m \times R^r \to R$ be, for fixed $y \in \{\{Y(t), \gamma_0\} \mid t \in [0,1]\}$, convex in s for all $s \in \{S(t) \mid t \in [0,1]\}$. Let $f: R^m \times R^r \to R^m$ be, for fixed $y \in R^m$, linear in s for all $s \in \{S(t) \mid t \in [0,1]\}$. Then CP_Q satisfies assumptions A5-A6.

<u>Proof.</u> By condition C1, there exists an $L \in R$ such that $\|u(t)\|_{R^r} \leq L$ for all $u \in Q_U$. Hence, Q_U is weakly sequentially compact (see Corollary 11 in [4]). Therefore, we have that

$$\Gamma^{-1}\{Q_X, \delta\} \cap Q_U \text{ is weakly sequentially compact for all } \delta \in [0, \gamma_0].$$

It follows from Satz 4.1 in [1] that weak convergence of a sequence $\{u^k\}_{k \in N'} \subset \Gamma^{-1}\{Q_X, \delta\} \cap Q_U$ ($\delta \in [0, \gamma_0]$) to $u \in U$ implies

$$\lim_{k \to \infty} \|\Gamma u - \Gamma u^k\|_X = 0. \tag{4.17}$$

Because of condition C1, Q_U is convex and closed. By Theorem 2.2. and Assertion 2.1., assumption A6 is satisfied. Further, we have that Q_U is weakly closed (see proof of Theorem 2.2.). Since $\{Q_X, \delta\}$ is closed for all $\delta \in [0, \gamma_0]$, we have by (4.17) that

$$\Gamma^{-1}\{Q_X, \delta\} \cap Q_U \text{ is weakly closed for all } \delta \in [0, \gamma_0].$$

Hence, assumption A5 is satisfied.

In order to construct a discretization method applicable to CP_Q, define for every $n \in N'$:

$$I_n := \{\tfrac{\nu}{n} \mid \nu = 0, 1, \ldots, n\},$$

$$X_n := \{x_n : I_n \to R^m \mid \frac{\nu}{n} \to x_n(\frac{\nu}{n})\} \text{ with } \|x_n\|_{X_n} := \max_{\nu=0,1,..,n} \|x_n(\frac{\nu}{n})\|_{R^m}, \tag{4.18}$$

$$\chi_n : X \to X_n \text{ by } \chi_n(x) : I_n \to R^m \text{ with } \chi_n(x)(\frac{\nu}{n}) := x(\frac{\nu}{n}),\ \nu=0,1,\ldots,n, \tag{4.19}$$

$$U_n := \{u_n : I_n \to R^r \mid \frac{\nu}{n} \to u_n(\frac{\nu}{n})\} \text{ with } \|u_n\|_{U_n} := \frac{1}{n}\sum_{\nu=0}^{n-1} \|u_n(\frac{\nu}{n})\|_{R^r}, \tag{4.20}$$

$$\partial_n : U \to U_n \text{ by } \partial_n(u) : I_n \to R^r \text{ with } \partial_n(u)(\frac{\nu}{n}) := u(\frac{\nu}{n}),\ \nu=0,1,\ldots,n, \tag{4.21}$$

$$P_n := \{p_n : I_n \to R^m \mid \frac{\nu}{n} \to p_n(\frac{\nu}{n}),\ p_0 := p_n(0)\}$$
$$\text{with } \|p_n\|_{P_n} := \|p_0\|_{R^m} + \max_{\nu=1,\ldots,n} \|p_n(\frac{\nu}{n})\|_{R^m}, \tag{4.22}$$

$$\pi_n : P \to P_n \text{ by } \pi_n(\hat{p}) : I_n \to R^m \text{ with } \pi_n(\hat{p})(\frac{\nu}{n}) := \begin{cases} p_0, & \nu=0, \\ p(\frac{\nu}{n}), & \nu=1,\ldots,n. \end{cases} \tag{4.23}$$

Further, define

$T_n := \tau_n(T) : X_n \times U_n \to P_n$ by

$$T_n(x_n(\frac{\nu}{n}), u_n(\frac{\nu}{n})) := \begin{cases} x_n(0) - x_0, & \nu=0, \\ n(x_n(\frac{\nu}{n}) - x_n(\frac{\nu-1}{n}) - \frac{1}{n} f(x_n(\frac{\nu-1}{n}), u_n(\frac{\nu-1}{n}))), & \nu=1,\ldots,n, \end{cases} \tag{4.24}$$

and

$$F_n := \phi_n(F) : X_n \times U_n \to R \text{ by } F_n(x_n, u_n) := \frac{1}{n}\sum_{\nu=0}^{n-1} g(x_n(\frac{\nu}{n}), u_n(\frac{\nu}{n})). \tag{4.25}$$

Because of (4.15) and (4.18)-(4.25), we have that $V = \{X_n \times U_n, \xi_n, \phi_n, \tau_n, P_n, \pi_n\}_{n \in N'}$ is applicable to CP_Q.

We obtain:

<u>Lemma 4.2.</u> The discretization method V applicable to CP_Q satisfies assumptions A8 and A9. Further, assumptions A10 and A11 hold.

<u>Proof.</u> Since ΓQ_U is compact for all $\delta \in [0, \gamma_0]$ (see [1], p. 22), we have, applying Lemma 5.3 in [6], that V satisfies A9 and A10. By Lemma 5.2 in [6], one obtains that V satisfies A8. A11 follows immediately from (4.24).

<u>Lemma 4.3.</u> Let V be the dicretization method applicable to CP_Q. Then assumptions A12 and A13 hold.

<u>Proof.</u> Assumption A12 follows directly from the conditions required for Y(t) and Z.
Now, let us show that A13 holds. Let $\delta > 0$ be arbitrarily given. Argueing as in the first part of the proof of Theorem 3.2. (see (3.23)), we conclude that there is an $n_1 \in N'$ implying that

$$T_n^{-1}(0)\cap(\chi_n(\{Q_X,\delta\})\times Q_{U_n})=\emptyset \quad \text{for all } n\geq n_1,\ n\in N'. \tag{4.26}$$

For every $n\geq n_1$, let $(x_n,u_n)\in T_n^{-1}(0)\cap(\chi_n(\{Q_X,\delta\})\times Q_{U_n})$. Then for every $n\in N'$, $n\geq n_1$, there is an $u^{(n)}\in Q_U(n)$ satisfying $u_n=\partial_n(u^{(n)})$. Define

$x^{(n)}:=\Gamma u^{(n)}$ for every $n\geq n_1$, $n\in N'$.

Let $\varepsilon>0$. By Lemma 2.1., there is an $n_2\in N'$, $n_2\geq n_1$, such that

$$\|(x_n,u_n)-(\chi_n(x^{(n)}),\partial_n(u^{(n)}))\|_{X_n\times U_n}=\|x_n-\chi_n(x^{(n)})\|_{X_n}<\frac{\varepsilon}{2}$$

holds for all $n\geq n_2$, $n\in N'$.

Now, A12 implies that

$$\chi_n(x^{(n)})\in\chi_n(\{\{Q_X,\delta\},\tfrac{\varepsilon}{2}\}) \text{ for all } n\geq n_2,\ n\in N', \tag{4.27}$$

and all $u^{(n)}\in Q_U(n)$ satisfying

$$\partial_n(u^{(n)})\in\{u_n\in Q_{U_n}\mid \exists\ x_n\in X_n \text{ with } (x_n,u_n)\in T_n^{-1}(0)\cap(\chi_n(\{Q_X,\delta\})\times Q_{U_n})\}.$$

Since ΓQ_U is compact, there is an $n_3\in N'$ such that, for all $n\geq n_3$,

$$\|\Gamma u^{(n)}(t)-\chi_n(\Gamma u^{(n)})(\tfrac{\nu}{n})\|_{R^m}<\frac{\varepsilon}{2} \tag{4.28}$$

holds for all $t\in[\frac{\nu-1}{n},\frac{\nu}{n}]$, $\nu=1,2,\ldots,n$, and all $u^{(n)}\in Q_U(n)$.

Defining $n_0:=\max(n_2,n_3)$, the assertion follows from (4.27),(4.28) and C1.

Applying the discretization method to CP_Q, we obtain problems $\{CP_{Q(\delta)_n}\}_{n\in N'}$ of the following type:

$$\text{Min!}\quad \sum_{\nu=0}^{n-1} g(x_n(\tfrac{\nu}{n}),u_n(\tfrac{\nu}{n}))$$

subject to

$$x_n(\tfrac{\nu}{n})=x_n(\tfrac{\nu-1}{n})+\tfrac{1}{n}f(x_n(\tfrac{\nu-1}{n}),u_n(\tfrac{\nu-1}{n})),\quad \nu=1,2,\ldots,n-1,$$

$$x_n(0)=0,$$

$$x_n(\tfrac{\nu}{n})\in\{Y(\tfrac{\nu}{n}),\delta\},\quad \nu=0,1,\ldots\ldots,n,$$

$$x_n(1)\in\{Z,\delta\},$$

$$u_n(\tfrac{\nu}{n})\in\bigcap_{t\in J_\nu^n}S(t),\quad \nu=0,1,\ldots\ldots,n-2.$$

References

[1] Bauer,H., Neumann,K.: Berechnung optimaler Steuerungen, Lecture Notes in Operations Research and Mathematical Systems 17, Springer Verlag 1969.

[2] Goldstein,A.: Constructive Real Analysis, Harper International 1967.

[3] Daniel,J.W.: The Approximate Minimization of Functionals, Prentice-Hall 1971.

[4] Dunford,N., Schwartz,J.: Linear Operators I: General Theory, Interscience 1962.

[5] Kantorowitsch,L.W., Akilov,G.P.: Funktionalanalysis in normierten Räumen, Akademie Verlag 1964.

[6] Köhler,M.: Approximation optimaler Prozesse unter Verwendung stabiler und konsistenter Diskretisierungsverfahren, Operations Research Verfahren-Methods of Operations Research, Verlag Anton Hain·Meisenheim, to appear.

[7] Stetter,H.J.: Analysis of Discretization Methods for Ordinary Differential Equations, Springer Tracts in Natural Philosophy 23, 1973.

Michael Köhler
Institut für Operations Research und
mathematische Methoden der Wirtschaftswissenschaften
der Universität Zürich
Weinbergstr. 59
CH-8006 Zürich

Un Algorithme Dual Pour le Calcul de la Distance Entre Deux Convexes

P.J. LAURENT
Université de Grenoble

RESUME : Un algorithme de type dual est proposé pour calculer la distance entre deux convexes de $\mathbb{R}^n$ qui sont définis chacun comme intersection d'une famille infinie de demi-espaces fermés. L'algorithme utilise le théorème d'échange de Stiefel, et, dans un cas particulier, une propriété d'alternance de même nature que celle qui intervient dans l'approximation au sens de Tchebycheff. Une généralisation est proposée pour la minimisation d'une fonction convexe ayant un noyau sur un convexe exprimé comme intersection infinie de demi-espaces fermés.

INTRODUCTION :

Si C_1 et C_2 sont deux convexes fermés disjoints de $\mathbb{R}^n$, on se propose de calculer la distance d entre ces deux convexes et des éléments $\bar{x}_1 \in C_1$ et $\bar{x}_2 \in C_2$ tels que $\|\bar{x}_1-\bar{x}_2\| = d$. Il est bien entendu équivalent de chercher la distance d de 0 à C_1-C_2 et l'élément $\bar{y} \in C_1-C_2$ tel que $\|\bar{y}\| = d$ (on aura $\bar{x}_1-\bar{x}_2 = \bar{y}$). Diverses méthodes directes ont été proposées pour résoudre ce problème ; cf [7] . Dans le cas où les convexes sont donnés par une infinité de contraintes, il semble intéressant d'utiliser un algorithme de type dual qui ne fasse intervenir à chaque itération qu'un nombre fini de contraintes. On a décrit en [5] un tel algorithme pour calculer, notamment, la distance d'un point à un convexe. Mais la difficulté est ici la suivante : si C_1 et C_2 sont définis sous la forme d'une intersection infinie de demi-espaces, il n'est pas évident a priori d'exprimer C_1-C_2 sous la même forme.

Si l'on désigne par C le produit $C_1 \times C_2$ dans l'espace $F = \mathbb{R}^n \times \mathbb{R}^n$ et si l'on note T l'application linéaire de F dans $G = \mathbb{R}^n$ définie par $T(x) = T(x_1,x_2) = x_1-x_2$, alors le problème de la distance entre C_1 et C_2 est équivalent au problème de la minimisation de $\|T(x)\|$ (ou $\frac{1}{2}\|T(x)\|^2$) pour $x \in C$. On reconnait la formulation abstraite du problème des fonctions-spline dans un convexe ; cf [3] . Un algorithme dual a déjà été proposé pour calculer une telle fonction-spline ; cf [4] , [6] . Si l'on considère donc le problème de la distance entre deux convexes comme un cas très particulier de problème de fonction-spline et si l'on particularise cet algorithme, on obtient un algorithme dual très simple que nous nous proposons de décrire directement.

Dans le paragraphe 1, nous donnons un théorème simple pour l'existence. Dans les paragraphes 2 et 3 nous exprimons C_1-C_2 comme une intersection de demi-espaces. Après avoir rappelé au paragraphe 4 l'algorithme dual pour le calcul de la distance d'un point à un convexe (avec un nouveau résultat de convergence), nous décrivons au paragraphe 5 l'algorithme pour la distance entre deux convexes. L'algorithme fait intervenir le théorème d'échange de Stiefel tel qu'il intervient dans l'algorithme de Rémès ; cf [1] , [3]. Lorsque les convexes C_1 et C_2 sont définis en faisant intervenir des polynomes (paragraphe 6) ou plus généralement un sous-espace vectoriel de Haar, on peut remplacer le théorème d'échange par une propriété d'alternance tout à fait analogue à celle qui intervient dans l'algorithme de Rémès pour l'approximation uniforme de fonctions continues par un polynome. Enfin, au paragraphe 7, on donne une extension de la méthode pour calculer le minimum d'une fonction de la forme $f(x) = g(T(x))$ (où T est linéaire et g est uniformément convexe) sur un convexe C qui est une intersection infinie de demi-espaces.

§1. PROBLEME DE LA DISTANCE ENTRE DEUX CONVEXES.

Soit $E = \mathbb{R}^n$, l'espace Euclidien de dimension n muni du produit scalaire usuel :

$$< x,y > = \sum_{i=1}^{n} x^i y^i ,$$

et de la norme associée

$$\|x\| = < x,x > ^{1/2} .$$

On considère deux convexes fermés C_1 et C_2 de E définis comme intersection d'une famille (en général infinie) de demi-espaces fermés :

$$(1.1)\quad \begin{aligned} C_1 &= \{x_1 \in E \mid < x_1,a_1(t) > \leq b_1(t) \text{ , pour tout } t \in T_1\} \\ C_2 &= \{x_2 \in E \mid < x_2,a_2(t)> \leq b_2(t), \text{ pour tout } t \in T_2\} \end{aligned}$$

Il est toujours possible, quitte à faire un changement de variable, de supposer que T_1 et T_2 sont disjoints. Pour la simplicité, on notera donc a l'application de $T = T_1 \cup T_2$ dans E définie par :

$$a(t) = \begin{cases} a_1(t) \text{ , pour } t \in T_1 \text{ ,} \\ a_2(t) \text{ , pour } t \in T_2 \text{ ,} \end{cases}$$

et de même b l'application de T dans $\mathbb{R}$ définie par :

$$b(t) = \begin{cases} b_1(t) \text{ , pour } t \in T_1 \text{ ,} \\ b_2(t) \text{ , pour } t \in T_2 \text{ .} \end{cases}$$

On suppose que $0 \notin a(T)$ et on note D_t le demi-espace fermé :

(1.2) $$D_t = \{x \in E \mid < x,a(t) > \leq b(t)\} .$$

On a alors :

(1.3) $$C_1 = \bigcap_{t \in T_1} D_t \text{ ,} \qquad C_2 = \bigcap_{t \in T_2} D_t \text{ .}$$

On suppose que les convexes fermés C_1 et C_2 sont disjoints. Posons :

(1.4) $$d = \inf_{\substack{x_1 \in C_1 \\ x_2 \in C_2}} \|x_1 - x_2\| .$$

On cherche à déterminer de façon approchée, d'une part la distance d entre C_1 et C_2 d'autre part deux éléments $\bar{x}_1$ et $\bar{x}_2$ (s'ils existent) tels que :

(1.5) $$\bar{x}_1 \in C_1 \text{ , } \bar{x}_2 \in C_2 \text{ , tels que } \|\bar{x}_1 - \bar{x}_2\| = d .$$

(1.6.) <u>REMARQUE</u> :

Posons :

(1.7) $$\Omega = C_1 - C_2 = \{x \in E \mid x = x_1 - x_2,\ x_1 \in C_1,\ x_2 \in C_2\} .$$

Comme C_1 et C_2 sont disjoints, on a $0 \notin \Omega$. Le problème revient à trouver $\bar{y} \in \Omega$ tel que :

(1.8) $$\|\bar{y}\| = \min_{y \in \Omega} \|y\| .$$

Si Ω est fermé, il existe $\bar{y} \in \Omega$ unique, solution de (1.8) et tout couple d'éléments $\bar{x}_1 \in C_1$, $\bar{x}_2 \in C_2$ tels que $\bar{x}_1 - \bar{x}_2 = \bar{y}$ est solution de (1.5). Diverses conditions classiques entrainent que $C_1 - C_2$ est fermé. Considérons par exemple l'hypothèse suivante :

(H1) L'ensemble $\{x \in E \mid < x,a(t) > \leq 0$, pour tout $t \in T\}$ est un sous-espace vectoriel.

On a le résultat suivant :

(1.9) THEOREME :

Si l'hypothèse (H1) est vérifiée, alors $\Omega = C_1 - C_2$ est fermé, il existe $\bar{y} \in \Omega$ unique vérifiant (1.8) et il existe au moins un couple $[\bar{x}_1, \bar{x}_2]$ vérifiant (1.5).

DEMONSTRATION :

Le cône asymptote $C_{1\infty}$ du convexe C_1 (cf [3]) est égal à :

$$C_{1\infty} = \{x_1 \in E \mid < x_1, a(t) > \leq 0 \text{ , pour tout } t \in T_1\},$$

et de même :

$$C_{2\infty} = \{x_2 \in E \mid < x_2, a(t) > \leq 0 \text{ , pour tout } t \in T_2\}.$$

D'après une extension d'un théorème de Dieudonné (cf [3], p. 490), si $C_{1\infty} \cap C_{2\infty}$ est un sous-espace vectoriel, alors $C_1 - C_2$ est fermé. (Ce résultat est aussi une conséquence de [8], cor. 9.11, p. 74). On a évidemment :

$$C_{1\infty} \cap C_{2\infty} = \{x \in E \mid < x, a(t) > \leq 0 \text{ , pour tout } t \in T\} .$$

(1.10) REMARQUE :

1°/ Si l'ensemble $\{x \in E \mid < x, a(t) > \leq 0 \text{ , pour tout } t \in T\}$ est un sous-espace vectoriel, alors il est égal à l'ensemble :

$$\{x \in E \mid < x, a(t) > = 0 \text{ , pour tout } t \in T\}.$$

2°/ Si les ensembles C_1 et C_2 sont disjoints et si (H1) est vérifiée, alors on a forcément $d > 0$.

(1.11) EXEMPLE :

On définit un polynome de degré n-1 par le vecteur de ses coefficients $x = [x_1, \ldots, x_n]$:

$$p(t) = \sum_{j=1}^{n} x_j p_j(t)$$

avec $p_j(t) = t^{j-1}$, $t \in [0,1]$, $j=1,\ldots,n$. Soit g_1 une fonction réelle continue définie sur $[0,1]$. On définit :

$$C_1 = \{x_1 \in E \mid \sum_{j=1}^{n} x_{1j} p_j(t) \leq g_1(t) \text{ , pour tout } t \in [0,1]\} . \tag{1.12}$$

Le convexe C_1 peut être identifié à l'ensemble des polynomes de degré n-1 dont le graphe se trouve au-dessous de celui de la fonction continue g_1. De la même façon, si g_2 est une fonction réelle continue définie sur $[0,1]$, on définit :

$$C_2 = \{x_2 \in E \mid \sum_{j=1}^{n} x_{2j} p_j(t) \geq g_2(t) \text{ , pour tout } t \in [0,1]\} . \tag{1.13}$$

On suppose que $C_1 \cap C_2 = \emptyset$. On cherche deux vecteurs $\bar{x}_1$ et $\bar{x}_2$ tels que :

$$(1.14) \qquad \bar{x}_1 \in C_1, \ \bar{x}_2 \in C_2 \ , \ \|\bar{x}_1-\bar{x}_2\| = \min_{\substack{x_1 \in C_1 \\ x_2 \in C_2}} \|x_1-x_2\|$$

Pour mettre le problème sous la forme proposée, il suffit de poser :

$$(1.15) \qquad \begin{aligned} a_1(t) &= [p_1(t),\ldots,p_n(t)] \in E \\ b_1(t) &= g_1(t) \quad , \quad \text{pour } t \in [0,1] = T_1 \end{aligned}$$

En ce qui concerne le convexe C_2, il est facile de faire un changement de variable de façon à avoir $T_2 \cap T_1 = \emptyset$:

$$(1.16) \qquad \begin{aligned} a_2(t) &= [-p_1(t-2),\ldots,-p_n(t-2)] \in E \ , \\ b_2(t) &= -g_2(t-2) \quad , \quad \text{pour } t \in [2,3] = T_2 \ . \end{aligned}$$

On verra que ce changement de variable n'est qu'une étape intermédiaire pour se placer dans les notations adoptées. Au contraire, il sera intéressant pour la description de l'algorithme, de conserver l'intervalle [0,1] commun aux deux ensembles de contrainte

§ 2. DISTANCE ENTRE DEUX CONVEXES ELEMENTAIRES

(2.1) DEFINITIONS :

Un sous-ensemble $\tilde{T} \subset T$ sera appelé annulateur si l'on a :

$$0 \in co(a(\tilde{T}))$$

où co(A) désigne l'enveloppe convexe de A.

Un annulateur $S \subset T$ sera dit minimal s'il n'existe pas d'annulateur qui soit strictement contenu dans S. On notera $\mathcal{S}$ l'ensemble des annulateurs minimaux.

Enfin, un annulateur minimal $R \subset T$ sera dit régulier si $R \cap T_1 \neq \emptyset$ et $R \cap T_2 \neq \emptyset$.

On notera $\mathcal{R}$ l'ensemble des annulateurs minimaux réguliers.

(2.2) PROPRIETES :

(i) D'après le théorème de Caratheodory, un annulateur minimal a au plus n+1 points.

(ii) Un ensemble $S \subset T$ comprenant k+1 points est un annulateur minimal si :

a) $0 \in ir\ co(a(S))$

b) a(S) engendre un sous-espace vectoriel de dimension k

(on note ir(A) l'intérieur relatif de A).

(iii) Tout annulateur $\tilde{T} \subset T$ contient au moins un annulateur minimal.

(iv) Si $S \in \mathcal{S}$, alors il existe des coefficients $\rho_S(s)$, $s \in S$ tels que :

(2.3) $$\rho_S(s) > 0 \ , \ \sum_{s\in S} \rho_S(s) = 1 \quad \text{et} \quad \sum_{s\in S} \rho_S(s)a(s) = 0.$$

Ces coefficients sont uniques.

Pour $S \in \mathcal{S}$, on pose $S_1 = S \cap T_1$ et $S_2 = S \cap T_2$ et on définit les deux polyèdres suivants :

(2.4) $$C_{S_1} = \{x_1 \in E \mid \langle x_1, a(s) \rangle \leq b(s) \ , \text{ pour tout } s \in S_1\}$$
$$C_{S_2} = \{x_2 \in E \mid \langle x_2, a(s) \rangle \leq b(s), \text{ pour tout } s \in S_2\}$$

On convient que $C_{S_1} = E$ lorsque $S_1 = \emptyset$, et de même pour C_{S_2}. Il est clair que $C_{S_1} \supset C_1$ et $C_{S_2} \supset C_2$. On pose alors :

(2.5) $$\alpha(S) = \sum_{s\in S_1} \rho_S(s)a(s) = - \sum_{s\in S_2} \rho_S(s)a(s)$$

(2.6) $$\beta(S) = \sum_{s\in S} \rho_S(s)b(s)$$

On convient que $\sum_{s\in S_1} \rho_S(s)a(s) = 0$, lorsque $S_1 = \emptyset$, et de même pour S_2.

On a alors le résultat suivant :

(2.7) <u>THEOREME</u> :

Pour tout $S \in \mathcal{S}$, on a :

$$C_{S_1} - C_{S_2} = D(S) \ ,$$

avec

$$D(S) = \{y \in E \mid \langle y, \alpha(S) \rangle \leq \beta(S)\} \ .$$

<u>DEMONSTRATION</u> :

a) Soit $y \in C_{S_1} - C_{S_2}$, c'est-à-dire $y = x_1 - x_2$, avec $x_1 \in C_{S_1}$ et $x_2 \in C_{S_2}$. On a :

$$\langle x_1, a(s) \rangle \leq b(s) \ , \text{ pour tout } s \in S_1 \ ,$$
$$\langle x_2, a(s) \rangle \leq b(s) \ , \text{ pour tout } s \in S_2 \ .$$

On a donc :

$$\sum_{s\in S_1} \rho_S(s) \langle x_1, a(s) \rangle + \sum_{s\in S_2} \rho_S(s) \langle x_2, a(s) \rangle \leq \sum_{s\in S} \rho_S(s)b(s),$$

et comme :

$$\sum_{s\in S_1} \rho_S(s)a(s) = - \sum_{s\in S_2} \rho_S(s)a(s) = \alpha(S) ,$$

on obtient finalement :

$$< x_1-x_2,\alpha(S) > \leq \beta(S) ,$$

c'est-à-dire $x_1-x_2 \in D(S)$.

b) Inversement, soit $y \in D(S)$, c'est-à-dire $< y,\alpha(S) > \leq \beta(S)$ et montrons que $y \in C_{S_1} - C_{S_2}$. Comme $C_{S_1} - C_{S_2}$ est un polyèdre, il est égal à l'intersection des demi-espaces fermés qui le contiennent. Il suffit donc de montrer que y appartient à tout demi-espace fermé contenant $C_{S_1} - C_{S_2}$. Soit $D = \{z \in E| < z,\gamma > \leq \delta\}$ un tel demi-espace, avec $\gamma \neq 0$. On a donc :

$$< x_1-x_2,\gamma > \leq \delta \quad , \text{pour tout } x_1 \in C_{S_1} \text{ et tout } x_2 \in C_{S_2},$$

et par conséquent :

$$<[x_1,x_2] \in E \times E| < x_1,\gamma > + < x_2,-\gamma > \leq \delta\} \supset C_{S_1} \times C_{S_2} .$$

Notons ℓ l'application de T dans E × E définie par :

$$(2.8) \qquad \ell(s) = \begin{cases} [a(s),0] & , \text{si } s \in T_1 , \\ [0,a(s)] & , \text{si } s \in T_2 . \end{cases}$$

Avec ces notations, on a :

$$C_{S_1} \times C_{S_2} = \{x \in E \times E| << x,\ell(s) >> \leq b(s) , \text{pour tout } s \in S\} ,$$

où $<<x,y>> = < x_1,y_1 > + < x_2,y_2 >$ désigne le produit scalaire dans E × E.
L'inclusion précédente s'écrit alors :

$$\{x \in E \times E| << x, \tilde{\gamma} >> > \delta \text{ et } << x,\ell(s) >> \leq b(s) , s \in S\} = \emptyset ,$$

avec $\tilde{\gamma} = [\gamma,-\gamma] \in E \times E$. On sait (cf [3], p. 22) qu'il existe alors des coefficients $\lambda_o \geq 0$, $\lambda(s) \geq 0$, $s \in S$, non tous nuls, tels que :

$$\lambda_o \tilde{\gamma} = \sum_{s\in S} \lambda(s)\ell(s) ,$$

$$\lambda_o \delta \geq \sum_{s\in S} \lambda(s)b(s) .$$

On a $\lambda_o \neq 0$; en effet, si l'on avait $\lambda_o = 0$, on aurait $C_{S_1} \times C_{S_2} = \emptyset$. Quitte à diviser par λ_o, on peut supposer que $\lambda_o = 1$. On a donc :

$$\gamma = \sum_{s \in S_1} \lambda(s)a(s) = - \sum_{s \in S_2} \lambda(s)a(s) ,$$

$$\delta \geq \sum_{s \in S} \lambda(s)b(s) .$$

D'autre part, comme S est un annulateur minimal, les coefficients $\lambda(s)$ sont proportionnels à $\rho_S(s)$:

$$\lambda(s) = k\, \rho_S(s) \quad , \quad \text{avec} \quad k = \sum_{s \in S} \lambda(s) > 0 .$$

Il en résulte que l'on a :

$$\gamma = k\, \alpha(S) ,$$
$$\delta \geq k\, \beta(S) .$$

Comme $y \in D(S)$, on a $< y,\gamma > \; = \; < y,k\alpha(S) > \; \leq \; k\beta(S) \; \leq \delta$, donc $y \in D$.

(2.10) REMARQUES :

(i) Lorsque $S \in \mathcal{R}$, alors S_1 et S_2 sont non vides. Dans ce cas $\alpha(S)$ est différent de zéro et $D(S)$ est un demi-espace fermé de E. Lorsque $S \in \mathcal{S} \setminus \mathcal{R}$, alors l'un des deux ensembles S_1 ou S_2 est vide. Supposons que ce soit S_1. On a alors $C_{S_1} = E$ et $C_{S_1} - C_{S_2} = E$. Dans ce cas $\alpha(S) = 0$ et $D(S) = E$.

(ii) Soit $S \in \mathcal{R}$ tel que $C_{S_1} \cap C_{S_2} = \emptyset$. Alors la distance entre C_{S_1} et C_{S_2} est égale à $d(S) = \dfrac{\beta(S)}{\|\alpha(S)\|}$ et la projection de 0 sur $D(S)$ est $\bar{y} = d(S)\, \dfrac{\alpha(S)}{\|\alpha(S)\|}$. L'ensemble des couples $[x_1,x_2] \in C_{S_1} \times C_{S_2}$ qui réalisent le minimum de la distance entre C_{S_1} et C_{S_2} est :

$$\{[x_1,x_2] \mid \; < x_1,a(s) > = b(s) \text{ , pour tout } s \in S_1$$
$$< x_2,a(s) > = b(s) \text{ , pour tout } s \in S_2$$
$$x_1 - x_2 = \bar{y} \; \} .$$

Lorsque S a exactement n+1 éléments, ce dernier ensemble est réduit à un seul élément $[\bar{x}_1,\bar{x}_2]$.

§3. EXPRESSION DE $C_1 - C_2$

On fait les deux hypothèses supplémentaires suivantes :

(H2) Les ensembles T_1 et T_2 sont compactes et les applications a et b sont continues.

(H3) Il existe $\tilde{x}_1 \in E$ et $\tilde{x}_2 \in E$ tels que :

$$< \tilde{x}_1, a(t) > < b(t) \text{ , pour tout } t \in T_1 \text{ ,}$$

$$< \tilde{x}_2, a(t) > < b(t) \text{ , pour tout } t \in T_2 \text{ ,}$$

(hypothèse de régularité des contraintes).

Posons :

$$(3.1) \qquad \begin{aligned} f_1(x_1) &= \max_{t \in T_1} (< x_1, a(t) > - b(t)) \text{ ,} \\ f_2(x_2) &= \max_{t \in T_2} (< x_2, a(t) > - b(t)) \text{ ,} \end{aligned}$$

et pour $x = [x_1, x_2] \in E \times E$:

$$(3.2) \qquad f(x) = \max(f_1(x_1), f_2(x_2)) = \max_{t \in T} (<<x, \ell(t)>> - b(t))$$

On a :

$$(3.3) \qquad C_1 = \{x_1 \in E | f_1(x_1) \leq 0\} \quad , \quad C_2 = \{x_2 \in E | f_2(x_2) \leq 0\} \quad \text{et}$$

$$(3.4) \qquad C = C_1 \times C_2 = \{x \in E \times E | f(x) \leq 0\} \text{ .}$$

(3.5) REMARQUES :

(i) L'hypothèse (H2) entraine que les fonctions f_1, f_2 et f sont des fonctions convexes continues et les domaines effectifs de leurs polaires sont des convexes compacts. On a notamment pour le domaine effectif de la polaire f^* de f :

$$\text{dom } (f^*) = \text{co}(\ell(T)) \text{ .}$$

(ii) L'hypothèse (H3) entraine qu'il existe $[\tilde{x}_1, \tilde{x}_2]$ tel que $f(\tilde{x}_1, \tilde{x}_2) < 0$.
On a donc $\inf_x f(x) = -f^*(0) < 0$.

Le convexe $C_1 - C_2$ est alors donné par le théorème suivant :

(3.6) THEOREME :

Avec les hypothèses (H1), (H2) et (H3) on a :

$$C_1 - C_2 = \bigcap_{S \in \mathcal{R}} D(S) \text{ .}$$

DEMONSTRATION :

a) Soit $y \in C_1 - C_2$, c'est-à-dire $y = x_1 - x_2$, avec $x_1 \in C_1$, $x_2 \in C_2$. Pour tout $S \in \mathcal{R}$, $x_1 \in C_{S_1}$ et $x_2 \in C_{S_2}$, on a donc $x_1 - x_2 \in C_{S_1} - C_{S_2} = D(S)$; cf th. (2.7).

On a donc $y \in \bigcap_{S \in \mathcal{R}} D(S)$.

b) Inversement, soit $y \in \bigcap_{S \in \mathcal{R}} D(S)$. Comme $D(S) = E$ lorsque $S \in \mathcal{S} \setminus \mathcal{R}$, on a aussi $y \in \bigcap_{S \in \mathcal{S}} D(S)$. Montrons que $y \in C_1 - C_2$. D'après le théorème (1.9), l'ensemble $C_1 - C_2$ est fermé. Il est donc égal à l'intersection des demi-espaces fermés qui le contiennent. Il suffit donc de montrer que y appartient à tout demi-espace fermé contenant $C_1 - C_2$. Soit $D = \{z \mid \langle z,\gamma\rangle \leq \delta\}$ un tel demi-espace ($\gamma \neq 0$). On a donc :

$$\{[x_1,x_2] \in E \times E \mid \langle x_1,\gamma\rangle + \langle x_2,-\gamma\rangle \leq \delta\} \supset C_1 \times C_2 = C$$

Soit χ_C la fonction indicatrice de C et χ_C^* sa fonction d'appui. Si l'on note $\tilde{\gamma} = [\gamma,-\gamma] \in E \times E$, l'inclusion précédente est équivalente à :

$$\chi_C^*(\tilde{\gamma}) \leq \delta .$$

Comme $C = \{x \in E \times E \mid f(x) \leq 0\}$ avec $f^*(0) > 0$, en désignant par f_∞^* la fonction asymptote de f^*, on a (cf [8], p. 118 et p. 79) :

$$\chi_C^*(\tilde{\gamma}) = \operatorname{Inf}\left(\mu f^*\left(\frac{\tilde{\gamma}}{\mu}\right), \quad \mu > 0 \;;\; f_\infty^*(\tilde{\gamma})\right)$$

et la borne inférieure est atteinte. Le domaine effectif de f^* étant borné, on a $f_\infty^*(y) = \chi_{\{0\}}(y)$. Comme $\tilde{\gamma} \neq 0$, on a donc :

$$\chi_C^*(\tilde{\gamma}) = \min_{\mu>0}\left(\mu f^*\left(\frac{\tilde{\gamma}}{\mu}\right)\right) .$$

Il existe donc $\mu > 0$ tel que $\chi_C^*(\tilde{\gamma}) = \mu\, f^*\left(\frac{\tilde{\gamma}}{\mu}\right)$. On a donc :

$$f^*\left(\frac{\tilde{\gamma}}{\mu}\right) = f^*\left(\frac{\gamma}{\mu}, -\frac{\gamma}{\mu}\right) \leq \frac{\delta}{\mu} .$$

Notons m l'application de T dans $E \times E \times \mathbb{R}$ définie par :

$$m(t) = [\ell(t),b(t)] .$$

L'hypothèse (H2) entraine que :

$$\text{épi}(f^*) = \text{co}(m(T)) + 0 \times \mathbb{R}^+ ,$$

(où 0 est l'origine de $E \times E$ et $\mathbb{R}^+$ l'ensemble des réels positifs ou nuls). Pour tout $y \in \text{dom}\,(f^*)$, on a $[y,f^*(y)] \in \text{co}(m(T))$. On a donc :

$$\left[\frac{\tilde{\gamma}}{\mu}, f^*\left(\frac{\tilde{\gamma}}{\mu}\right)\right] \in \text{co}(m(T)) .$$

Par conséquent, il existe un sous-ensemble fini $\tilde{T} \subset T$ et $\rho(t) > 0$, $t \in \tilde{T}$, $\sum_{t\in\tilde{T}} \rho(t) = 1$ tel que :

$$\frac{\tilde{\gamma}}{\mu} = \sum_{t\in\tilde{T}} \rho(t)\ell(t) = \left[\sum_{t\in\tilde{T}_1} \rho(t)a(t), \sum_{t\in\tilde{T}_2} \rho(t)a(t)\right] ,$$

$$f^*\left(\frac{\tilde{\gamma}}{\mu}\right) = \sum_{t\in\tilde{T}} \rho(t)b(t) ,$$

où $\tilde{T}_1 = \tilde{T} \cap T_1$ et $\tilde{T}_2 = \tilde{T} \cap T_2$. On a donc :

$$\gamma = \mu \sum_{t\in\tilde{T}_1} \rho(t)a(t) = -\mu \sum_{t\in\tilde{T}_2} \rho(t)a(t) ,$$

$$\delta \geq \mu \sum_{t\in\tilde{T}} \rho(t)b(t).$$

Montrons maintenant qu'il existe des annulateurs minimaux $S^j \in \mathcal{S}$, $j=1,\dots,m$ et des réels $\mu_j > 0$, $j=1,\dots,m$, tels que :

$$[\gamma,-\gamma] = \sum_{j=1}^{m} \mu_j [\sum_{s\in S_1^j} \rho_{S^j}(s)a(s) \,,\, \sum_{s\in S_2^j} \rho_{S^j}(s)a(s)]$$

$$\mu \sum_{t\in\tilde{T}} \rho(t)b(t) = \sum_{j=1}^{m} \mu_j [\sum_{s\in S^j} \rho_{S^j}(s)b(s)]$$

En effet, $\tilde{T}$ est un annulateur, et l'on a :

$$\mu \sum_{t\in\tilde{T}} \rho(t)\ell(t) = [\gamma,-\gamma] .$$

Soit $S^1 \in \mathcal{S}$ un annulateur minimal contenu dans $\tilde{T}$ et posons :

$$\sum_{s\in S^1} \rho_{S^1}(s)\ell(s) = [\gamma_1,-\gamma_1] .$$

Formons :

$$\mu \sum_{t\in\tilde{T}} \rho(t)\ell(t) - \mu_1 \sum_{s\in S^1} \rho_{S^1}(s)\ell(s) = \sum_{t\in\tilde{T}} \theta(t)\ell(t) ,$$

avec :

$$\theta(t) = \begin{cases} \mu\, \rho(t) - \mu_1\, \rho_{S^1}(t) , & \text{si } t \in S^1 , \\ \mu\, \rho(t) & , \text{si } t \in \tilde{T} \setminus S^1 . \end{cases}$$

Si l'on choisit $\mu_1 > 0$ correctement ($\mu_1 = \min_{t\in S^1} \frac{\mu\, \rho(t)}{\rho_{S^1}(t)} = \frac{\mu\, \rho(t_1)}{\rho_{S^1}(t_1)}$) on aura $\theta(t) \geq 0$, pour tout $t \in \tilde{T}$ et $\theta(t_1) = 0$. On a donc :

$$\mu \sum_{t\in\tilde{T}} \rho(t)\ell(t) = \mu_1 \sum_{s\in S^1} \rho_{S^1}(s)\ell(s) + \sum_{t\in\tilde{T}_1} \theta(t)\ell(t) ,$$

où $\tilde{T}_1 \subset \tilde{T}$ est un annulateur qui a un nombre d'éléments strictement inférieur à celui de $\tilde{T}$. On a aussi :

$$\mu \sum_{t\in\tilde{T}} \rho(t)b(t) = \mu_1 \sum_{s\in S^1} \rho_{S^1}(s)b(s) + \sum_{t\in\tilde{T}_1} \theta(t)b(t).$$

On répète cette opération un nombre fini de fois (inférieur au nombre d'éléments de $\tilde{T}$) pour obtenir la propriété.

Pour tout S^j , $j=1,\dots,m$, on a :

$$< y,\alpha(S^j) > \; \leq \; \beta(S^j) ,$$

donc :

$$< y, \sum_{s \in S_1^j} \rho_{S^j}(s)a(s) > \leq \sum_{s \in S^j} \rho_{S^j}(s)b(s) ,$$

et par conséquent :

$$< y, \sum_{j=1}^{m} \mu_j \sum_{s \in S_1^j} \rho_{S^j}(s)a(s) > \leq \sum_{j=1}^{m} \mu_j \sum_{s \in S^j} \rho_{S^j}(s)b(s) ,$$

soit :

$$< y, \gamma > \leq \mu \sum_{t \in \tilde{T}} \rho(t)b(t) \leq \delta ,$$

ce qui termine la démonstration.

§4. ALGORITHME POUR LA DISTANCE D'UN POINT A UN CONVEXE.

On a vu que le problème de trouver la distance entre deux convexes C_1 et C_2 est équivalent au problème de trouver la distance de 0 au convexe C_1-C_2, et que, avec des hypothèses convenables, ce convexe s'exprime comme intersection d'une famille de demi-espaces fermés. On va rappeler dans ce paragraphe un algorithme dual pour calculer la distance de 0 à un convexe. Cet algorithme a été décrit dans [2] dans le cas d'un nombre fini de contraintes puis généralisé dans [5] pour la minimisation d'une fonction uniformément convexe avec une infinité de contraintes. Nous donnons ici un résultat de convergence nouveau qui est utile pour la suite.

Soit K un convexe fermé de E défini par :

(4.1) $\qquad K = \{y \in E| < y,z(j) > \leq c(j) , \text{ pour tout } j \in J \} ,$

où J est un ensemble quelconque et z, c des applications bornées de J dans E, $\mathbb{R}$ respectivemnet. Notons mz et mc leurs bornes supérieures respectives :

(4.2) $\qquad mz = \operatorname{Sup}_{j \in J} \|z(j)\| \quad \text{et} \quad mc = \operatorname{Sup}_{j \in J} |c(j)| .$

Si l'on note :

(4.3) $\qquad K_j = \{y \in E| < y,z(j) > \leq c(j)\} , \quad j \in J ,$

on a :

(4.4) $\qquad K = \bigcap_{j \in J} K_j .$

On suppose que $0 \notin K$ et on note $\bar{y}$ l'élément de K tel que :

(4.5) $\qquad \|\bar{y}\| = \min_{y \in K} \|y\| .$

Supposons qu'a l'itération ν , on ait un élément $y^\nu \in E$ tel que le demi-espace fermé associé :

(4.6) $$D^{\nu} = \{y \in E \mid \langle y^{\nu}, y - y^{\nu} \rangle \geq 0\}$$

contiennent K. On remarque que si $y^{\nu} \in K$, alors c'est la solution $\bar{y}$. Supposons donc que $y^{\nu} \notin K$ et soit $j^{\nu} \in J$ tel que $y^{\nu} \notin K_{j^{\nu}}$. On définit alors l'élément $y^{\nu+1}$ comme étant la projection de O sur $D^{\nu} \cap K_{j^{\nu}}$:

(4.7) $$y^{\nu+1} \in D^{\nu} \cap K_{j^{\nu}} \text{ et } \|y^{\nu+1}\| = \min_{y \in D^{\nu} \cap K_{j^{\nu}}} \|y\|$$

Il est bien connu que le demi-espace associé :

(4.8) $$D^{\nu+1} = \{y \in E \mid \langle y^{\nu+1}, y - y^{\nu+1} \rangle \geq 0\}$$

contiendra alors $D^{\nu} \cap K_{j^{\nu}}$, donc aussi K .

(4.9) <u>REMARQUE</u> :

La résolution du sous-problème (4.7) est très facile ; comme la projection y^{ν} de O sur D^{ν} n'appartient pas à $K_{j^{\nu}}$, on peut procéder en deux étapes :

1°/ On calcule la projection $\tilde{y}^{\nu+1}$ de O sur $K_{j^{\nu}}$ qui est égale, lorsque $O \notin K_{j^{\nu}}$, à $\dfrac{c(j^{\nu})}{\|z(j^{\nu})\|^2} z(j^{\nu})$. Si $\tilde{y}^{\nu+1} \in D^{\nu}$ alors on a $y^{\nu+1} = \tilde{y}^{\nu+1}$.

2°/ Dans le cas contraire, $y^{\nu+1}$ est égale à la projection de O sur l'intersection des hyperplans frontières de D^{ν} et $K_{j^{\nu}}$; on a alors $y^{\nu+1} = \alpha z(j^{\nu}) + \beta y^{\nu}$ où α et β sont solution de :

$$\begin{aligned} \alpha \langle z(j^{\nu}), z(j^{\nu}) \rangle + \beta \langle y^{\nu}, z(j^{\nu}) \rangle &= c(j^{\nu}) \\ \alpha \langle z(j^{\nu}), y^{\nu} \rangle + \beta \langle y^{\nu}, y^{\nu} \rangle &= \langle y^{\nu}, y^{\nu} \rangle \end{aligned}$$

<u>PROPRIETES DE LA SUITE</u> y^{ν} :

Voyons d'abord quelles sont les propriétés de la suite y^{ν} sans préciser davantage le choix de l'indice j^{ν} à chaque itération. Comme $D^{\nu} \supset D^{\nu} \cap K_{j^{\nu}} \supset K$, on a :

(4.10) $$\|y^{\nu}\| \leq \|y^{\nu+1}\| \leq \|\bar{y}\| ,$$

et par conséquent :

(4.11) $$\lim_{\nu \to \infty} \|y^{\nu}\| = \ell \leq \|\bar{y}\| ,$$

donc :

(4.12) $$\lim_{\nu \to \infty} (\|y^{\nu+1}\| - \|y^{\nu}\|) = 0 .$$

Comme $y^{\nu+1} \in D^{\nu}$, c'est-à-dire $\langle y^{\nu}, y^{\nu+1} - y^{\nu} \rangle \geq 0$, on a :

$$\|y^{\nu+1} - y^{\nu}\|^2 = \|y^{\nu+1}\|^2 - \|y^{\nu}\|^2 - 2 \langle y^{\nu+1} - y^{\nu}, y^{\nu} \rangle \leq \|y^{\nu+1}\|^2 - \|y^{\nu}\|^2 .$$

Il en résulte que :

(4.13) $$\lim_{\nu\to\infty} \|y^{\nu+1} - y^\nu\| = 0 .$$

CHOIX DE L'INDICE j^ν :

Comme $y^\nu \notin K_{j^\nu}$, il existe $\theta^\nu \in]0,1]$ tel que :

(4.14) $$0 < (< y^\nu,z(j^\nu) > - c(j^\nu)) = \theta^\nu \operatorname{Sup}_{j\in J} (< y^\nu,z(j) > - c(j)).$$

On a montré dans [5] que si l'on choisit pour K_{j^ν} la contrainte "la plus mal satisfaite" c'est-à-dire $j^\nu \in J$ tel que :

$$(< y^\nu,z(j^\nu) > - c(j^\nu)) = \operatorname{Sup}_{j\in J} (< y^\nu,z(j) > - c(j))$$

(ce qui correspond à $\theta^\nu = 1$), ou encore (ce dernier choix étant difficile en pratique, et même impossible si la borne supérieure n'est pas atteinte) si l'on choisit $j^\nu \in J$ tel que :

$$(< y^\nu,z(j^\nu) > - c(j^\nu)) \geq m \operatorname{Sup}_{j\in J} (< y^\nu,z(j) > - c(j))$$

(ce qui correspond à $\theta^\nu \geq m$, pour tout ν) où m est une constante appartenant à $]0,1[$, alors la suite y^ν converge vers $\bar{y}$. Le théorème suivant montre qu'il en est de même lorsque la condition $\theta^\nu \geq m$ est satisfaite pour au moins une sous-suite.

(4.15) THEOREME :

S'il existe une sous-suite, définie par $\mathbb{N}_1 \subset \mathbb{N}$ et une constante $m > 0$ telle que θ^ν défini dans (4.14) vérifie $\theta^\nu \geq m$, pour tout $\nu \in \mathbb{N}_1$, alors $\lim_{\nu\to\infty} \|y^\nu\| = \|\bar{y}\|$ et $\lim_{\nu\to\infty} y^\nu = \bar{y}$.

DEMONSTRATION :

Comme $y^{\nu+1} \in K_{j^\nu}$, on a :

$$\|y^\nu-y^{\nu+1}\| \geq d(y^\nu;K_{j^\nu}) = \operatorname{Inf}_{y\in K_{j^\nu}} \|y^\nu-y\| = \frac{< y^\nu,z(j^\nu) > - c(j^\nu)}{\|z(j^\nu)\|}$$

On en déduit que :

$$\operatorname{Sup}_{j\in J} (< y^\nu,z(j) > - c(j)) \leq \frac{mz}{m} \|y^\nu-y^{\nu+1}\| \quad , \text{ pour tout } \nu \in \mathbb{N}_1$$

et donc que l'on a :

$$\limsup_{\substack{\nu\to\infty \\ \nu\in\mathbb{N}_1}} (\operatorname{Sup}_{j\in J} (< y^\nu,z(j) > - c(j))) \leq 0 .$$

La suite y^ν étant bornée, il existe $\mathbb{N}_2 \subset \mathbb{N}_1$ tel que $\lim\limits_{\substack{\nu\to\infty \\ \nu\in\mathbb{N}_2}} y^\nu = \tilde{y}$.

L'élément $\tilde{y}$ vérifie d'une part $\|\tilde{y}\| = \ell \leq \|\bar{y}\|$ et d'autre part $\sup\limits_{j\in J}(<\tilde{y},z(j)> - c(j)) \leq 0$ donc $\tilde{y} \in K$. On en déduit que $\tilde{y} = \bar{y}$ et $\ell = \|\bar{y}\|$. La suite $\|y^\nu\|$ étant monotone, on a donc :

$$\lim_{\nu\to\infty} \|y^\nu\| = \ell = \|\bar{y}\| .$$

Enfin, l'élément $\bar{y} \in K \subset D^\nu$ vérifie $< y^\nu,\bar{y}-y^\nu > \geq 0$, ce qui entraine, comme on l'a vu plus haut :

$$\|\bar{y}-y^\nu\|^2 \leq \|\bar{y}\|^2 - \|y^\nu\|^2 .$$

Il en résulte que :

$$\lim_{\nu\to\infty} \|\bar{y}-y^\nu\| = 0 .$$

(4.16) REMARQUE :

La même amélioration du résultat de convergence peut être faite dans le cas plus général de la minimisation d'une fonctionnelle uniformément convexe dans un espace de Banach réflexif tel qu'il est traité en [5] .

§5. ALGORITHME POUR LA DISTANCE ENTRE DEUX CONVEXES

Dans le paragraphe 3, on a exprimé le convexe $\Omega = C_1-C_2$ comme une intersection de demi-espaces fermés :

(5.1) $$\Omega = \bigcap_{S\in\mathcal{R}} D(S) .$$

L'ensemble $\mathcal{R}$ des annulateurs minimaux réguliers joue donc le rôle de l'ensemble d'indices J dans le paragraphe précédent. Supposons que l'on ait obtenu l'itéré y^ν tel que le demi-espace associé D^ν contienne Ω . L'application directe de l'algorithme précédent nous conduirait à choisir la contrainte la plus mal satisfaite en y^ν, c'est-à-dire en fait l'annulateur minimal régulier $S^\nu \in \mathcal{R}$ tel que :

(5.2) $$(< y^\nu,\alpha(S^\nu) > - \beta(S^\nu)) = \sup_{S\in\mathcal{R}} (< y^\nu,\alpha(S) > - \beta(S)) .$$

Un tel S^ν n'existe pas forcément (si la borne supérieure n'est pas atteinte) et même dans le cas où il existe, sa détermination pratique constitue un problème plus difficile que le problème initial. De même, si l'on se donne a priori $\theta \in]0,1[$, il serait difficile de déterminer $S^\nu \in \mathcal{R}$ tel que :

(5.3) $$(< y^\nu,\alpha(S^\nu) > - \beta(S^\nu)) \geq \theta \sup_{S\in\mathcal{R}} (< y^\nu,\alpha(S) > - \beta(S)) .$$

L'algorithme que nous allons décrire consiste à être moins exigeant pour le choix de la contrainte la plus mal satisfaite. Si $S^{\nu-1} \in \mathcal{R}$ est l'annulateur minimal régulier associé à $y^{\nu-1}$ (correspondant à la contrainte $D(S^{\nu-1})$ qui a été considéré à l'itération ν-1), on choisira pour S^{ν} un élément de $\mathcal{R}$ qui contient seulement un nouveau point $t^{\nu} \in T$ par rapport à $S^{\nu-1}$. A l'élément y^{ν} sont associés deux éléments x_1^{ν} et x_2^{ν} tels que $x_1^{\nu} - x_2^{\nu} = y^{\nu}$. Le point nouveau t^{ν} correspondra à la contrainte la plus mal satisfaite en x_1^{ν} ou x_2^{ν}.

a) Description d'une itération de l'algorithme

A l'itération ν on suppose que l'on a un itéré $y^{\nu} \in E$ tel que :

(5.4) $$D^{\nu} = \{y \in E | < y^{\nu}, y-y^{\nu} > \geq 0\} \supset \Omega \ .$$

On suppose que l'on a aussi un annulateur minimal régulier $S^{\nu-1} \in \mathcal{R}$ tel que :

(5.5) $$< y^{\nu}, \alpha(S^{\nu-1}) > = \beta(S^{\nu-1}) \ ,$$

c'est-à-dire $y^{\nu} \in Fr(D(S^{\nu-1}))$. Nous supposerons que cet annulateur se compose exactement de n+1 points. Dans ces conditions il existe deux éléments (uniques) x_1^{ν} et x_2^{ν} tels que :

(5.6) $$x_1^{\nu} \in C_{S_1^{\nu-1}} \ , \quad x_2^{\nu} \in C_{S_2^{\nu-1}} \ , \quad x_1^{\nu} - x_2^{\nu} = y^{\nu} \ .$$

On montre qu'en fait, x_1^{ν} appartient à la frontière de tous les demi-espaces qui constituent $C_{S_1^{\nu-1}}$ et de même pour x_2^{ν} . Ces deux éléments vérifient donc les équations :

(5.7) $$\begin{aligned} &< x_1^{\nu}, a(s) > = b(s) \ , \text{ pour tout } s \in S_1^{\nu-1} \\ &< x_2^{\nu}, a(s) > = b(s) \ , \text{ pour tout } s \in S_2^{\nu-1} \\ &x_1^{\nu} - x_2^{\nu} = y^{\nu} \end{aligned}$$

(et ces équations les déterminent de façon unique).
On montre facilement que si $x_1^{\nu} \in C_1$ et $x_2^{\nu} \in C_2$ alors c'est une solution du problème. Supposons que ce ne soit pas le cas. On choisit alors la contrainte la plus mal satisfaite pour le couple $x^{\nu} = [x_1^{\nu}, x_2^{\nu}]$, c'est-à-dire le point $t^{\nu} \in T$ tel que :

(5.8) $$<< x^{\nu}, \ell(t^{\nu}) >> - b(t^{\nu}) = \max_{t \in T} (<< x^{\nu}, \ell(t) >> - b(t)) = f(x^{\nu})$$

On pourrait en fait se contenter de prendre $t^{\nu} \in T$ tel que :

$$<< x^{\nu}, \ell(t^{\nu}) >> - b(t^{\nu}) \geq \theta \ f(t^{\nu})$$

avec θ fixe appartenant à $]0,1[$. On note que l'équation (5.8) n'a pas forcément une solution unique : s'il y a plusieurs solutions on choisit simplement l'une quelconque d'entre elles. Le point t^{ν} peut appartenir soit à T_1, correspondant ainsi à une contrainte relative à C_1, soit à T_2, correspondant à une contrainte relative à C_2. On a forcément $t^{\nu} \notin S^{\nu-1}$. On sait (théorème d'échange généralisé, cf [3],[1]) qu'il existe un annulateur minimal (unique) S^{ν} contenu dans $S^{\nu-1} \cup \{t^{\nu}\}$ et différent de $S^{\nu-1}$ (l'annulateur S^{ν} comprend donc effectivement t^{ν}). Nous supposerons à nouveau que S^{ν} se compose exactement de n+1 éléments. Ainsi S^ν est obtenu en remplaçant un élément de $S^{\nu-1}$ par t^ν ; il existe une méthode constructive simple pour déterminer l'élément de $S^{\nu-1}$ qui disparait ; cf [3] , p. 118.

On considère l'ensemble $D(S^\nu)$ associé à S^ν . On a bien sûr $y^\nu \notin D(S^\nu)$. On remarque que l'annulateur minimal S^ν est forcément régulier (s'il ne l'était pas on aurait $D(S^\nu)=E$). Soit $y^{\nu+1}$ la projection de 0 sur $D^\nu \cap D(S^\nu)$:

(5.9) $$y^{\nu+1} \in D^\nu \cap D(S^\nu) \;, \; \|y^{\nu+1}\| = \min_{y \in D^\nu \cap D(S^\nu)} \|y\|$$

Il est évident que le demi-espace associé à $y^{\nu+1}$:

(5.10) $$D^{\nu+1} = \{y \in E | < y^{\nu+1}, y-y^{\nu+1} > \geq 0\}$$

contient $D^\nu \cap D(S^\nu)$, donc aussi Ω . Pour la détermination effective de $y^{\nu+1}$, nous renvoyons à la remarque (4.9). Comme la projection y^ν de 0 sur D^ν n'appartient pas à $D(S^\nu)$ l'élément $y^{\nu+1}$ appartient à la frontière de $D(S^\nu)$, c'est-à-dire vérifie :

(5.11) $$< y^{\nu+1}, \alpha(S^\nu) > = \beta(S^\nu) .$$

Comme S^ν se compose exactement de n+1 points, il existe deux éléments (uniques) $x_1^{\nu+1}$ et $x_2^{\nu+1}$ tels que :

(5.12) $$x_1^{\nu+1} \in C_{S_1^\nu} \;, \; x_2^{\nu+1} \in C_{S_2^\nu} \;, \; x_1^{\nu+1} - x_2^{\nu+1} = y^{\nu+1} .$$

Comme précédemment ces deux éléments vérifient les équations :

(5.13) $$\begin{aligned} &< x_1^{\nu+1}, a(s) > = b(s) \;, \text{ pour tout } s \in S_1^\nu \;, \\ &< x_2^{\nu+2}, a(s) > = b(s) \;, \text{ pour tout } s \in S_2^\nu \;, \\ &x_1^{\nu+1} - x_2^{\nu+1} = y^{\nu+1} \;, \end{aligned}$$

qui les déterminent complètement.

b) Démarrage de l'algorithme :

A l'itération $\nu = 1$, on choisit un annulateur minimal régulier $S°$ tel que $\beta(S°) < 0$. Le demi-espace $D(S°)$ associé contient Ω mais ne contient pas l'origine. Un tel annulateur existe puisque $0 \notin \Omega$. On suppose que $S°$ comporte exactement n+1 éléments. On pose $D^1 = D(S°)$. Soit y^1 la projection de 0 sur $D(S°)$. On a $y^1 \neq 0$ et D^1 s'écrit encore

(5.14) $$D^1 = \{y \in E | < y^1, y-y^1 > \geq 0\}$$

et l'on a :

(5.15) $$< y^1, \alpha(S°) > = \beta(S°).$$

Il existe alors deux éléments (uniques) x_1^1 et x_2^1 tels que :

$$(5.16)\quad \begin{aligned} &< x_1^1,a(s) > = b(s) \text{ , pour tout } s \in S_1^\circ \text{ ,}\\ &< x_2^1,a(s) > = b(s) \text{ , pour tout } s \in S_2^\circ \text{ ,}\\ &x_1^1 - x_2^1 = y^1 \text{ .}\end{aligned}$$

(5.17) REMARQUES :

(i) Comme $y^\nu \notin D(S^\nu)$, on a :

$$< y^\nu,\alpha(S^\nu) > - \beta(S^\nu) > 0$$

Comme dans le paragraphe 4, soit $\theta^\nu \in]0,1]$ tel que :

$$(5.18)\quad < y^\nu,\alpha(S^\nu) > - \beta(S^\nu) = \theta^\nu \operatorname*{Sup}_{S\in \mathcal{R}} (< y^\nu,\alpha(S) > - \beta(S)).$$

Evaluons les deux côtés de cette égalité. Pour tout $S \in \mathcal{R}$, on a :

$$< y^\nu,\alpha(S) > - \beta(S) = < x_1^\nu-x_2^\nu, \sum_{s\in S_1} \rho_S(s)a(s) > - \sum_{s\in S} \rho_S(s)b(s)$$

$$= \sum_{s\in S_1} \rho_S(s)(< x_1^\nu,a(s)> - b(s)) + \sum_{s\in S_2} \rho_S(s)(<x_2^\nu,a(s)>-b(s))$$

On a donc :

$$(5.19)\quad \operatorname*{Sup}_{S\in \mathcal{R}} (< y^\nu,\alpha(S) > - \beta(S)) \leq f(x_1^\nu,x_2^\nu) \text{ .}$$

Considérons maintenant le cas de $S = S^\nu$; On peut encore écrire :

$$(5.20)\quad < y^\nu,\alpha(S) > - \beta(S) = \sum_{s\in S} \rho_S(s)(<< x^\nu,\ell(s) >> - b(s)) \text{ ,}$$

avec $x^\nu= [x_1^\nu,x_2^\nu]$. On a noté t^ν le point de S^ν qui a été introduit à l'itération ν . Les autres points de S^ν appartiennent en fait à $S^{\nu-1}$. On a donc :

$$(5.21)\quad << x^\nu,\ell(s) >> - b(s) = \begin{cases} f(x_1^\nu,x_2^\nu) \text{ , si } s = t^\nu \text{ ,}\\ 0 \qquad\qquad \text{ , si } s \in S^\nu \text{ , } s \neq t^\nu \text{ .}\end{cases}$$

On obtient donc :

$$(5.22)\quad < y^\nu,\alpha(S^\nu) > - \beta(S^\nu) = \rho_{S^\nu}(t^\nu)f(x_1^\nu,x_2^\nu) \text{ .}$$

De (5.18), (5.19) et (5.22), on déduit :

$$(5.23)\quad \rho_{S^\nu}(t^\nu) \leq \theta^\nu.$$

Il suffit donc que, au moins pour une sous-suite, les coefficients $\rho_{S^\nu}(t^\nu)$ soit bornés inférieurement par un nombre strictement positif, pour que l'algorithme converge (cf théorème (4.15)), c'est-à-dire que les contraintes soient satisfaites à la limite pour y^ν et que y^ν converge vers $\bar{y}$.

(ii) Comme dans la démonstration du théorème (4.15), on peut écrire :

$$(5.24) \qquad \|y^\nu - y^{\nu+1}\| \leq d(y^\nu;D(S^\nu)) = \frac{< y^\nu,\alpha(S^\nu) > -\beta(S^\nu)}{\|\alpha(S^\nu)\|} = \frac{\rho_{S^\nu}(t^\nu) f(x_1^\nu,x_2^\nu)}{\|\alpha(S^\nu)\|} .$$

L'ensemble $a(T_1)$ étant borné, il existe $k \in \mathbb{R}$ tel que $\|a(t)\| \leq k$, pour tout $t \in T_1$, et on a alors aussi :

$$\|\alpha(S^\nu)\| \leq k \quad , \quad \text{pour tout } \nu .$$

On obtient donc :

$$f(x_1^\nu,x_2^\nu) \leq \frac{k}{\rho_{S^\nu}(t^\nu)} \|y^\nu - y^{\nu+1}\| .$$

On voit donc que si $\rho_{S^\nu}(t^\nu)$ est borné inférieurement par une constante positive, alors les contraintes seront aussi vérifiées à la limite pour $[x_1^\nu,x_2^\nu]$.

La convergence de cet algorithme sera étudiée de façon plus approfondie dans un travail ultérieur.

§6. EXEMPLE

Nous reprenons ici l'exemple (1.11). Il est facile de voir que les hypothèses (H1),(H2) et (H3) sont vérifiées. Comme dans le cas de l'approximation au sens de Tchebycheff, il est intéressant de ne pas distinguer (par un changement de variable comme en (1.16)) les ensembles T_1 et T_2. Un annulateur minimal S est défini ici par k+1 couples distincts d'abscisses et de signes $[\varepsilon_i,t_i]$, $i=1,\dots,k+1$, $(k \leq n)$, avec $\varepsilon_i = \pm 1$ et $t_i \in [0,1]$ tels qu'il existe des coefficients positifs ρ_i , $\sum_{i=1}^{k+1} \rho_i = 1$, pour lesquels :

$$(6.1) \qquad \sum_{i=1}^{k+1} \rho_i \varepsilon_i \, p(t_i) = 0 \text{ , pour tout polynome de degré } n-1 .$$

Lorsque $\varepsilon_i = +1$, le point t_i est interprété comme appartenant à T_1 (contrainte C_1) et lorsque $\varepsilon_i = -1$, le point t_i est interprété comme appartenant à T_2, au changement de variable près, (contrainte C_2). Il est bien connu que l'on a forcément :

(6.2) soit $k = n$: toutes les abscisses sont distinctes, et si l'on suppose $t_1 < t_2 < \dots < t_{n+1}$, on a $\varepsilon_i = -\varepsilon_{i+1}$, $i=1,\dots,k$ et les $(\rho_i\varepsilon_i)$ sont proportionnels aux coefficients de la différence divisée basée sur ces abscisses,

(6.3) soit $k = 1$: $t_1 = t_2$, $\varepsilon_1 = -\varepsilon_2$, $\rho_1 = \rho_2 = \frac{1}{2}$.

L'hypothèse que nous avons faite pour décrire l'algorithme correspond à supposer que l'on ne rencontre pas d'annulateurs minimaux de deuxième type. On a donc toujours n+1 abscisses distinctes et si elles sont rangées dans l'ordre croissant, elles concernent alternativement l'ensemble C_1 et l'ensemble C_2. La règle d'échange de Stiefel peut alors être simplifiée , comme dans le cas de l'approximation sur un intervalle avec condition de Haar : la nouvelle abscisse (qui concerne soit C_1 soit C_2) est échangée avec l'une des abscisses précédentes de façon à garder l'alternance des abscisses de type C_1 et C_2. L'algorithme se simplifie alors de la façon suivante :

a) <u>Description d'une itération de l'algorithme</u> :

Supposons qu'à l'itération ν on ait obtenu :

$$
(6.4)\qquad
\begin{aligned}
q^\nu(t) &= \sum_{j=1}^{n} y_j^\nu \, p_j(t) ,\\
p_1^\nu(t) &= \sum_{j=1}^{n} x_{1j}^\nu \, p_j(t) ,\\
p_2^\nu(t) &= \sum_{j=1}^{n} x_{2j}^\nu \, p_j(t) ,
\end{aligned}
$$

tels que :

$$
(6.5)\qquad
\begin{aligned}
&p_1^\nu(t_i^{\nu-1}) = g_1(t_i^{\nu-1}) \text{ , pour tout i tel que } \varepsilon_i^{\nu-1} = 1 ,\\
&p_2^\nu(t_i^{\nu-1}) = g_2(t_i^{\nu-1}) \text{ , pour tout i tel que } \varepsilon_i^{\nu-1} = -1 ,\\
&p_1^\nu(t) - p_2^\nu(t) = q^\nu(t) \text{ , pour tout t ,}
\end{aligned}
$$

où $t_1^{\nu-1} < t_2^{\nu-1} < \ldots < t_{n+1}^{\nu-1}$ et $\varepsilon_{i+1}^{\nu-1} = -\varepsilon_i^{\nu-1}$, $i=1,\ldots,n$. On calcule alors :

$$
(6.6)\qquad
\begin{aligned}
\eta_1^\nu &= \max_{t\in[0,1]} (p_1^\nu(t) - g_1(t)) ,\\
\eta_2^\nu &= \max_{t\in[0,1]} (g_2(t) - p_2^\nu(t)) ,\\
\eta^\nu &= \max\,(\eta_1^\nu \,;\, \eta_2^\nu) .
\end{aligned}
$$

Si $\eta^\nu \leq 0$, alors p_1^ν et p_2^ν sont solutions du problème. Supposons donc que $\eta^\nu > 0$. Si $\eta_1^\nu \geq \eta_2^\nu$, on pose $\hat{\varepsilon}^\nu = 1$ et on désigne par $\hat{t}^\nu$ un point de [0,1] tel que $\eta_1^\nu = p_1^\nu(\hat{t}^\nu) - g_1(\hat{t}^\nu)$. La contrainte nouvelle, caractérisée par $[\hat{\varepsilon}^\nu,\hat{t}^\nu]$ sera relative à C_1. Si $\eta_1^\nu < \eta_2^\nu$, on pose $\hat{\varepsilon}^\nu = -1$ et on désigne par $\hat{t}^\nu$ un point de [0,1] tel que $\eta_2^\nu = g_2(\hat{t}^\nu) - p_2^\nu(\hat{t}^\nu)$. La contrainte nouvelle, caractérisée par $[\hat{\varepsilon}^\nu,\hat{t}^\nu]$ sera alors relative à C_2.

On suppose que $\hat{t}^\nu \neq t_i^{\nu-1}$, i=1,...,n+1. On échange alors le couple $[\hat{\varepsilon}^\nu,\hat{t}^\nu]$ avec l'un des couples $[\varepsilon_i^{\nu-1},t_i^{\nu-1}]$, i=1,...,n+1, de façon à ce que le nouvel ensemble obtenu

$[\varepsilon_i^\nu, t_i^\nu]$, qui caractérise S^ν , vérifie

$$(6.7)\qquad \begin{aligned} & t_1^\nu < t_2^\nu < \ldots\ldots < t_{n+1}^\nu , \\ & \varepsilon_{i+1}^\nu = -\varepsilon_i^\nu \quad , \quad i=1,\ldots,n . \end{aligned}$$

Cet échange est exactement le même que celui que l'on pratique dans l'algorithme de Rémès pour l'approximation au sens de Tchebycheff d'une fonction continue par un polynome. Soit $\rho_i^\nu > 0$, $\sum_{i=1}^{n+1} \rho_i^\nu = 1$, les coefficients (uniques) tels que

$$(6.8)\qquad \sum_{i=1}^{n+1} \rho_i^\nu \varepsilon_i^\nu \, p(t_i^\nu) = 0 \text{ , pour tout polynome p de degré n-1.}$$

On forme alors :

$$(6.9)\qquad \alpha^\nu \text{ de composantes } \alpha_j^\nu = \sum_{\substack{i=1 \\ \varepsilon_i^\nu=1}}^{n+1} \rho_i^\nu \, p_j(t_i^\nu) ,$$

$$(6.10)\qquad \beta^\nu = \sum_{\substack{i=1 \\ \varepsilon_i^\nu=1}}^{n+1} \rho_i^\nu \, g_1(t_i^\nu) - \sum_{\substack{i=1 \\ \varepsilon_i^\nu=-1}}^{n+1} \rho_i^\nu \, g_2(t_i^\nu) ,$$

et le demi-espace :

$$(6.11)\qquad D(S^\nu) = \{y \in E | < y, \alpha^\nu > \, \leq \beta^\nu\} .$$

On a par ailleurs :

$$(6.12)\qquad D^\nu = \{y \in E | < y^\nu, y-y^\nu > \, \geq 0\} ,$$

qui est supposé contenir Ω . Soit $y^{\nu+1}$ la projection de 0 sur $D^\nu \cap D(S^\nu)$:

$$(6.13)\qquad y^{\nu+1} \in D^\nu \cap D(S^\nu) , \quad \|y^{\nu+1}\| = \min_{y \in D^\nu \cap D(S^\nu)} \|y\|$$

et $q^{\nu+1}(t) = \sum_{j=1}^{n} y_j^{\nu+1} p_j(t)$ le polynome associé. Il existe alors $p_1^{\nu+1}(t) = \sum_{j=1}^{n} x_{1j}^{\nu+1} p_j(t)$ et $p_2^{\nu+1}(t) = \sum_{j=1}^{n} x_{2j}^{\nu+1} p_j(t)$, uniques, tels que :

$$(6.14)\qquad \begin{aligned} & p_1^{\nu+1}(t_i^\nu) = g_1(t_i^\nu) \text{ , pour tout i tel que } \varepsilon_i^\nu = 1 , \\ & p_2^{\nu+1}(t_i^\nu) = g_2(t_i^\nu) \text{ , pour tout i tel que } \varepsilon_i^\nu = -1 , \\ & p_1^{\nu+1}(t) - p_2^{\nu+1}(t) = q^{\nu+1}(t) \text{ , pour tout t .} \end{aligned}$$

On remarque que le demi-espace $D^{\nu+1}$ associé à $y^{\nu+1}$ contiendra automatiquement $D^{\nu} \cap D(S^{\nu})$ donc Ω .

b) Démarrage de l'algorithme :

A l'itération $\nu = 1$, on choisit un ensemble de n+1 abscisses de [0,1] : $t_1^o < t_2^o < \ldots < t_{n+1}^o$. Il existe des coefficients ρ_i^o uniques positifs, $\sum_{i=1}^{n+1} \rho_i^o = 1$ tels que :

(6.15) $$\sum_{i=1}^{n+1} \rho_i^o \, \varepsilon_i^o \, p(t_i^o) = 0 \text{ , pour tout polynome p de degré n-1 ,}$$

où les ε_i^o sont égaux à $\varepsilon(-1)^i$, avec ε est égal à +1 ou -1. L'annulateur S^o est défini par les n+1 couples $[\varepsilon_i^o, t_i^o]$, i=1,...,n+1. On forme :

(6.16) $$\alpha^o = \alpha(S^o) \text{ , de composantes } \alpha_j^o = \sum_{\substack{i=1 \\ \varepsilon_i^o=1}}^{n+1} \rho_i^o \, p_j(t_i^o) \text{ ,}$$

(6.17) $$\beta^o = \beta(S^o) = \sum_{\substack{i=1 \\ \varepsilon_i^o=1}}^{n+1} \rho_i^o \, g_1(t_i^o) - \sum_{\substack{i=1 \\ \varepsilon_i^o=-1}}^{n+1} \rho_i^o \, g_2(t_i^o) \text{ ,}$$

et le demi-espace :

(6.18) $$D(S^o) = \{y \in E | < y,\alpha^o > \leq \beta^o > \text{ .}$$

Comme $0 \notin C_1 - C_2$, il est toujours possible de choisir S^o de sorte que $\beta^o < 0$, c'est-à-dire $0 \notin D(S^o)$. Si l'on a, en particulier, $g_1(t_i^o) < g_2(t_i^o)$, i=1,...,n+1 , on montre que ce sera forcément le cas pour l'une au moins des deux valeurs $\varepsilon = +1$ ou $\varepsilon = -1$ plus haut. On pose $D^1 = D(S^o)$ et soit y^1 la projection de 0 sur D^1. Comme $0 \notin D^1$, on a $y^1 \neq 0$ et D^1 s'écrit encore :

(6.19) $$D^1 = \{y \in E | < y^1, y-y^1 > \geq 0\} \text{ .}$$

On considère le polynome associé $q^1(t) = \sum_{j=1}^{n} y_j^1 \, p_j(t)$ et on détermine les polynomes $p_1^1(t) = \sum_{j=1}^{n} x_{1j}^1 \, p_j(t)$ et $p_2^1(t) = \sum_{j=1}^{n} x_{2j}^1 \, p_j(t)$ comme en (6.14).

(6.20) REMARQUE :

On peut appliquer le même algorithme si l'on définit C_1 et C_2 à l'aide de fonctions plus générales $p_1,\ldots,p_n$ (non nécessairement polynominales) mais qui engendrent un sous-espace vectoriel vérifiant la condition de Haar sur l'intervalle considérée.

§7. UNE EXTENSION DE LA METHODE

La méthode numérique que nous avons décrite pour le calcul de la distance entre deux convexes est en fait un cas particulier d'une méthode plus générale que nous allons exposer brièvement sans donner les démonstrations.
Soient $F = \mathbb{R}^m$ et $G = \mathbb{R}^n$ munis de produits scalaires quelconques ($< .,. >_F$ et $< .,. >_G$) et des normes associées. Soit T une application linéaire de F sur G dont le noyau N est de dimension $q = m-n$. Enfin g désigne une fonction uniformément convexe définie sur G à valeurs dans $\mathbb{R}$. On définit alors la fonction f par :

(7.1) $\quad f(x) = g(T(x))$, pour tout $x \in F$.

On considère un ensemble convexe fermé non vide de F défini par :

(7.2) $\quad C = \{x \in F \mid < x,\ell(t) >_F \leq b(t)$, pour tout $t \in T\}$,

où ℓ et b sont des applications de T dans F et $\mathbb{R}$ respectivement. On étudie le problème de la minimisation de f sur C. On recherche donc un élément $\bar{x}$ (s'il existe)tel que :

(7.3) $\quad \bar{x} \in C \, , \, f(\bar{x}) = \min_{x \in C} f(x)$.

Si l'on pose $\Omega = T(C)$, le problème revient à chercher $\bar{y}$ tel que :

(7.4) $\quad \bar{y} \in \Omega \, , \, g(\bar{y}) = \min_{y \in \Omega} g(y)$.

On fait l'hypothèse suivante :

(K1) L'ensemble $C_\infty \cap N$ est un sous-espace vectoriel,
où $C_\infty = \{x \in F \mid < x,\ell(t) >_F \leq 0$, pour tout $t \in T\}$ est le cône asymptote associé à C.

On a le résultat suivant :

(7.5) THEOREME :

Si l'hypothèse (K1) est vérifiée, alors $\Omega = T(C)$ est fermé, il existe $\bar{y}$ unique vérifiant (7.4) et il existe au moins un élément $\bar{x}$ vérifiant (7.3).

On suppose dans la suite que :

$$\min_{y \in G} g(y) < \min_{y \in \Omega} g(y)$$

c'est-à-dire que l'élément $\tilde{y} \in Y$ tel que $g(\tilde{y}) = \min_{y \in G} g(y)$ (qui existe et est unique) n'appartient pas à Ω .

(7.6) REMARQUE :

On retrouve le problème de la distance entre deux convexes en posant $F = E \times E$, $C = C_1 \times C_2$, $G = E$, $T(x) = T(x_1,x_2) = x_1-x_2$ et $g(y) = \frac{1}{2}\|y\|^2$.

(7.7) DEFINITIONS :

Un sous-ensemble $\tilde{T} \subset T$ sera dit un annulateur de N si l'on a $co(\ell(\tilde{T})) \cap N^{\perp} \neq \emptyset$ et un annulateur S de N sera dit minimal s'il n'existe pas d'annulateur de N qui soit strictement contenu dans S. On notera encore $\mathcal{S}$ l'ensemble des annulateurs minimaux de N. Enfin un annulateur minimal $R \subset T$ sera dit régulier si $0 \notin co(\ell(R))$. On notera $\mathcal{R}$ l'ensemble des annulateurs minimaux réguliers de N.

(7.8) REMARQUE :

On retrouve les définitions données en (2.1) si l'on prend pour ℓ l'application définie en (2.8) et pour N le sous-espace :

$$\{[x_1,x_2] \in E \times E | x_1 = x_2\} .$$

On montre qu'un annulateur minimal a au plus q+1 éléments. Si $S \in \mathcal{S}$, alors il existe des coefficients $\rho_S(s)$, $s \in S$ tels que :

(7.9) $$\rho_S(s) > 0 , \sum_{s\in S} \rho_S(s) = 1 \quad \text{et} \quad x_S = \sum_{s\in S} \rho_S(s)\ell(s) \in N^{\perp} .$$

Les coefficients $\rho_S(s)$ sont uniques. A $S \in \mathcal{S}$, on associe le convexe :

(7.10) $$C_S = \{x \in F| < x,\ell(s) >_F \leq b(s) , \text{ pour tout } s \in S\} .$$

Désignons par T' l'application adjointe de T. Comme T est surjectif, T' est une application injective de G dans F et l'on a $T'(G) = N^{\perp}$.
Pour tout $S \in \mathcal{S}$, il existe donc un élément unique $\alpha(S)$ tel que :

(7.11) $$T'(\alpha(S)) = x_S .$$

Lorsque $S \in \mathcal{R}$, alors $x_S \neq 0$ et $\alpha(S) \neq 0$. On note encore :

(7.12) $$\beta(S) = \sum_{s\in S} \rho_S(s)b(s).$$

On démontre alors le théorème suivant qui généralise (2.7) :

(7.13) THEOREME :

Pour tout $s \in \mathcal{S}$, on a :

$$T(C_S) = D(S)$$

avec : $D(S) = \{y \in G| < y,\alpha(S) >_G \leq \beta(S)\}$
(on remarque que si $S \in \mathcal{R}$, alors $\alpha(S) \neq 0$ et D(S) est un demi-espace).

On fait les hypothèses suivantes :

(K2) L'ensemble T est compact et les applications ℓ et b sont continues.

(K3) Il existe $\tilde{x} \in F$ tel que $< \tilde{x},\ell(t) >_F < b(t)$, pour tout $t \in T$.

On a alors le théorème suivant qui généralise (3.6) :

(7.14) THEOREME :

Avec les hypothèses (K1), (K2) et (K3) , on a :

$$\Omega = T(C) = \bigcap_{S \in \mathcal{R}} D(S) .$$

L'algorithme décrit au paragraphe 5 se généralise alors de la façon suivante :

a) Description d'une itération de l'algorithme :

Supposons que l'on ait, à l'itération ν , un élément $y^\nu \in G$ tel qu'il existe $z^\nu \in \partial g(y^\nu)$ (le sous différentiel de g en y^ν) pour lequel on ait :

(7.15) $D^\nu = \{y \in G \mid < z^\nu,y-y^\nu >_G \geq 0\} \supset \Omega$,

et un annulateur minimal régulier $S^{\nu-1} \in \mathcal{R}$ tel que :

(7.16) $< y^\nu,\alpha(S^{\nu-1}) >_G = \beta(S^{\nu-1})$,

(on remarque que g est alors minimum sur D^ν en y^ν). Si l'on suppose que cet annulateur comporte exactement q+1 éléments, alors il existe un unique élément x^ν tel que $x^\nu \in C_{S^{\nu-1}}$ et $T(x^\nu) = y^\nu$. En fait cet élément x^ν vérifie les équations suivantes qui le déterminent complètement :

(7.17)
$$< x^\nu,\ell(s) >_F = b(s) \text{ , pour tout } s \in S^{\nu-1} \text{ ,}$$
$$T(x^\nu) = y^\nu \text{ .}$$

Si $x^\nu \in C$, c'est une solution du problème. Si ce n'est pas le cas, on choisit $t^\nu \in T$ tel que :

(7.18) $< x^\nu,\ell(t^\nu) >_F - b(t^\nu) = \max\limits_{t \in T} (< x^\nu,\ell(t) >_F - b(t))$.

Il existe un annulateur minimal (unique) S^ν contenu dans $S^{\nu-1} \cup \{t^\nu\}$ et différent de $S^{\nu-1}$. Nous supposerons que cet annulateur comporte à nouveau exactement q+1 éléments. On remarque que S^ν est forcément régulier. On considère le demi-espace $D(S^\nu)$ associé et on détermine l'élément $y^{\nu+1}$ tel que :

(7.19) $y^{\nu+1} \in D^\nu \cap D(S^\nu)$, $g(y^{\nu+1}) = \min\limits_{y \in D^\nu \cap D(S^\nu)} g(y)$.

On sait qu'il existe alors $z^{\nu+1} \in \partial g(y^{\nu+1})$ tel que :

(7.20) $< z^{\nu+1},y-y^{\nu+1} >_G \geq 0$, pour tout $y \in D^\nu \cap D(S^\nu)$.

Comme $D^{\nu} \cap D(S^{\nu}) \supset \Omega$, le demi-espace :

(7.21) $$D^{\nu+1} = \{y \in G \mid \langle z^{\nu+1}, y-y^{\nu+1} \rangle_G \geq 0\}$$

contient Ω . D'autre part, comme $y^{\nu} \notin D(S^{\nu})$, l'élément $y^{\nu+1}$ appartient à la frontière de $D(S^{\nu})$, c'est-à-dire vérifie :

(7.22) $$\langle y^{\nu+1}, \alpha(S^{\nu}) \rangle_G = \beta(S^{\nu}) .$$

Comme S^{ν} se compose de q+1 éléments, il existe $x^{\nu+1}$ (unique) tel que :

(7.23) $$\begin{aligned} &\langle x^{\nu+1}, \ell(s) \rangle_F = b(s) \text{ , pour tout } s \in S^{\nu} \text{ ,} \\ &T(x^{\nu+1}) = y^{\nu+1} . \end{aligned}$$

b) Démarrage de l'algorithme :

Le démarrage de l'algorithme se fait de façon analogue à ce qui a été décrit au paragraphe 5. On choisit un annulateur minimal régulier S° tel que $\tilde{y} \notin D(S^{\circ})$. Il existe $y^1 \in D(S^{\circ})$ (unique) tel que $g(y^1) = \min\limits_{y \in D(S^{\circ})} g(y)$ et il existe alors $z^1 \in \partial g(y^1)$, $z^1 \neq 0$, tel que :

(7.24) $$D(S^{\circ}) = \{y \in G \mid \langle z^1, y-y^1 \rangle_G \geq 0\} .$$

On choisit alors $D^1 = D(S^{\circ})$. On a évidemment $\langle y^1, \alpha(S^{\circ}) \rangle_G = \beta(S^{\circ})$.

7.25) REMARQUE :

L'algorithme peut être décrit dans des espaces de dimension infinie. On peut supposer par exemple que F et G sont des espaces de Hilbert et que T est une application linéaire continue de F sur G dont le noyau N est de dimension q. C'est en fait ce qui a été fait dans [4] , [6] , pour les fonctions-spline, mais avec $g(y) = \frac{1}{2} \|y\|^2$ et avec pour T un ensemble fini.

REFERENCES

[1] Carasso,C. ; and Laurent, P.J. : An exchange algorithme for the minimization of convex functionals (to appear).

[2] Haugazeau, Y. : Sur les inéquations variationnelles et la minimisation de fonctionnelles convexes. Thèse, Paris, 7 juin 1967.

[3] Laurent, P.J. : Approximation et Optimisation. Hermann, Paris,(1972).

[4] Laurent, P.J. : Construction of spline functions in a convex set. In "Approximation with special emphasis on spline-functions". I.J. Schoenberg Ed., Acad. Press, (1969), 415-446.

[5] Laurent,P.J. ; et Martinet, B. : Méthodes duales pour le calcul du minimum d'une fonction convexe sur une intersection de convexes. Symposium on Optimization, Nice, June 29th-July 5th, (1969), Lecture Notes in Math., 132, Springer-Verlag, (1970), 159-179.

[6] Morin, M. : Méthodes de calcul des fonctions-spline dans un convexe. Thèse, Grenoble, (1969).

[7] Pierra, G. : Sur le croisement de méthodes de descente et l'intersection ou la distance de deux convexes. Colloque d'Analyse Numérique du C.N.R.S., La Colle Sur Loup, mai-juin 1973.

[8] Rockafellar, R.T. : Convex Analysis. Princeton, Univ. Press, (1970).

P.J. LAURENT
Mathématiques Appliquées
B.P. 53
38041 GRENOBLE (FRANCE)

The Design of a Nonlinear Optimization Programme for Solving Technological Problems

F.A. Lootsma

University of Technology
Delft, the Netherlands

Abstract

This paper presents the design of the programme MINIFUN for solving non-linear optimization problems arising in research, development and engineering laboratories. The underlying algorithm is based on the penalty function approach whereby constrained optimization problems are solved via sequential unconstrained optimization. The paper shows the performance of several unconstrained optimization algorithms incorporated in MINIFUN, the significance of numerical differentiation, and finally a comparison of MINIFUN with other well-known programmes for non-linear optimization.

1. Introduction

Although it is dangerous to generalize from a limited number of observations, we feel that the applications of non-linear optimization in industry fall into three distinct categories.

a) Technological applications. There is a variety of design problems in research, development and engineering departments (Bracken and McCormick (1968)). These problems have a highly non-linear nature and can be formulated with a relatively small number of variables (two to twenty). Larger models (fifty to hundred variables) are frequently studied in electricity companies to solve dispatching problems. A significant proportion (70%) of the problems reduces to the minimization of a sum of squares arising from curve fitting. The number of research members and engineers concerned with a particular problem is relatively small (two to five), thus reducing the difficulties of communication and coordination.

b) Business applications. In this area, non-linear optimization is mostly a refinement of a linear-programming study. Applications of linear programming are mainly found in the area of production planning (long term or medium term capacity planning, medium term or short term production allocation and production scheduling). The size of the problems tends to be very large, both in the number of variables and constraints (hundreds, or even thousands) and in the number of people involved. Mostly, linear programming can only be carried out after a favourable high-level management decision has been taken. In addition data collection and the implementation of a solution calculated by linear programming techniques depend heavily on the continuous support by management. In many of these problems one finds a number of non-linearities (due to economies of scale), although it is often extremely difficult to precisely determine these non-linearities in a given practical situation. If they cannot be neglected, they can often be handled via local linearization and repeated application of linear programming (method of approximation programming), or via separable programming (Beale (1968)).

c) Discretized problems. There is a widespread interest in the attempts to solve problems arising from the discretization of continuous problems by means of non-linear optimization techniques. This is a feasible approach to solve optimal control problems, for instance (Tabak and Kuo (1971)). The discretized problems tend to have hundreds of variables, each connected with a grid point (specification point) of the discretization. This is in sharp contrast with model formulation in the first-named category of technological applications where the majority of the variables have a distinct physical meaning (temperature, pressure, dimensions of devices, -----). There is a similar distinction with the second category where the majority of the variables are related to quantities of certain items (batch sizes, for instance) so that they have a managerial significance.

In this paper, we shall mainly be concerned with an optimization programme for solving problems of the first category. The size of the problems, and the vagueness of the non-linearities seem to indicate that it is preferable to consider the second category merely as an extension of linear-programming applications. For the time being, we hesitate to deal with the third category. We feel that our experience with large non-linear problems is still too limited to draw any valid conclusions.

The programme to be discussed is the ALGOL 60 procedure MINIFUN (minimization of functions) originally developed by the author (1972c) on the Electrologica X8 computer in Philips Research Laboratories, Eindhoven, the Netherlands. Several unconstrained minimization methods have been incorporated in it, and constrained minimization problems are solved via unconstrained minimization of a mixed interior-exterior penalty function. If the user is not willing to supply the first-order or the second-order derivatives of the problem functions, they are automatically generated by numerical differentiation of the functions and their first-order derivatives respectively. The Gauss-Newton algorithm is available if the problem reduces to the minimization of a sum of squares.

The programme has been operational in several industrial laboratories and universities and it is also available in the library of the Numerical Algorithms Group (a joint effort of the Computer Laboratories in several British Universities). Certification and validation of the programme have been completed. Hence, this paper presents some details of the performance of MINIFUN. Preliminary results have been published earlier by the author (1972a), and the present paper may accordingly be considered as a continuation. Particular attention will be given to the requirements of the user and to the significance of numerical differentiation. Finally, we shall briefly indicate the performance of MINIFUN with respect to some other, well-known non-linear optimization codes.

2. General requirements for unconstrained minimization

Today, there is a confusing variety of algorithms for unconstrained minimization of a twice differentiable function f: straightforward search methods, a large class of gradient methods (ranging from steepest descent to the sophisticated variable-metric or quasi-Newton methods) operating with the first derivatives of f, and lastly Newton's method (with several variants) using first and second derivatives of f. For recent surveys we may refer to Powell (1970, 1971) and to Sargent (1973). Some of these methods have been programmed, comparisons between methods have been made, and some computer procedures are published in well-known journals or available on request. Nevertheless, as soon as one starts to think about the desirable properties of a procedure for unconstrained minimization, several questions arise which are almost never discussed in the literature.

The first problem arising in practice is the choice of a method. The variety of methods proposed in the last ten years shows in fact that we miss the unique, powerful method for solving any problem in this field. It is undesirable, however, that a computing centre should have all the proposed methods available in the software library. The documentation would grow beyond reasonable limits. Furthermore, one would have the intolerable task of finding the most appropriate method for each minimization problem which is presented. Thus, one has to choose a restricted set of methods, and the choice must be such that one can roughly indicate, for each particular problem, which of the selected methods is the most suitable one to use.

Another requirement is flexibility, in the sense that the user must be able to try other methods on his problem if the method which he initially used happens to fail. To meet this requirement one could design a procedure incorporating several methods; the choice of a particular method would depend on the setting of a parameter when the procedure is called. Such a design is also suggested by the unconstrained methods themselves, since many of them do operate along the same lines. One may, for instance, compare the gradient methods, where each iteration consists of two major operations:

find the search direction to the next iteration point by a transformation of the gradient of f at the current iteration point, and find a minimum of f along this direction by application of a method for one-dimensional minimization.

The literature on unconstrained minimization is largely concerned with the choice of search directions. Little attention was paid to the numerical question of how to find a minimum of f along a line. The efficiency of the procedure, however, depends critically on the efficiency and the accuracy of such a one-dimensional search (alternatively referred to as univariate search, line search or line minimization). There are several methods for finding a line minimum: golden section, repeated quadratic interpolation, and repeated cubic interpolation, for instance (Kowalik and Osborne (1968), Box, Davies and Swann (1969)). However, the question is not that of choosing the most efficient one, but of choosing the most efficient combination of unconstrained minimization method (choice of search directions) and univariate search. The question is by no means trivial This is clearly illustrated by the recent developments in the variable-metric methods: attempts are made to find methods which only need a rough approximation to a line minimum, or no line search at all (Fletcher (1970)).

Many descriptions of methods and computer procedures are rather vague about the accuracy which can be obtained. In fact, many stopping rules are obscure. What does it mean, for instance, that the iterative process terminates as soon as a point is obtained where $|| \nabla f ||$ is below a certain threshold preset by the user? What the user needs is a procedure which produces a solution with the previously required accuracy, either in function value, or in position, or possibly in both.

It is urgent that a collection of comprehensible and difficult test problems should be selected. The problems which appear in the literature are sometimes rather simple, and mostly they are not difficult to solve. Hence, success of a procedure on these problems is by no means a guarantee of success in general.

Comprehensible output and protection against errors in the problem formulation seem to be somewhat neglected. On many occasions, however, it is useful to compare different problem formulations in order to see how the computations proceed and where they terminate. Thus, elaborate output facilities must be available in minimization procedures. When first (and second) derivatives of f are supplied by the user, it is necessary that they should be checked by differencing the function (and the first derivatives) at some iteration points (or by calling a formula manipulator) whereafter the deviations between the user-supplied derivatives and their numerical approximations must be printed.

This paper describes how the procedure MINIFUN was developed in an attempt to meet the requirements of flexibility and desired accuracy for an unconstrained minimization programme. This has been achieved by the utilization of test problems arising in constrained minimization via penalty functions. This approach enabled us at the same time to develop MINIFUN as a general, flexible programme for constrained minimization. It is beyond the scope of the present paper to discuss the output facilities of MINIFUN. We may note, however, that the check of the analytical derivatives (by numerical differentiation at the starting point only) appears to be extremely helpful in the initial phase of an optimization project.

3. The role of penalty-function techniques in constrained and unconstrained minimization

Recently, several authors have studied a class of methods which reduce the computational process for solving the problem

$$\text{minimize } f(x_1, \ldots, x_n) \text{ subject to}$$

$$g_i(x_1, \ldots, x_n) \geq 0;\ i = 1, \ldots, m,$$

$$h_j(x_1, \ldots, x_n) = 0;\ j = 1, \ldots, p,$$

to sequential unconstrained minimization of a so-called penalty function: a function P combining in a particular way the objective function f, the constraint functions $g_1, \ldots, g_m, h_1, \ldots, h_p$, and a controlling parameter. A survey of these developments was given by Fiacco and McCormick (1968) and the author (1972b). Obviously, a broad new field of applications was opened to unconstrained minimization techniques: the solution of constrained-minimization or non-linear-programming problems.

Computational success with these penalty-function techniques rests on the power of unconstrained-minimization methods, and indeed, some of them were powerful enough to give striking results. Particularly Newton's method and the quasi-Newton or variable-metric methods have been applied with a good deal of success.

There are two aspects of the penalty-function techniques studied by the author (1970) that will be discussed here in more detail. First, the condition number of the Hessian matrix of a penalty function at a point where the function attains its minimum varies with the inverse of the controlling parameter if the constrained minimum is a boundary point of the constraint set (which is true for almost any practical problem). Then the constrained minimum is approached via a trajectory where the condition number tends to infinity. Accordingly, there are difficulties in minimizing a penalty function due to ill-conditioning of its Hessian matrix. Second, a considerable acceleration of the convergence is obtained by extrapolation. This is possible since the trajectory to the constrained minimum can be expanded in a Taylor series in terms of the controlling parameter. Thus, greater successes with penalty-function methods can be obtained by extrapolation on accurate approximations of penalty-function minima. Loss of accuracy will immediately lead to deterioration of the extrapolation scheme.

Accurate minimization of penalty-functions with ill-conditioned Hessian matrices: a precarious approach, but presumably it provides an excellent class of problems to test and to compare the algorithms for unconstrained minimization. These ideas arose mainly in the course of many fruitful discussions with J.D. Pearson (ISA-Research, N.V. Philips, Eindhoven, Netherlands, at present Research Analysis Corporation, McLean,

Virginia, U.S.A.). Our concern was a possible extension of the SUMT programme written by McCormick, Mylander and Fiacco (1965), and the continuing development of the procedure MINIMIZE published by the author (1970). True, they are designed for solving constrained-minimization problems, but they rest on unconstrained-minimization methods, and the experiences with penalty functions will possibly yield a significant impression of the power of the available algorithms in the field of unconstrained minimization.

4. The choice of algorithms in MINIFUN

In this section we give a brief description of how the algorithms incorporated in MINIFUN have been selected from the variety of methods proposed in the last few years.

There are two criteria by which unconstrained-minimization methods may be classified: (a) quadratic convergence or quadratic termination, and (b) the order of the method (the highest order of the derivatives required to generate the search directions).

A method for unconstrained minimization is said to have the property of quadratic convergence (Fletcher and Powell (1963)) if it minimizes a quadratic function with a positive-definite Hessian matrix in a finite number of exact linear searches. Numerically, these methods are highly efficient in a neighbourhood of a minimum. It is possible that they waste a lot of effort if the starting point of the minimization process is "far away" from the minimum. This suggests that a hybrid strategy could successfully be employed. First, a "cheap" method could be used (to find a point which is "close" to the minimum), and thereafter a method with quadratic convergence. The terms "far away" and "close" have never been precisely defined, however, and hence we prefer the methods which do have the property of quadratic convergence, and accordingly fast ultimate convergence towards a minimum (see also Fletcher (1965, 1970), and Powell (1970, 1971)). These methods do mostly provide a more accurate approximation of a minimum.

Methods for unconstrained minimization differ in the highest order of the derivatives required to find the successive search directions. There are zeroth-order methods using function values only; there is a large class of first-order methods (including the variable-metric methods); and lastly, one finds the second-order methods (Newton's method with several variants). In most cases, higher-order methods seem to be faster and more reliable than lower-order methods (with some notorious exceptions like the steepest-descent method which is probably inferior to the direct search algorithms of Hooke and Jeeves (1961) and Nelder and Mead (1964)). One may take advantage of this when higher-order derivatives can easily be supplied. Hence, we decided that MINIFUN should contain a quadratically convergent method of order 0, order 1, and order 2.

One of the best methods for unconstrained minimization without calculating derivatives seems to be the P64 method of Powell (1964). It is a method with quadratic convergence, and it has accordingly been chosen as the zeroth-order method to be incorporated in MINIFUN.

A frequently cited first-order method is the DFP algorithm due to Davidon (1959), Fletcher and Powell (1963). Although a powerful method, it is exceeded by the BFS algorithm independently developed by Broyden (1970), Fletcher (1970) and Shanno (1970).

In its classical form, the second-order method of Newton has a number of drawbacks if the function to be minimized is non-convex, or if its Hessian matrix is singular at some iteration points. These disadvantages are possibly removed by the modification of Fiacco and McCormick (1968). We have extensively tested this promising variant, the NFM algorithm, of Newton's method.

For many problems it is difficult or practically impossible to supply the first and/or second derivatives of the problem functions. This would impose a serious limitation on the choice of methods for unconstrained minimization. Therefore, MINIFUN has been written to generate derivatives by numerical differentiation so that the user can easily proceed to a higher-order method. In this context, the order of information is understood to be the highest order of the derivatives supplied by the user. With the facilities of numerical differentiation, MINIFUN has been designed in such a way that the first-order BFS method can also be employed with zeroth-order information (no derivatives), and the second-order NFM algorithm with first-order information. The first and second derivatives are obtained by differencing function values and first derivatives respectively. We have omitted the possibility of using the second-order NFM algorithm with information of order 0 only.

The choice of the linear search method (quadratic interpolation with a strong convergence criterion) has been motivated by the author (1972a) in a comparison with some other well-known methods (golden section algorithm, cubic interpolation, ...).

Finally, the choice of the interior-exterior penalty function (with a logarithmic-barrier term and a quadratic-loss term) rests upon arguments brought forward by Fiacco and McCormick (1968) and by the author (1970, 1972b). For more details the reader is referred to the manual of MINIFUN (Lootsma (1972c)).

5. Performance of unconstrained-minimization methods in MINIFUN

The problems that have been used to test and to compare the aforenamed algorithms are widely known, and frequently cited in the literature. We have also given a description in a previous paper (Lootsma (1972a)). The complete list of test problems is as follows:

Test functions for unconstrained minimization

1. The test function of Rosenbrock (2 variables)
2. The test function of Beale (2 variables)
3. The test function of Powell (4 variables)
4. The test function of Wood (4 variables)

Test problems for constrained minimization

5 . The cubic example of Fiacco and McCormick with a feasible starting point (3 non-negative variables, 1 linear and 2 quadratic constraints, and a cubic objective function).

6 . The cubic example with an unfeasible starting point.

7 . The test problem of Rosen and Suzuki (4 variables, 3 quadratic constraints and a quadratic objective function).

8 . The distance between a given sphere and a given cylinder (6 variables, 2 linear and 2 quadratic constraints, and a quadratic objective function).

9 . Mylander's largest hexagon with unit diameter (9 variables, 13 quadratic constraints, and a quadratic objective function).

10. Colville's primal Shell problem (5 non-negative variables, 10 linear constraints, and a cubic objective function).

11. Colville's dual Shell problem (15 non-negative variables, 5 quadratic constraints, and a cubic objective function).

The performance of MINIFUN is shown in Table I presenting the execution times in seconds on the Electrologica X8 computer (with a so-called Colville Standard time of 720 seconds). The heading INFO of the second column indicates the order of the available information. Thus, the first-order BFS algorithm is used in two different manners: with INFO = 0 (first derivatives not supplied by the user, hence approximated via numerical differentiation of the problem functions), and with INFO = 1 (first derivatives available). Similarly, the second-order NFM algorithm is used with INFO = 1 (first derivatives available, second derivatives approximated via numerical differentiation of the first derivatives), and with INFO = 2 (first and second derivatives available).

We have only recorded the results for the constrained-minimization problems 5-11. Measurement of the elapsed time on the Electrologica X8 computer was not accurate enough to record the short execution times for the unconstrained problems 1-4. It is worth noting, however, that these problems have been successfully solved via each of the unconstrained algorithms in MINIFUN with one exception only: the BFS algorithm with numerical approximations to the first derivatives (INFO = 0) failed on Rosenbrock's test function.

Table II presents an overall comparison of the unconstrained algorithms on the basis of scaled execution times. The entries in table II are easily derived from table I. In each column, a weight of 100 is assigned to the method with the smallest execution time. The remaining entries in that column are the ratios of the smallest and the actual execution time, multiplied by 100. A zero indicates a failure of the algorithm for that particular test problem. The last column contains the row averages, thus providing a rough measure for the relative performance of the algorithm in general.

Obviously, numerical differentiation is an important feature of MINIFUN. If first derivatives are not available (INFO = 0), then the BFS algorithm with numerical approximationsto the first derivatives is mostly superior to the P64 algorithm (which suffers from a number of failures). Similarly, if only the first derivatives are available (INFO = 1), the NFM algorithm with numerical approximations to the second derivatives is mostly to be preferred above the BFS algorithm. True, these observations are based on experiences with the minimization of penalty functions, but Goffin, de Beer, and Kilsdonk (1974) obtained similar results for unconstrained problems. Moreover, they found that the NFM algorithm with numerical approximations to the first <u>and</u> second derivatives is competitive with many gradient methods, even if the derivatives are supplied by the user.

Finally, we note that the desired accuracy has been obtained for all the test problems. Hopefully, MINIFUN does provide such solutions for a large class of properly scaled problems. Moreover, if the Gauss-Newton method fails on least-squares problems, the user has several unconstrained-minimization techniques available as a last resort.

UNC. MIN.	INFO	Results for problem						
		5	6	7	8	9	10	11
P64	0	12	23	--	44	--	265	--
BFS	0	10	19	34	32	287	120	--
	1	9	16	24	26	152	71	793
NFM	1	5	11	19	13	270	39	313
	2	5	8	14	12	199	39	234

Table I

Execution times in seconds of unconstrained-minimization algorithms in MINIFUN used to solve test problems 5-11 on an Electrologica X8 computer (Colville standard time of 720 seconds). The parameter INFO indicates whether no derivatives (INFO = 0), first derivatives (INFO = 1), or first and second derivatives (INFO = 2) are available via user-supplied procedures.

UNC. MIN.	INFO	Results for problem							Average
		5	6	7	8	9	10	11	
P64	0	42	35	0	27	0	15	0	17
BFS	0	50	42	41	38	53	33	0	37
	1	56	50	58	46	100	55	30	56
NFM	1	100	73	74	92	56	100	75	81
	2	100	100	100	100	76	100	100	97

Table II

Scaled execution times and ranking of unconstrained-minimization algorithms in MINIFUN. The parameter INFO indicates whether no derivatives (INFO = 0), first derivatives (INFO = 1), or first and second derivatives (INFO = 2) are available via user-supplied procedures.

6. Comparison with other non-linear optimization programmes

In a recent paper, Staha and Himmelblau (1973) presented an extensive comparison (on a Control Data 6600, with 25 test problems of up to 100 variables) of several readily available FORTRAN programmes for non-linear optimization.

a) The GREG programme documented by Guigou (1971), and based upon Wolfe's generalized reduced gradient method as described by Abadie and Guigou (1969).

b) The COMET programme of Staha (1973) for constrained minimization via moving exterior truncations (a penalty-function method).

c) The GPMNLC programme of Kreuser and Rosen (1971) using a gradient-projection method for non-linear constraints.

d) The FORTRAN version MINI of the ALGOL 60 procedure MINIMIZE (Lootsma (1970)) using a mixed interior-exterior penalty function and the DFP algorithm of Davidon (1959), Fletcher and Powell (1963) for unconstrained minimization. MINIFUN is an improved and extended version of MINIMIZE.

e) Several other non-linear optimization programmes. These were rejected during the experiments as being too unreliable, too slow, or too difficult to manage for an unsophisticated user.

The GREG, COMET, GPMNLC and MINI programmes have been compared on the basis of scaled execution times generated in the same manner as we described in the previous section.

The results (the row averages over 4 unconstrained problems, 5 linearly constrained problems, 16 problems with non-linear constraints, and over the total set of 25 test problems) may be found in table III. They reveal many significant properties of the underlying algorithms.

If the first derivatives are available (INFO = 1), then COMET is very fast for unconstrained problems; GREG and GPMNLC are more powerful for linearly constrained problems than the penalty-function programmes COMET and MINI; GREG, closely followed by COMET, shows an excellent behaviour for problems with non-linear constraints.

It is interesting to observe the performance of COMET and MINI in the important cases where the first derivatives are not available (INFO = 0). COMET slows down remarkably, and this is probably due to the linear search using gradients along the search direction and cubic interpolation. Facilities for numerical differentiation are not available in GREG and GPMNLC.

Staha and Himmelblau (1973) note that even the GREG and the GPMNLC programme are unsuitable for the large category of unsophisticated users. The leading GREG programme is lengthy (over 2000 lines of coding), compared with COMET and MINI (each approximately 600 lines).

Although it is difficult to compare the FORTRAN programme MINI and the ALGOL 60 procedure MINIFUN (approximately 900 lines of coding) on different computers, our most pessimistic estimate of the average scaled execution times for MINIFUN would read as follows:

MINIFUN	BFS	INFO = 0	16
		INFO = 1	24
	NFM	INFO = 1	35
		INFO = 2	40

These numbers roughly indicate the position of MINIFUN on the scale of Staha and Himmelblau.

It is a matter of course that these numbers should be considered with a great deal of care. The relative performance of programmed algorithms depends on the computer used. Moreover, there are many arbitrary steps in the programme, such as the setting of threshold values to terminate the linear search or the manner in which the provision of function values and derivatives is organized, which greatly affect the efficiency of the programme. These numbers should therefore be used as a rough estimate of the relative performance of the underlying algorithms.

Programme	INFO	UNC.MIN.	LIN. CONSTR.	NON-L. CONSTR.	OVERALL
GREG	1	38	71	77	69
COMET	1	91	52	61	64
GPMNLC	1	59	71	17	35
MINI	1	18	18	25	22
COMET	0	25	2	23	19
MINI	0	13	6	17	14
Number of problems	---	4	5	16	25

Table III

Ranking of constrained-minimization programmes according to Staha and Himmelblau (1973), on the basis of 4 unconstrained test problems, 5 linearly constrained problems, and 16 problems with non-linear constraints. The parameter INFO indicates whether no derivatives (INFO = 0), or first derivatives (INFO = 1) are available via user-supplied procedures.

7. Optimization in practice

Mostly, the solution of an optimization problem (whether it is a linear-programming, integer-programming, non-linear-programming, or unconstrained-minimization problem) seems to go through the following phases.

a) A practical problem is formulated in mathematical terms, and attempts are made to solve the mathematical problem. The answers are inspected, and possibly several alternative formulations are used to find the most appropriate mathematical model for the problem at hand. One or more methods for solving the problem are employed, and comparisons in efficiency, accuracy etc. are made.

b) As soon as a model and a method are chosen, the mathematical problem is solved for several values of certain parameters in the formulation. Alternatively, a special programme is written incorporating an optimization procedure which is frequently called to solve the problem for different values of certain parameters.

In our opinion the design presented in this paper is particularly useful in the first phase. It provides easy facilities for experiments with several methods and with higher-order derivatives. This was also emphasized in the previous sections. We have been less concerned with the second phase. Here, successes depend largely on the mode of operation in the first phase. As soon as a method is chosen for solving a given problem, it is possible to streamline the minimization procedure omitting so much that it can only operate according to this particular method.

Again, we emphasize the importance of output facilities. A modern linear-programming procedure contains many output procedures, providing prints of the input data, a matrix picture showing the matrix structure and the order of magnitude of its elements, the course of the iterative process, the final solution and the associated dual solution, indications for sensitivity analysis, etc. The underlying idea is, that it helps the user to analyse his problem. This idea is also true for unconstrained and constrained minimization. Therefore, it is important to find out which data is relevant for the user, particularly in the first phase of the problem solution. This is the critical phase, and the design of MINIFUN was almost entirely devoted to it.

Acknowledgement

It is a pleasure to thank John Pearson (Research Analysis Corporation, McLean, Virginia, USA) for the inspiring discussions about the subject of this paper. I am also greatly indebted to Shirley Lill (University of Liverpool) and to Heather Liddell (Queen Mary College, London) for the careful certification and validation of MINIFUN.

Delft, January 1975

References

1. J. Abadie and J. Guigou (1969. Gradient réduit généralisé. Note HI 069/02. Electricité de France, Paris.
2. E.M.L. Beale (1968). Mathematical Programming in Practice. Pitman & Sons, London.
3. M.J. Box (1966). A comparison of several current optimization methods and the use of transformations in constrained problems. The Comp. J. 9, 67-77.
4. M.J. Box, D. Davies and W.H. Swann (1969). Nonlinear Optimization Techniques. Oliver & Boyd, London.
5. J. Bracken and G.P. McCormick (1968). Selected Applications of Nonlinear Programming. Wiley, New York.
6. C.G. Broyden (1970). The convergence of a class of double-rank minimization algorithms. 1. General considerations. J.I.M.A., 6, 76-90. 2. The new algorithm. J.I.M.A., 222-231.
7. W.C. Davidon (1959). Variable metric method for minimization. A.E.C. Research and Development Report ANL-5990.
8. A.V. Fiacco and G.P. McCormick (1968). Nonlinear Programming, Sequential Unconstrained Minimization Techniques. Wiley, New York.
9. R. Fletcher (1965). Function minimization without evaluating derivatives - review. The Comp. J., 8, 33-41.
10. R. Fletcher (1970). A new approach to variable metric algorithms. The Comp. J., 13, 317-322.
11. R. Fletcher and M.J.D. Powell (1963). A rapidly convergent descent method for minimization. The Comp. J. 6, 163-168.
12. R. Goffin, A. de Beer, A.C.M. Kilsdonk (1974). Non-linear Optimization Package OPTAC-II. Information Systems and Automation Department, N.V. Philips, Eindhoven.
13. J. Guigou (1971). Présentation et utilisation du code GREG. Note HI 582/2. Electricité de France, Paris.
14. R. Hooke and T.A. Jeeves (1961). Direct search solution of numerical and statistical problems. J.A.C.M. 8, 212-229.
15. J. Kowalik and M.R. Osborne (1968). Methods for Unconstrained Optimization Problems. American Elsevier, New York.
16. J.L. Kreuser and J.B. Rosen (1971). GPM/GPMNLC extended gradient projection method for non-linear constraints. University of Wisconsin, Madison.
17. F.A. Lootsma (1970). Boundary properties of penalty functions for constrained minimization. Philips Res. Repts. Suppl. 3.
18. F.A. Lootsma (1972a). Penalty function performance of several unconstrained minimization techniques. Philips Res. Repts. 27, 358-385.
19. F.A. Lootsma (1972b). A survey of methods for solving constrained-minimization problems via unconstrained minimization. In F.A. Lootsma (ed.), Numerical Methods for Non-linear Optimization. Academic Press, London, 313-348.
20. F.A. Lootsma (1972c). The Algol 60 procedure minifun for solving non-linear optimiz-

ation problems. Report 4761, Philips Research Laboratories, Eindhoven, the Netherlands.

21. G.P. McCormick, W.C. Mylander, and A.V. Fiacco (1965). Computer program implementing the sequential unconstrained minimization technique for nonlinear programming. Technical paper RAC-TP-151, Research Analysis Corporation, McLean, Virginia, USA.
22. J.A. Nelder and R. Mead (1964). A simplex method for function minimization. The Comp. J. 7, 308-313.
23. M.J.D. Powell (1964). An efficient method for finding the minimum of a function of several variables without calculating derivatives. The Comp. J. 7, 155-162.
24. M.J.D. Powell (1970). A survey of numerical methods for unconstrained optimization. SIAM Review 12, 79-97.
25. M.J.D. Powell (1971). Recent advances in unconstrained optimization. Math. Progr. 1, 26-57.
26. R.W.H. Sargent (1973). Minimization without Constraints. In M. Avriet, M.J. Rijckaert and D.J. Wilde (eds.), Optimization and Design. Prentice-Hall, Englewood Cliffs, New Jersey, 37-75.
27. D.F. Shanno (1970). Conditioning of quasi-Newton Methods for function minimization. Math. of Comp. 24, 647-656.
28. R.L. Staha (1973). Constrained optimization via Moving Exterior Truncations. Thesis, University of Texas, Austin.
29. R.L. Staha and D.M. Himmelblau (1973). Evaluation of constrained nonlinear programming techniques. Working paper, University of Texas, Austin.
30. D. Tabak and B.C. Kuo (1971). Optimal Control by Mathematical Programming. Prentice-Hall, Englewood Cliffs, New Jersey.

Optimale Steuerprozesse mit Zustandsbeschränkungen

von

H.Maurer und U.Heidemann

1. Einleitung

In dieser Arbeit soll das qualitative Verhalten der Lösungen optimaler Steuerprozesse mit Zustandsbechränkungen untersucht werden. Dabei stellen sich die beiden folgenden Probleme.

(1) Hat die optimale Trajektorie nur Kontaktpunkte bzw. Berührpunkte mit der Zustandsbegrenzung oder verläuft die optimale Trajektorie ein Stück auf der Zustandsbegrenzung?

(2) Ist die optimale Trajektorie glatt an den Eintritts- bzw. Austrittspunkten der Zustandsbegrenzung oder treten Ecken auf?

Die Beantwortung von Problem (1) hängt wesentlich von der Ordnung p der Zustandsbeschränkung ab. Beim Problem (2) müssen zusätzlich die beiden Fälle unterschieden werden, ob die Steuervariable nichtlinear (genauer: ob die HAMILTON-Funktion regulär ist) oder linear auftritt.

Der Fall regulärer HAMILTON-Funktion ist in Jacobson et al [8] und Hamilton Jr. [5] untersucht worden. Die Autoren beweisen die Glattheit der Steuerung bis zur (p-2)-ten zeitlichen Ableitung und die wichtige Aussage, daß für p ungerade, $p \geq 3$, i.a. nur Berührpunkte mit der Begrenzung auftreten. Die Ergebnisse von [5],[8],[14] sind in Abschnitt 4 unter Auslassung der Beweise zusammengestellt.

Tritt die Steuervariable linear auf, so ist die Steuerung i.a. unstetig in den Eintritts- bzw. Austrittspunkten der Zustandsbegrenzung. Statt der Glattheit der Steuerung erhält man eine entsprechende Glattheit der Schaltfunktion. Die sich daran anschließende Theorie ist in Maurer [9] entwickelt worden und ist dual zur Theorie singulärer Steuerungen in McDanell, Powers [12]. Abschnitt 5 faßt die Ergebnisse von [9] unter Verzicht auf Beweise so zusammen, daß auch die Ähnlichkeit zu den Resultaten in Abschnitt 4 sichtbar wird.

2. Problemstellung und Definitionen

Gesucht ist eine skalare, stückweis stetige Steuerfunktion $u(t)$, $t \in [0,T]$, welche das Funktional

(2.1) $$J(u) = G(x(T))$$

maximiert unter den Nebenbedingungen

(2.2) $$\dot{x} = f(x,u) \; , \; 0 \leq t \leq T$$

(2.3) $$x(0) = x_0 \; , \; \psi(x(T)) = 0$$

(2.4) $$u(t) \in U \subset \mathbb{R}$$

und der Zustandsbeschränkung der Ordnung p

(2.5) $$S(x) \leq \alpha \quad \text{für} \quad 0 \leq t \leq T \; , \; \alpha \in \mathbb{R} \; .$$

Hierbei ist $x \in \mathbb{R}^n$ der Zustandsvektor und $\psi: \mathbb{R}^n \to \mathbb{R}^k$. Die skalaren Funktionen G,S und die Funktionen f (n-Vektor), ψ (k-Vektor, $k \leq n$) seien hinreichend oft differenzierbar, sodaß die im folgenden auftretenden Ausdrücke sinnvoll sind. Der Parameter α in (2.5) ist aus numerischen Gründen als ein Homotopie-Parameter zu verstehen, d.h. man möchte die Lösungsschar von (2.1)-(2.5) in Abhängigkeit von α studieren.

Bekanntlich lassen sich nichtautonome Steuerungsprobleme, Probleme mit freier Endzeit, etc., auf das obige Problem reduzieren.

Eine optimale Lösung von (2.1)-(2.4) ohne die Zustandsbeschränkung (2.5) heißt <u>unbeschränkte Extremale</u>, während eine optimale Lösung von (2.1)-(2.5) <u>beschränkte Extremale</u> heißt.

Entlang einer Trajektorie $x(t)$ von (2.2) werde die i-te zeitliche Ableitung von $S(x(t))$ mit S^i $(i \geq 0)$ bezeichnet, wobei $S^0 = S$. Dann gilt

(2.6) $$S^i = S^i(x) \; , \; i=0,\ldots,p-1 \; , \; S^p = S^p(x,u) \; ,$$

da nach Definition der Ordnung p der Zustandsbeschränkung (2.5) S^p die erste zeitliche Ableitung ist, welche die Steuerung u explizit enthält.

Teilstücke von $x(t)$ mit $S(x(t)) < \alpha$ heißen <u>innere Teilstücke</u>; Teilstücke von $x(t)$ mit $S(x(t)) \equiv \alpha$ in $[t_1,t_2]$, $0 \leq t_1 < t_2 \leq T$, heißen <u>Randstücke</u>. Der Zeitpunkt t_1 bzw. t_2 heißt <u>Eintritts</u>- bzw. <u>Austrittspunkt</u> des Randstückes. Die Trajektorie $x(t)$ hat einen <u>Kontaktpunkt</u> mit der Begrenzung in $t_1 \in \,]0,T[$, wenn $S(x(t_1)) = \alpha$ und $S(x(t)) < \alpha$ für $t \neq t_1$ in einer Umgebung von t_1 . Für $p \geq 2$ ist jeder Kontaktpunkt schon ein <u>Berührpunkt</u>, da hier zusätzlich die Bedingung $S^1(x(t_1)) = 0$ folgt. Durch Differentation von $S(x(t)) \equiv \alpha$, $t_1 \leq t \leq t_2$, erhält man die <u>Eintrittsbedingungen</u> eines Randstückes

(2.7) $$S^i(x(t_1)) = \delta_{i,o}\,\alpha\ ,\ i=0,\ldots,p-1\ .$$

Hierbei bedeutet $\delta_{i,o}$ das KRONECKER-Symbol. Die Randsteuerung $u=u(x)$ ist bestimmt durch $S^p(x,u) \equiv 0$, wobei noch $\frac{\partial}{\partial u} S^p(x,u) \neq 0$ entlang eines Randstückes vorausgesetzt sei.

3. Notwendige Bedingungen: Maximumprinzip

Die HAMILTON-Funktion wird in [8] definiert durch

(3.1) $$H(x,u,\lambda,\eta) = \lambda^T f(x,u) + \eta\cdot S(x)\ .$$

Dabei ist $\lambda \in \mathbb{R}^n$ und T bedeutet die Transponierung. Der skalare Multiplikator $\eta(t)$ genügt $\eta(t) \leq 0$ und $\eta(t)S(x(t)) \equiv 0$ in $[0,T]$. Die notwendigen Bedingungen in [8],[14] sind die folgenden.

1) Die adjungierten Gleichungen für $\lambda(t) \in \mathbb{R}^n$ lauten

(3.2) $$\dot{\lambda} = -H_x = -\lambda^T f_x - \eta S_x\ .$$

2) Es gelten die Sprungbedingungen im Eintritts- bzw. Austrittspunkt t_1 bzw. t_2

(3.3) $$\lambda(t_i^+) = \lambda(t_i^-) + \nu_i S_x(x(t_i))\ ,\ \nu_i \geq 0\ \ (i=1,2)\ .$$

Die Sprungbedingungen für einen Kontaktpunkt t_1 sind hierin enthalten mit $t_1 = t_2$.

3) Die optimale Steuerung $u(t)$ maximiert die HAMILTON-Funktion, d.h.

(3.4) $$H(x(t),u(t),\lambda(t),\eta(t)) = \max_{u\in U} H(x(t),u,\lambda(t),\eta(t))\ .$$

Die HAMILTON-Funktion H heißt regulär, wenn H ein eindeutig bestimmtes Maximum $u(t)$ in (3.4) hat für alle $t \in [0,T]$.

Bemerkung: Beim Beweis dieser notwendigen Bedingungen erhält man zunächst nur eine Funktion $\eta^*(t)$ von beschränkter Variation, die mit $\eta(t)$ in (3.1) durch $\eta = \eta^{*\prime}$ (vgl. [14, Corollary 3.4]) zusammenhängt, d.h. $\eta^*(t)$ muß zusätzlich als stückweise differenzierbar angenommen werden. Es wäre interessant, die Voraussetzungen zu ermitteln, unter denen dies gilt.

4. Verhalten der Extremalen bei nichtlinear auftretender Steuerung (reguläre HAMILTON-Funktion)

Der Steuerbereich U in (2.4) sei ein offenes Intervall. Dann ist die optimale Steuerung durch $H_u \equiv 0$, $t \in [0,T]$, festgelegt. Im folgenden gelte außerdem die strenge LEGENDRE-CLEBSCH-Bedingung $H_{uu} < 0$. Sei nun t_1 ein Kontaktpunkt oder ein Eintritts- bzw. ein Austrittspunkt. Zu den Bedingungen (3.1)-(3.4) tritt dann noch die Stetigkeit von H in t_1 hinzu

$$(4.1) \qquad H(t_1^+) = H(t_1^-) .$$

SATZ 4.1 [8]: *Sei* H *regulär.*

(i) $\underline{p=1}$: *Die Steuerung* $u(t)$ *ist stetig in* t_1 *und es gilt* $\nu_1=0$ *in (3.3) und* $S^1(x(t_1),u(t_1)) = 0$.

(ii) $\underline{p\geq 2}$: *Die Steuerung* $u(t)$ *und ihre ersten* $p-2$ *zeitlichen Ableitungen sind stetig in* t_1 .

(iii) Ist $\underline{p \text{ ungerade}}$, $\underline{p\geq 3}$, *und ist die* $(p-1)$*-te zeitliche Ableitung von* $u(t)$ *unstetig in* t_1, *so kann die beschränkte Extremale die Begrenzung in* t_1 *nur berühren.*

Der Beweis von (i) für einen Kontaktpunkt t_1 findet sich nicht in der Literatur und werde daher hier skizziert. In t_1 muß für S^1 als Funktion von t gelten: $S^1(t_1^-) \geq 0$ und $S^1(t_1^+) \leq 0$. Wäre $u(t)$ unstetig in t_1, so können nicht beide Ausdrücke verschwinden und es sei etwa $S^1(t_1^+) < 0$. Aus (4.1) folgt dann mit (3.3) durch eine einfache Rechnung

$$(4.2) \qquad \nu_1 = \lambda(t_1^-)^T\{f(x(t_1),u(t_1^-)) - f(x(t_1),u(t_1^+))\}/ S^1(t_1^+) .$$

Hierin ist der Zähler positiv, da H regulär ist und also $u(t_1^-)$ das eindeutige Maximum von $H(t_1^-)$ ist. Wegen $S^1(t_1^+) < 0$ gilt dann $\nu_1 < 0$ im Widerspruch zu $\nu_1 \geq 0$ in (3.3). Daher ist $u(t)$ stetig in t_1 und dies impliziert $S^1(x(t_1),u(t_1)) = 0$. Weiterhin folgt aus der Stetigkeit von $u(t)$ in t_1 die Beziehung $\nu_1 = 0$ aufgrund von $H_u(t_1^+) = H_u(t_1^-) = 0$ und $p = 1$, vgl. [14, Corollary 3.2].

Der Beweis von 4.1(iii) beruht auf der für $p \geq 2$ geltenden Beziehung

$$(4.3) \qquad \nu_1 = (-1)^p H_{uu}(t_1)\{u^{(p-1)}(t_1^+) - u^{(p-1)}(t_1^-)\}/ S_u^p(t_1) \geq 0 .$$

Die Unstetigkeit von $u^{(p-1)}(t)$ in t_1 ist daher äquivalent zu $\nu_1>0$ für $p\geq 2$. Der Fall $\nu_1 = 0$ wird in [5] für $p\geq 1$ untersucht.

SATZ 4.2 [5]: *Sei t_1 ein Eintrittspunkt und sei $u^{(p+m)}(t)$ $(m\geq 0)$ die niedrigste Ableitung von $u(t)$, welche unstetig in t_1 ist. Dann ist $p + m$ eine ungerade Zahl.*

Die Voraussetzung dieses Satzes läßt sich auch ausdrücken durch die Beziehungen

$$(4.4) \qquad \eta^{(i)}(t_1^+) = 0 \quad (i=0,\dots,m-1) \;,\; \eta^{(m)}(t_1^+) < 0$$

mit $\eta(t)$ aus (3.1). Für $p = 1$ zeigen die numerischen Beispiele, daß die beschränkte Extremale in der Regel keine Kontaktpunkte, sondern nur Randstücke besitzt, welche vom Typ $m = 0$ in 4.2 sind (also $\dot{u}(t)$ unstetig in t_1).

LEMMA 4.3: *Sei $p = 1$. Ist die unbeschränkte Extremale eindeutig durch die notwendigen Bedingungen des Maximumprinzipes bestimmt, so kann jede beschränkte Extremale nicht nur Kontaktpunkte enthalten, sondern enthält mindestens ein Randstück.*

Hätte nämlich die beschränkte Extremale nur Kontaktpunkte, so wäre die adjungierte Variable $\lambda(t)$ stetig auf $[0,T]$ wegen $\nu_1 = 0$ nach 4.1(i) und (3.3). Also würde auch die beschränkte Extremale dem Maximumprinzip des unbeschränkten Problems genügen, was der vorausgesetzten Eindeutigkeit widerspricht.

Aus (2.7) und Satz 4.1 ergeben sich zusammenfassend die folgenden Bedingungen in einem Eintrittspunkt t_1 eines Randstückes. Für das Weitere sei der 'Normalfall' $\nu_1 > 0$ für $p \geq 2$ angenommen.

$$(4.5) \quad \underline{p = 1} \quad : \quad S(x(t_1)) = \alpha \;,\; S^1(x(t_1),u(t_1)) = 0$$

$$(4.6) \quad \underline{p \text{ gerade}} : \quad S^i(x(t_1)) = \delta_{i,0}\,\alpha \;,\; i = 0,\dots,p-1$$

$$S^i(x(t_1),u(t_1)) = 0 \;,\; i = p,\dots,2p-2$$

Wir halten fest, daß nur für $p = 1$ die Kontaktbedingungen gleichzeitig die Eintrittsbedingungen sind.

Für die unbeschränkte Extremale $x^0(t)$ werde nun definiert

$$(4.7) \qquad \alpha_o = \max_{t\in[0,T]} S(x^o(t)) = S(x^o(t_1)) .$$

Ist $p = 1$, so gilt (4.5) für $\alpha = \alpha_o$ schon in allen Punkten t_1 definiert durch (4.7). Für $p \geq 2$ hingegen ist die Eintrittsbedingung (4.6) i.a. nicht erfüllt in t_1 . Dies ist der Grund für das folgende Verhalten der beschränkten Extremalen $x(t;\alpha)$ in Abhängigkeit von $\alpha < \alpha_o$ und der Ordnung p , sofern $x(t;\alpha)$ existiert.

p=1: $x(t;\alpha)$ enthält mindestens ein Randstück für alle $\alpha < \alpha_o$. Falls der Zeitpunkt t_1 in (4.7) eindeutig bestimmt ist, dann enthält $x(t;\alpha)$ genau ein Randstück für $\alpha_o - \varepsilon \leq \alpha < \alpha_o$, $\varepsilon > 0$ klein; vgl. Fig.1 für einen festen Endpunkt $x(T)$.

p gerade: Es existiert ein Parameter $\alpha_1 < \alpha_o$, sodaß $x(t;\alpha)$ für alle $\alpha_1 \leq \alpha < \alpha_o$ die Begrenzung nur berührt. Sind dann bzgl. α zuerst in α_1 die Eintrittsbedingungen (4.6) in einem Berührpunkt t_1 erfüllt, so wird dieser Punkt t_1 sich in ein Randstück von $x(t;\alpha)$ für $\alpha < \alpha_1$ aufspalten; vgl. Fig.2 . Ein solcher Parameter α_1 , für den (4.6) gilt, braucht nicht zu existieren, sodaß dann die beschränkte Extremale die Begrenzung nur berührt für alle $\alpha < \alpha_o$ [5, example 2].

p ungerade: $x(t;\alpha)$ berührt die Begrenzung nur für alle $\alpha < \alpha_o$.

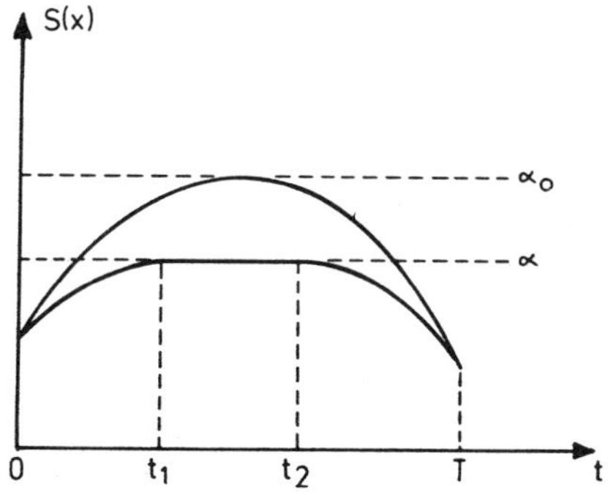

Figur 1

Typisches Verhalten der Extremalen für p=1

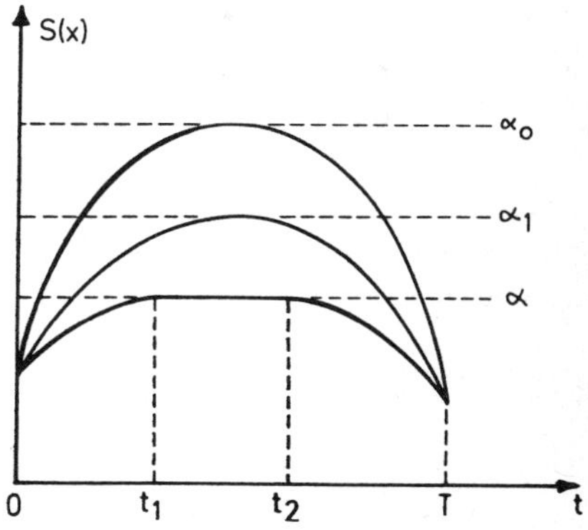

Figur 2

Typisches Verhalten der Extremalen für p gerade

Den Betrachtungen für $p = 1$ und p gerade liegt die Annahme zugrunde, daß $x(t;\alpha)$ und die zugehörige adjungierte Variable $\lambda(t;\alpha)$ stetig von α abhängen. Diese Annahme läßt sich wegen $H_{uu} < 0$ unter gewissen Voraussetzungen beweisen.

Es ist bisher noch kein Beispiel mit gerader Ordnung $p \geq 4$ bekannt, wo die beschränkte Extremale ein Randstück enthält. Für $p = 2$ ist die obige Folge - Berührung für $\alpha_1 \leq \alpha < \alpha_o$, Randstück für $\alpha < \alpha_1$ - das typische Verhalten der Lösung; für ein analytisches Beispiel vgl. [2, S.121]. Algorithmen zur Lösung der betrachteten Steuerungsprobleme, welche das geschilderte Verhalten der Extremalen in Abhängigkeit von p und α berücksichtigen, sind in Maurer, Gillessen [10] unter Verwendung der Mehrzielmethode von Bulirsch, Deuflhard, Stoer [3] entwickelt worden. Die mit diesen Methoden gerechneten numerischen Beispiele finden sich in [10], Gillessen [4] und Wick [15]. Für andere numerische Verfahren verweisen wir auf Miele et al [13] und die dort aufgeführte Literatur.

5. Verhalten der Extremalen bei linear auftretender Steuerung

Die Dynamik (2.2) hat bei linear auftretender Steuervariable die Form

$$\dot{x} = f(x,u) = f_1(x) + f_2(x) \cdot u \,. \tag{5.1}$$

Der Steuerbereich U in (2.4) sei ein kompaktes Intervall $U = [a,b]$. Die Funktion $S^p(x,u)$ in (2.6) ist hier linear in u, d.h.

$$S^p(x,u) = \alpha(x) + \beta(x) \cdot u \,, \quad \beta(x) \neq 0 \,. \tag{5.2}$$

Die Randsteuerung ist daher $u(x) = -\alpha(x)/\beta(x)$ und es gelte längs eines Randstückes in $[t_1,t_2]$

$$a < u(x(t)) < b \quad \text{für} \quad t_1 < t < t_2 \,, \tag{5.3}$$

d.h. die Randsteuerung $u(x(t))$ soll im Inneren des Steuerbereiches U liegen mit Ausnahme der Punkte t_1 und t_2.

Die HAMILTON-Funktion (3.1) ist ebenfalls linear in u

$$H(x,u,\lambda,\eta) = \lambda^T f_1(x) + \lambda^T f_2(x) \cdot u + \eta S(x) \,. \tag{5.4}$$

Der Koeffizient von u in (5.4) heißt die Schaltfunktion

$$(5.5) \qquad \Phi(t) = \lambda^T(t) f_2(x(t))$$

Die HAMILTON-Funktion H ist genau in den Punkten nicht regulär, in denen $\Phi(t)$ eine Nullstelle hat.

Optimale Steuerung für innere Teilstücke: Die Schaltfunktion $\Phi(t)$ habe nur isolierte Nullstellen, d.h. der singuläre Fall, wo $\Phi(t)$ identisch auf einem Teilintervall verschwindet, sei ausgeschlossen. Die Maximierung von H in (5.4) über $U = [a,b]$ gibt dann die bang-bang Steuerung

$$(5.6) \qquad u(t) = \begin{cases} b & \text{für } \Phi(t) > 0 \\ a & \text{für } \Phi(t) < 0 \end{cases}$$

Optimale Steuerung für Randstücke: Die optimale Steuerung ist die Randsteuerung $u(x) = -\alpha(x)/\beta(x)$, welche verträglich sein muß mit der Maximierung von H über $U = [a,b]$. Die Annahme (5.3) impliziert daher

$$(5.7) \qquad H_u = \Phi(t) \equiv 0 \quad \text{für} \quad t_1^+ \leq t \leq t_2^- .$$

Die Randsteuerung verhält sich demnach wie eine singuläre Steuerung in Steuerungsproblemen ohne Zustandsbeschränkungen. Dies weist auf eine Dualität dieser beiden Arten von Steuerungsproblemen hin. Die folgenden Untersuchungen sollen diese Dualität genauer erläutern.

Sei t_1 ein Kontaktpunkt oder ein Eintrittspunkt eines Randstückes. Mit Hilfe der Formeln in [5, (16)-(18)] für die i-te zeitliche Ableitung $\Phi^{(i)}(t)$ der Schaltfunktion ergeben sich in einfacher Weise die folgenden Beziehungen

$$(5.8) \qquad \Phi^{(i)}(t_1^+) = \Phi^{(i)}(t_1^-) \quad , \; i = 0,\ldots,p-2 \quad (\text{leer für } p=1)$$

$$(5.9) \qquad \Phi^{(p-1)}(t_1^+) = \Phi^{(p-1)}(t_1^-) + \nu_1(-1)^{p-1}\beta(x(t_1)) \quad , \; \nu_1 \geq 0 .$$

Hier ist $\nu_1 \geq 0$ aus (3.3). Bei der Herleitung dieser Formeln muß $p \leq 2q + r$ vorausgesetzt werden, wobei $\Phi^{(2q)}$ die niedrigste Ableitung ist, welche die Steuerung u explizit enthält (s. [12]), und $u^{(r)}(t)$ die niedrigste Ableitung ist, welche unstetig in t_1 ist. Für $p \geq 2$ gilt stets $p \leq 2q + r$, da $q \geq 1$. Die Gleichungen (5.8) bzw. (5.9) sind das Gegenstück zu Satz 4.1(ii) bzw. (4.3). Ferner ist die

Bedingung $\nu_1 > 0$ äquivalent zur Unstetigkeit von $\Phi^{(p-1)}(t)$ in t_1 .

Ist insbesondere t_1 ein Eintrittspunkt, so folgt $\Phi^{(i)}(t_1^+) = 0$ ($i \geq 0$) wegen (5.7) und man erhält die Eintrittsbedingungen

(5.10) $$\Phi^{(i)}(t_1^-) = 0 \quad , \; i = 0,\dots,p-2 \quad ,$$

(5.11) $$\nu_1 = (-1)^p \, \Phi^{(p-1)}(t_1^-)/\beta(x(t_1)) \geq 0 \quad .$$

(5.10),(5.11) gelten auch in einem Austrittspunkt t_2 , wenn t_1^- durch t_2^+ und ν_1 durch $-\nu_2$ ersetzt wird.

Wir wollen nun die möglichen Nahtstellen zwischen inneren Teilstücken und Randstücken studieren. Dabei sei die Steuerung als stückweis analytisch angenommen, um die beim Beweise benötigten TAYLOR-Entwicklungen zu rechtfertigen. Alle folgenden Sätze gelten auch im Austrittspunkt t_2 . Der nächste Satz und der zugehörige Beweis sind dual zu einem Resultat für singuläre Steuerungsprobleme [12, Theorem 1] .

SATZ 5.1: *Sei* t_1 *ein Eintrittspunkt und sei* $u^{(r)}(t)$ ($r \geq 0$) *die niedrigste Ableitung von* $u(t)$ *, welche unstetig in* t_1 *ist. Gilt* $\nu_1 > 0$ *, d.h. gilt* $\Phi^{(p-1)}(t_1^-) \neq 0$ *, so ist* $p + r$ *eine gerade Zahl.*

Aus Satz 5.1 folgt unmittelbar.

KOROLLAR 5.2:

(i) Wenn $p + r$ *ungerade ist und* $\nu_1 > 0$ *in* t_1 *gilt, dann kann* t_1 *nur ein Kontaktpunkt sein.*

(ii) Sei $p + r$ *ungerade. Wenn* t_1 *ein Eintrittspunkt ist, dann ist notwendig* $\nu_1 = 0$ *.*

In den bisher bekannten numerischen Beispielen ist stets $u(t)$ unstetig in einem (möglichen) Eintrittspunkt t_1 , also $r = 0$. Im Hinblick auf Lemma 5.4(i) sagt dann Teil (i) aus, daß für p ungerade, $p \geq 3$ nur eine Berührung der Begrenzung möglich ist. Dies ist dasgleiche Ergebnis wie in Satz 4.1(iii). Teil (ii) des Korollars kommt zur Anwendung im Falle $p = 1$, wo die beschränkte Extremale in der Regel nur Randstücke enthält, für die dann $\nu_1 = 0$ bzw. $\nu_2 = 0$ gelten muß. Wegen (5.11) zieht dies die folgenden Bedingungen für die Schaltfunktion nach sich

(5.12) $$\Phi(t_1^-) = 0 \; , \quad \Phi(t_2^+) = 0 \; .$$

Diese Relationen sind wichtig für die numerische Berechnung der adjungierten Variablen $\lambda(t)$.

Allgemein kann die Bedingung $\nu_1 = 0$, d.h. $\Phi^{(p-1)}(t_1^-) = 0$ für einen Eintrittspunkt t_1 , wiederum dual zu [12, Theorem 2] oder analog wie in Satz 4.2 untersucht werden.

<u>SATZ 5.3</u>: *Sei* t_1 *Eintrittspunkt. Sei* $\Phi^{(p+m)}(t_1^-)$ $(m \geq 0)$ *die niedrigste nichtverschwindende Ableitung von* $\Phi(t)$*, sei* $u^{(r)}(t)$ $(r \geq 0)$ *die niedrigste Ableitung von* $u(t)$ *, welche unstetig in* t_1 *ist, und es gelte* $p + m \leq 2q + r$ *. Dann ist* $p + r + m$ *eine ungerade Zahl.*

Die Voraussetzung $\Phi^{(p+i)}(t_1^-) = 0 \ (i = 0,\dots,m-1)$, $\Phi^{(p+m)}(t_1^-) \neq 0$ ist übrigens äquivalent zu (vgl. (4.4))

$$(5.13) \qquad \eta^{(i)}(t_1^+) = 0 \ (i = 0,\dots,m-1) \ , \ \eta^{(m)}(t_1^+) < 0 \ .$$

mit $\eta(t) \leq 0$ aus (3.1). Die Beweise von Satz 5.1 und Satz 5.3 motivieren außerdem, daß die Bedingungen $\nu_1 \geq 0$ und $\eta(t) \leq 0$ eine <u>duale</u> Rolle zur verallgemeinerten LEGENDRE-CLEBSCH-Bedingung $(-1)^q \frac{\partial}{\partial u} \Phi^{(2q)} \leq 0$ bei singulären Steuerungsproblemen spielen.

Für $p = 1$ tritt in allen bisherigen Beispielen der 'Normalfall' $r = 0$, $m = 0$ in Satz 5.3 auf. Weiterhin gilt für $p = 1$.

<u>LEMMA 5.4</u>: *Sei* $p = 1$.

(i) Ist t_1 *ein Kontaktpunkt und ist* $u(t)$ *unstetig in* t_1 *, so gilt* $\nu_1 = 0$ *und* $\Phi(t_1) = 0$.

(ii) Ist die unbeschränkte Extremale durch die notwendigen Bedingungen des Maximumprinzipes eindeutig bestimmt, so kann die beschränkte Extremale nicht nur Kontaktpunkte enthalten, sondern sie enthält mindestens ein Randstück.

Die Aussage (i) (vgl. hierzu auch Satz 4.1(i)) wird mittels der Beziehung (5.9) bewiesen und (ii) folgt mit dergleichen Überlegung wie bei Lemma 4.3.

Entsprechend (4.5),(4.6) seien hier für einen Eintrittspunkt t_1 die <u>Eintrittsbedingungen</u> zusammengefaßt. Es wird dabei nur der 'Normalfall' $r = 0$ und $\nu_1 > 0$ für $p \geq 2$ betrachtet.

(5.14) p = 1 : $S(x(t_1)) = \alpha$, $\Phi(t_1) = 0$

(5.15) p gerade : $S^i(x(t_1)) = \delta_{i,o}\,\alpha$, $i = 0,\ldots,p-1$

$\Phi^{(i)}(t_1) = 0$, $i = 0,\ldots,p-2$

Auch hier sind nur für $p = 1$ die Kontaktbedingungen gleichzeitig die Eintrittsbedingungen.

Man definiere für die unbeschränkte Extremale $x^o(t)$ wie in (4.7)

(5.16) $$\alpha_o = \max_{t\in[0,T]} S(x^o(t)) = S(x^o(t_1)) .$$

Für $p = 1$ sind die Punkte t_1 definiert durch (5.16) Schaltpunkte und (5.14) gilt für $\alpha = \alpha_o$. Hingegen ist für p gerade und $\alpha = \alpha_o$ die Gleichung (5.15) i.a. nicht erfüllt. Die Steuerung ist nämlich i.a. stetig in t_1 , d.h. t_1 ist kein Schaltpunkt, und daher hat man $\Phi(t_1) \neq 0$. Beispiele hierfür werden in [9] und in Abschnitt 6. für $p = 2$ diskutiert.

ANNAHME 5.5: Die beschränkten Extremalen $x(t;\alpha)$ und die adjungierten Variablen $\lambda(t;\alpha)$ hängen stetig von α ab.

Unter dieser Annahme wird man auf dasgleiche qualitative Verhalten von $x(t;\alpha)$ in Abhängigkeit von α geführt wie Abschnitt 4. für die Ordnungen $p = 1$, p gerade, p ungerade und $p \geq 3$. Fig.1 muß im vorliegenden Fall dahingehend abgeändert werden, daß die Funktion $S(x(t))$ in den Unstetigkeitsstellen von $u(t)$ Ecken hat, also insbesondere in t_1 für $\alpha = \alpha_o$ und in t_1 und t_2 für $\alpha < \alpha_o$. Fig.2 bleibt unverändert, da trotz der Unstetigkeiten in $u(t)$ die Funktion $S(x(t))$ glatt ist wegen $p \geq 2$.

Algorithmen und numerische Beispiele werden in Bock [1], Heidemann [7], Maurer, Gillessen [10] und Wick [15] behandelt. Diese Algorithmen sind dual zu Algorithmen für singuläre Steuerungsprobleme in Maurer [11]. Ein Beispiel mit $p = 2$, wo die beschränkte Extremale die Begrenzung berührt, wird in [9] gegeben. Hingegen ist noch kein Beispiel mit p gerade bekannt, bei dem die Abfolge - Berührung der Begrenzung für $\alpha_1 \leq \alpha < \alpha_o$ - Randstück für $\alpha < \alpha_1$ - auch tatsächlich auftritt.

In diesem Zusammenhang ist nun zu beachten, daß die Annahme 5.5 kritisch ist und nicht immer zutrifft. In der Regel ist zwar $x(t;\alpha)$ stetig bzgl. α , nicht jedoch $\lambda(t;\alpha)$. Bock [1] behandelt ein Problem der

Ordnung $p = 1$, in dem $\lambda(t;\alpha)$ unstetig in abzählbar vielen Parametern $\alpha_i < \alpha_o$, $i \in \mathbb{N}_+$, ist. Im nächsten Abschnitt wird ein Problem der Ordnung $p = 2$ aus [7] diskutiert, bei welchem die Berührungsphase für das Intervall $[\alpha_1,\alpha_o]$ übersprungen wird und $x(t;\alpha)$ ein Randstück schon für $\alpha < \alpha_o$ enthält. Dort ist dann notwendig $\lambda(t;\alpha)$ unstetig in $\alpha = \alpha_o$. Ein Grund für das Nichtzutreffen der Annahme 5.5 ist darin zu suchen, daß sich die Extremalen bei einer linear auftretenden Steuerung nicht immer in eine hinreichend große Schar von Extremalen einbetten lassen, da zu wenige 'Freiheitsgrade' (Schaltpunkte, Eintritts- und Austrittspunkte, etc.) zur Verfügung stehen, vgl. [9]. Dem Leser sei als ein offenes Problem überlassen, die genauen Voraussetzungen zu ermitteln, unter denen Annahme 5.5 gilt oder nicht gilt.

6. Numerisches Beispiel: Zeitoptimale Steuerung eines Nuklearreaktors

In Hassan et al [6] werden einige numerische Ergebnisse für die zeitoptimale Steuerung eines Nuklearreaktors angegeben. Eine vollständige numerische Lösung dieses Problems hat Heidemann [7] mit der Mehrzielmethode erhalten unter Verwendung numerischer Techniken aus [10]. Das Problem besteht darin, die Endzeit

$$J(u) = T \tag{6.1}$$

zu minimieren unter den Nebenbedingungen

$$\begin{aligned}
&\dot{x}_1 = k_1(x_3-1)x_1 + k_2x_2, && x_1(0) = n_o, && x_1(T) = n_T\\
&\dot{x}_2 = k_1x_1 - k_2x_2, && x_2(0) = n_ok_1/k_2, && x_2(T) = n_Tk_1/k_2\\
&\dot{x}_3 = u, && x_3(0) = 0, && x_2(T) = 0\\
&|u| \le 0.2
\end{aligned} \tag{6.2}$$

Es bedeuten x_1: Neutronendichte, x_2: verzögerte Neutronen-Konzentration, x_3: Reaktivität, und in Abänderung zu [6] sind die Zahlenwerte $k_1 = 5.$, $k_2 = 0.1$, $n_o = 0.04$, $n_T = 0.06$. Der Steuerprozeß ist außerdem unterworfen

(I) der Zustandsbeschränkung der Ordnung $p = 1$

$$S(x) = x_3 \le \rho, \quad 0 \le t \le T \tag{6.3}$$

(II) oder der Zustandsbeschränkung der Ordnung $p = 2$

$$S(x) = x_1 \le \alpha, \quad 0 \le t \le T \tag{6.4}$$

In [7] wird auch der Fall behandelt, daß beide Zustandsbeschränkungen gleichzeitig wirksam sind.

Unbeschränkte Extremale: Die optimale Steuerung ist

$$(6.5) \qquad u^o(t) = \begin{cases} 0.2 \ , & 0 \leq t \leq t^* \\ -0.2 \ , & t^* < t \leq t^{**} \\ 0.2 \ , & t^{**} < t \leq T \end{cases}$$

mit $T = 7.047806$, $t^* = 0.4798784 \cdot T$, $t^{**} = 0.9798784 \cdot T$. Die adjungierten Variablen $\lambda(t)$ für das Maximumprinzip (5.6) sind durch $\lambda(0) = (\ 2.970144\ ,\ 2.845469\ ,\ 5.\ )$ über die adjungierten Gleichungen festgelegt. Die drei Parameter t^*, t^{**}, T in (6.5) sind übrigens eindeutig durch die drei Bedingungen für den Endzustand $x(T)$ bestimmt.

(I) Zustandsbeschränkung (6.3) der Ordnung $p = 1$: Für die unbeschränkte optimale Trajektorie $x^o(t)$ gilt $\rho_o = \max x_3^o(t) = x_3^o(t^*) = 0.6764179$. Die Steuerung $u^o(t)$ hat einen Schaltpunkt in t^* und daher $\Phi(t^*) = 0$. Mit den Diskussionen in Abschnitt 5. (bzw. 4.) für $p = 1$ folgt dann, daß die beschränkte Extremale genau ein Randstück für $\rho_o - \varepsilon \leq \rho < \rho_o$ enthält, $\varepsilon > 0$ geeignet. Im vorliegenden Problem gilt diese Aussage sogar für alle $\rho < \rho_o$. Die Randsteuerung ist $u = S^1(x,u) = 0$. Dann hat die optimale Steuerung die folgende Gestalt für $\rho < \rho_o$

$$(6.6) \qquad u(t) = \begin{cases} 0.2 \ , & 0 \leq t < t_1 = 5\rho \\ 0. \ , & t_1 \leq t \leq t_2 \\ -0.2 \ , & t_2 < t \leq t^{**} \\ 0.2 \ , & t^{**} < t \leq T \end{cases}$$

Die drei Parameter t_2, t^{**}, T sind wiederum eindeutig bestimmt durch den Endzustand $x(T)$. Die numerischen Resultate für $x(t)$, $\lambda(t)$ und verschiedenen Parametern $\rho < \rho_o$ in [7] zeigen, daß stets Satz 5.3 mit $p = 1$, $r = 0$, $m = 0$, d.h. $\dot{\Phi}(t_1^-) \neq 0$, erfüllt ist.

(II) Zustandsbeschränkung (6.4) der Ordnung $p = 2$:

Sei $x^o(t)$ die unbeschränkte optimale Trajektorie und definiere $\alpha_o = \max x_1^o(t) = x_1^o(t_1) = 0.12...$, $t_1 = 0.54.. \cdot T$. Der Punkt t_1 ist kein Schaltpunkt der Steuerung (6.5) und es gilt $\Phi(t_1) \neq 0$ für die unbeschränkte Schaltfunktion, s. Fig.3 . Wäre Annahme 5.5 richtig, so könnte wegen der Diskussionen in Abschnitt 5. bzw. 4. die beschränkte Extremale die Begrenzung nur berühren für $\alpha_o - \varepsilon \leq \alpha < \alpha_o$ $\varepsilon > 0$

geeignet. Eine zu einer solchen Extremalen gehörende Steuerung hätte dann aber (o.E. im gleichen Bereich für α) diegleiche Struktur wie in (6.5). In (6.5) sind jedoch alle freien Parameter durch $x(T)$ bestimmt und es bleibt kein Freiheitsgrad für die Bedingung $x_1 \leq \alpha$ übrig. Daher kann die beschränkte Extremale die Begrenzung nicht berühren und es folgt weiter: wenn eine beschränkte Extremale mit einem Randstück für $\alpha < \alpha_0$ existiert, so ist Annahme 5.5 nicht erfüllt und die adjungierten Variablen $\lambda(t;\alpha)$ sind unstetig in α_0 als Funktion von α .

Eine beschränkte Extremale mit einem Randstück kann nun tatsächlich konstruiert werden für $\alpha < \alpha_0$. Die optimale Steuerung ist

$$(6.7) \qquad u(t) = \begin{cases} 0.2 , & 0 \leq t \leq t^* \\ -0.2 , & t^* < t < t_1 \\ -k_2x_3 , & t_1 \leq t \leq t_2 \\ -0.2 , & t_2 < t \leq t^{**} \\ 0.2 , & t^{**} < t \leq T \end{cases}$$

Die Randsteuerung $u = -k_2x_3$ erhält man aus $S^2(x,u) = 0$. Die fünf Parameter t^*, t_1, t_2, t^{**}, T in (6.7) und die adjungierten Variablen $\lambda(t)$ werden numerisch über die Bedingungen für $x(T)$ und (5.15), d.h. $x_3(t_1) = \alpha$, $\dot{x}_3(t_1) = 0$, $\Phi(t_1) = \Phi(t_2) = 0$, und die Transversalitätsbedingung $H(t) \equiv 0$ durch ein geeignetes Randwertproblem berechnet. Für $\alpha = \alpha_0$ gilt $t_1 = t_2$ in (6.7). Um einen numerischen Vergleich zu ermöglichen, seien die Ergebnisse angegeben für

$\underline{\alpha = 0.105}$: $T = 7.344445$, $t^* = 0.4253868 \cdot T$, $t_1 = 0.4793580 \cdot T$,
$t_2 = 0.6285059 \cdot T$, $t^{**} = 0.9806927 \cdot T$,
$\lambda(0) = (4.052238,\ 3.882192,\ 5.)$, $\nu_1 = -\lambda_1(t_1^-) = 3.838474 > 0$

und für $t_1 \leq t \leq t_2$: $\lambda_1 = \lambda_3 \equiv 0$, $\eta(t) = -k_1\lambda_2(t) = -k_1 c \exp(k_2 t) < 0$ wegen $c > 0$.

Die Beziehung $\nu_1 = -\lambda_1(t_1^-)$ folgt aus (5.11). Die Steuerung $u(t)$ in (6.7) ist aufgrund der Zahlenwerte für k_2x_3 immer unstetig in t_1 und t_2 . Die notwendigen Bedingungen von Satz 5.1 für ein Randstück sind deshalb hier erfüllt mit $\nu_1 > 0$, $p = 2$, $r = 0$, also $p + r$ gerade. Die zugehörige Schaltfunktion $\Phi(t)$ ist in Fig.4 dargestellt.

In (6.7) betrachte man den Grenzfall $\alpha \uparrow \alpha_0$, d.h. $t_1 - t_2 \to 0$, und man erhält die in Fig.3 gestrichelt gezeichnete Schaltfunktion, welche die Unstetigkeit der Schaltfunktion in $\alpha = \alpha_0$ deutlich macht. Die Schaltfunktion für $\alpha = \alpha_0^-$ ist nicht zulässig im Sinne des Maximumprinzipes, da offenbar $\nu_1 > 0$ nach (5.9) wegen der Unstetigkeit von $\dot{\Phi}(t)$ in t_1. Die Schaltfunktionen in Fig.3, Fig.4 sind auf das Einheitsintervall $[0,1]$ transformiert.

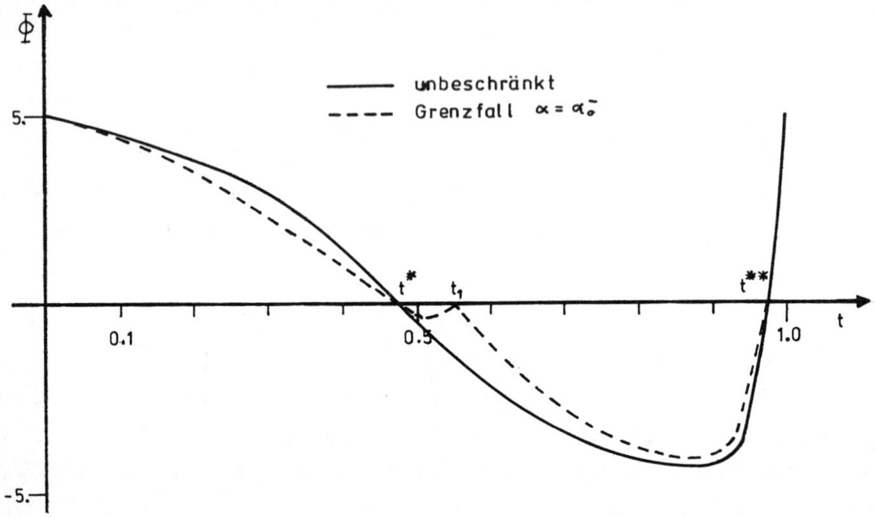

Figur 3 : Unstetigkeit der Schaltfunktion in $\alpha = \alpha_0$

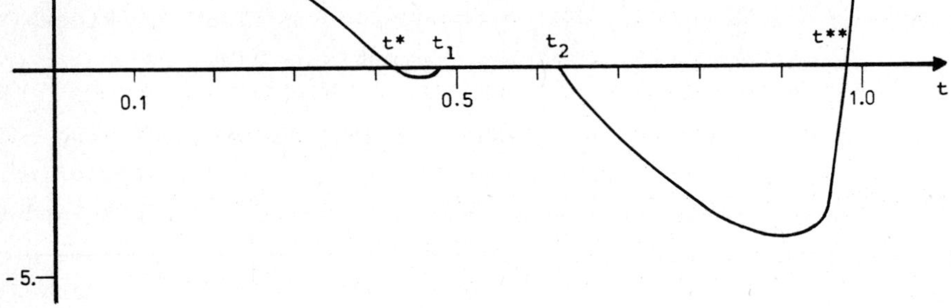

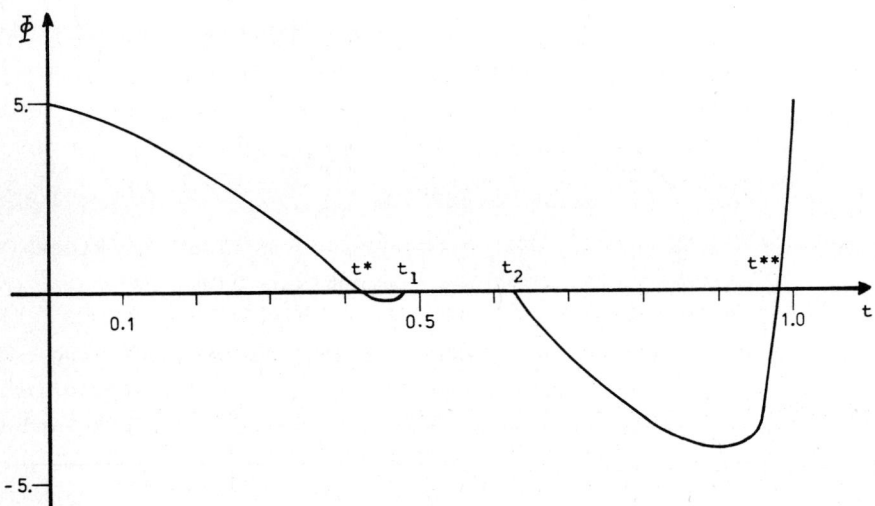

Figur 4 : Schaltfunktion für $\alpha = 0.105$

Literatur

[1] Bock, G.: Numerische Optimierung zustandsbeschränkter parameterabhängiger Prozesse mit linear auftretender Steuerung unter Anwendung der Mehrzielmethode, Diplomarbeit, Mathematisches Institut der Universität Köln, 1974.

[2] Bryson, A.E., Ho, Y.C.: Applied Optimal Control, Blaisdell Publishing Company, Waltham, Massachusetts, 1969.

[3] Bulirsch, R., Deuflhard, P., Stoer, J.: Numerical solution of nonlinear two-point boundary value problems I, erscheint in Numerische Mathematik, Handbook Series Approximation.

[4] Gillessen, W.: Optimale Steuerung bei einer Beschränkung im Phasenraum und deren numerische Berechnung unter Verwendung der Mehrzielmethode dokumentiert an zwei Beispielen aus der Flugbahnoptimierung für Ordnungen der Zustandsbeschränkung $q=0$ und $q=2$, Diplomarbeit, Mathematisches Institut der Universität Köln, 1973/74.

[5] Hamilton Jr., W.E.: On Nonexistence of Boundary Arcs in Control Problems with Bounded State Variables, IEEE Transactions on Automatic Control, AC-17, No.3, 338 -343 (1972).

[6] Hassan, M.A., Ghonaimy, M.A.R., Abdel Malek, N.R.: Computational Solution of the Nuclear Reactor Minimum Time Start-Up Problem with State Space Constraints, 2nd IFAC Symposium on multivariable technical control systems, Düsseldorf, 11-13 Oct. 1971, North Holland Publishing Company, Amsterdam, 1971.

[7] Heidemann, U.: Numerische Berechnungen optimaler Steuerungen mit Hilfe der Mehrzielmethode bei Beschränkungen im Zustandsraum mit Beispielen aus der Physik, Chemie und Raumfahrt, Diplomarbeit, Mathematisches Institut der Universität Köln, 1974.

[8] Jacobson, D.H., Lele, M.M., Speyer, J.L.: New Necessary Conditions of Optimality for Control Problems with State-Variable Inequality Constraints, J. of Math. Analysis and Appl. 35, 255 - 284 (1971).

[9] Maurer, H.: On optimal control problems with bounded state variables and control appearing linearly, eingereicht an SIAM Journal on Control.

[10] Maurer, H., Gillessen, W.: Application of multiple shooting to the numerical solution of optimal control problems with bounded state variables, erscheint in COMPUTING.

[11] Maurer, H.: Numerical solution of singular control problems using multiple shooting techniques, erscheint in J. of Optim. Theory and Appl..

[12] McDanell, J.P., Powers, W.F.: Necessary Conditions for Joining Optimal Singular and Nonsingular Subarcs, SIAM Journal on Control 9, 161 - 173 (1971).

[13] Miele, A., Well, K.H., Tietze, J.L.: Modified Quasilinearization Algorithms for Optimal Control Problems with Bounded State, J. of Optim. Theory and Appl. 12, 285 - 319 (1973).

[14] Norris, D.O.: Nonlinear Programming Applied to State- Constrained Optimization Problems, J. of Math. Analysis and Appl. 43, 261 - 272 (1973).

[15] Wick, R.: Numerische Lösung volkswirtschaftlicher Variationsprobleme mit Zustandsbeschränkungen unter Anwendung der Mehrzielmethode, Diplomarbeit, Mathematisches Institut der Universität Köln, 1973/74.

Algorithms for Rational Discrete Least Squares Approximation

Part II: Optimal Polefree Solution

Peter Spellucci, Mainz

Summary: In this paper an algorithm for the computation of a locally optimal polefree solution to the discrete rational least squares problem under a mild regularity condition is presented. It is based on an adaptation of projection methods [8], [12], [13], [14], [18], [19] to the modified Gauß-Newton method [4], [10]. A special device makes possible the direct handling of the infinitely many linear constraints present in this problem.

1. Introduction

Let $(\xi_1,\eta_1),\dots,(\xi_n,\eta_n)$, $\xi_1\leqq\dots\leqq\xi_n$, $\eta_i\in\mathbb{R}$, $i=1(1)n$ be given data points and

$$R_{l,m}(\xi;a,b)\equiv\sum_{i=0}^{l}\alpha_i\xi^i/\sum_{j=0}^{m}\beta_j\xi^j, \tag{1.1}$$

where $a=(\alpha_0,\dots,\alpha_n)^T\in\mathbb{R}^{l+1}$, $b=(\beta_0,\dots,\beta_m)^T\in\mathbb{R}^{m+1}$. l and m are held fixed throughout. The case of multiple ξ's is explicitly permitted. For reasons which become understandable in connection with the convergence theorem in 4. we require

$$|\{\xi_1,\dots,\xi_n\}|\geqq l+m+1=:p. \tag{1.2}$$

This paper is concerned with the problem:

Minimize

$$F(a,b)\equiv\sum_{i=1}^{n}\left(R_{l,m}(\xi_i;a,b)-\eta_i\right)^2$$

for $(a,b)\in\mathcal{L}(\varepsilon)$, where

$$\mathcal{L}(\varepsilon):=\{(a,b)\,|\,|\alpha_i|\leqq 1,\ i=0(1)n,\ |\beta_j|\leqq 1,\ j=0(1)m,\ \sum_{j=0}^{m}\beta_j\xi^j\geqq\varepsilon>0 \text{ for } \xi\in[\xi_1,\xi_n]\}.$$

(Admitting $\varepsilon\leqq 0$ and requiring $|R_{l,m}(\xi;a,b)|<\infty$ instead would yield a noncompact $\mathcal{L}(\varepsilon)$, where F is non-semicontinuous and where a minimum may not exist [11].
Moreover, if a local minimum exists on $\partial\mathcal{L}(0)$, there exists a corresponding local minimum in $\mathring{\mathcal{L}}(\varepsilon)$, for some $\varepsilon>0$ [11].)

Here and in the following we denote by $\partial\mathfrak{A}$ the boundary and by $\mathring{\mathfrak{A}}$ the interior of a set $\mathfrak{A}$.
Now without loss of generality we may assume

$$\xi_1=-1,\quad \xi_n=1,\quad \max\{|\eta_i|\ \ |i=1(1)n\}=1. \tag{1.3}$$

Then for any $\delta>0$ there exists a ε, $0<\varepsilon<1$, and for $(a,b)\in\mathcal{L}(\delta)$ a $g=(g_D^T,g_N^T)^T$, where $g_D=(\gamma_1,\dots,\gamma_m)^T$, $g_N=(\gamma_{m+1},\dots,\gamma_p)^T$, such that

$$R_{l,m}(\xi;a,b)\equiv f(\xi;g):=P_l(\xi;g_N)/Q_m(\xi;g_D) \tag{1.4}$$

$$P_l(\xi;g_N):=\sum_{i=0}^{l}\gamma_{m+i+1}T_i(\xi),\qquad Q_m(\xi;g_D):=\sum_{j=1}^{m}\gamma_jT_j(\xi)+1$$

and

$$Q_m(\xi;g_D)\geqq\varepsilon\quad\text{for } \xi\in[-1,1]. \tag{1.5}$$

Here T_i is the i-th Chebyshev polynomial of first kind. Moreover, there exists a $\bar\gamma$, dependent on δ, l and m only, such that $|\gamma_i|\leqq\bar\gamma$, $i=1(1)p$, uniformly on $\mathcal{L}(\delta)$. Therefore we consider the equivalent problem:

For $0<\varepsilon<1$ fixed, ξ_i, η_i as in (1.3), minimize

$$E(g):=\sum_{i=1}^{n}\left(f(\xi_i;g)-\eta_i\right)^2 \tag{1.6}$$

subject to

$$g_D\in\mathfrak{D}=\mathfrak{D}(\varepsilon):=\{\, c \mid c\in\mathbb{R}^m:\ Q_m(\xi;c)\geq\varepsilon \text{ for } \xi\in[-1,1]\,\}. \tag{1.7}$$

$\mathfrak{D}$ is convex and compact [16]. For $g_D\in\mathfrak{D}$ fixed there exists a unique $g_N\in\mathbb{R}^{l+1}$ such that

$$E((g_D^T,g_N^T)^T)=\min\{\, E(g) \mid g=(g_D^T,c^T)^T,\ c\in\mathbb{R}^{l+1}\}.$$

Therefore problem (1.6),(1.7) is well defined. Using the methods described by Golub and Pereyra in [9] it would be possible to reduce our problem to a minimization problem over $\mathfrak{D}$ alone. This approach is not taken here because of its complexity. However the algorithm to be described in 3. uses that fact in a quite simple fashion. Since

$$\min\{E(g) \mid g=(g_D^T,g_N^T)^T,\ g_D\in\mathfrak{D},\ g_N\in\mathbb{R}^{l+1}\}\leq E(0)$$

and $E(g)>2E(0)$ for $g_N\notin\mathfrak{N}$, $g_D\in\mathfrak{D}$, where

$$\mathfrak{N}=\{\, c \mid c\in\mathbb{R}^{l+1},\ |P_l(\xi_i;c)|\leq\sqrt{2}(\sqrt{n}+2)\max\{\, Q_m(\xi;d) \mid \xi\in[-1,1],\ d\in\mathfrak{D}\} \text{ for } i=1(1)n\},$$

only the convex compact set

$$\mathfrak{G}:=\mathfrak{D}\oplus\mathfrak{N} \tag{1.8}$$

is of interest for the solution of (1.6),(1.7). Therefore no additional restrictions have to be added to (1.6),(1.7) and we may consider this problem as equivalent to problem $(E,\mathfrak{G})$:

$$\text{Minimize } E(g)=\|f(g)-y\|^2 \text{ subject to } g\in\mathfrak{G}, \tag{1.9}$$

where we have made use of the notation

$$y=(\eta_1,\ldots,\eta_n)^T\ ,\quad f(g)=(f(\xi_1;g),\ldots,f(\xi_n;g))^T. \tag{1.10}$$

$\|.\|$ denotes the Euclidean norm throughout. As a matrix-norm we use the subordinate lub. Since E is not strict quasiconvex on $\mathfrak{G}$ in general (see [1]), we have to content **us** with the determination of a local minimum of E on $\mathfrak{G}$. This simplified problem will be dealt with in the following.
For general m it is not possible to give $\mathfrak{D}$ an explicit representation of the form

$$\mathfrak{D}=\{c \mid c\in\mathbb{R}^m,\ \phi_i(c)\geq 0 \text{ for } i=1(1)k,\ k\geq 1,\ \phi_i:\ \mathbb{R}^m\to\mathbb{R}\}.$$

In the cases m=1 and m=2 such a representation can be found trivially, but for m=2 it turns out that $\phi_i\notin C^2(\mathbb{R}^m)$ or that $\nabla\phi_i$ not linearly independent on $\partial\mathfrak{D}$, which is very undesirable from the view of both theory and practice.
In the first part of this paper [15] we described a simple algorithm for the minimization of E not subject to (1.7). Unfortunately the solutions obtained thereby may have poles within [-1,1] and therefore may be useless in practice. This especially occurs with high-degree approximations to "empirical" data. Therefore the infinitely many linear constraints (1.7) have to be introduced. Now for any $\varepsilon>0$ there exists a $\delta(\varepsilon)>\varepsilon$ and a finite partition $\zeta_1\leq\ldots\leq\zeta_{k(\varepsilon)}$ of [-1,1] such that

$$Q_m(\zeta_i;g)\geq\delta(\varepsilon) \text{ for } i=1(1)k(\varepsilon) \text{ implies } Q_m(\xi;g)\geq\varepsilon \text{ for } \xi\in[-1,1],$$

[11]. Therefore we may simplify problem $(E,\mathfrak{G})$ to a problem with only finitely many li-

near constraints, which then may be handled by one of the well known methods of nonlinear optimization. Pomentale [11] used such a discretization to solve the discrete rational least squares problem by a sequential unconstrained method. Unfortunately

$$k(\varepsilon)=\mathcal{O}(\varepsilon^{-1}) \text{ for } m \geq 2.$$

Therefore it will not be possible to decrease ε only quite modestly without increasing the computing effort beyond any sensible bound. Here we give a method which does not lean on such a discretization. The basic idea is to combine the modified Gauß-Newton method, chosen for its good numerical performance and good convergence properties [5] with a projection method. Moving "downhill" along a ray $g+\lambda z$ in a supporting hyperplane of $\mathcal{G}$ we occasionally will leave the feasible domain $\mathcal{G}$, which is not a polyhedron. An additional step is required to correct back with E sufficiently decreased, if g is not a constrained minimum. A new, simple and efficient algorithm for doing so will be described. The method given here generalizes to nonlinearly constrained nonlinear least squares and general nonlinear programming problems. The details will be given in a subsequent publication. Its main advantage is that it makes unnecessary knowlegde of the global behaviour of nonlinear projectors or constraining hyperfaces, in contrast to the methods in [12], [14] and yet has a global convergence property.

2. Characterization theorem

The following theorem gives a necessary and sufficient characterization of a local constrained minimum of E on $\mathcal{G}$. The necessity-part is constructive and gives an immediate basis for the algorithm to be described in the following section. For similar results in the case of only finitely many linear constraints compare [8], [12], [13].

Let g_o be given such that $E(g_o) \leqq E(0)$ and let

$$\hat{\mathcal{L}}(g):=\{c \mid c \in \mathbb{R}^p,\ E(c) \leqq E(g)\},\ \mathcal{L}(g):=\{c \mid c \in \hat{\mathcal{L}}(g),\ c \text{ path-connected to } g\} \quad (2.1)$$

$$\mathcal{G}_o := \mathcal{G} \cap \mathcal{L}(g_o) \quad (2.2)$$

$$\mathcal{G}_{o,r} := \partial\mathcal{D} \oplus \mathfrak{N} \cap \mathcal{L}(g_o) \quad . \quad (2.3)$$

Define the integer valued mapping $q: \mathcal{G} \to \{0,1,\dots,[m/2]+1\}$ by

$$\begin{aligned} &Q_m(\xi;g_D)=\varepsilon \quad \text{for } \xi \in \{\xi_1^*,\dots,\xi_q^*\}\ ,\ |\xi_i^*| \leqq 1 \text{ for } i=1(1)q \\ &Q_m(\xi;g_D)>\varepsilon \quad \text{for } \xi \in [-1,1] \setminus \{\xi_1^*,\dots,\xi_q^*\}\ . \end{aligned} \quad (2.4)$$

For any $g \in \mathcal{G}$ exactly q of the infinitely many constraints (1.7) are active. Let N_j denote the matrix whose columns are the inward normals corresponding to the first j of these constraints, namely

$$N_j^T = \begin{pmatrix} n_1^T \\ \vdots \\ n_j^T \end{pmatrix} = \begin{pmatrix} T_1(\xi_1^*),\dots,T_m(\xi_1^*),0,\dots,0 \\ \\ T_1(\xi_j^*),\dots,T_m(\xi_j^*),0,\dots,0 \end{pmatrix},\quad 1 \leqq j \leqq q\ ,\ n_i \in \mathbb{R}^p\ . \quad (2.5)$$

Elementary calculus shows that n_i, $i=1(1)q$, are linearly independent. Finally let denote

$$h=h(g):=\nabla E(g), \qquad J=J(g):=\frac{1}{2}(\nabla\nabla^T)E(g). \quad (2.6)$$

In the following, by writing $b=(b_D^T,b_N^T)^T$ for $b\in\mathbb{R}^p$ we always imply $b_D\in\mathbb{R}^m$ and $b_N\in\mathbb{R}^{l+1}$.

Theorem 1: Let $g\in\mathcal{G}_0$. g is a strong local minimum of E on $\mathcal{G}$ iff

$$(A):\quad H_q h=0 \text{ and } a_q=a_q(g):=(L^TN_q)^{\dagger}L^Th\geqq 0, \qquad (2.7)$$

supposed there holds the regularity condition $(a_q=(\alpha_1,\dots,\alpha_q)^T)$

$$z^TJz>0 \text{ for } z\in\mathcal{R}(g), \qquad (2.8)$$

where $\mathcal{R}(g):=\{y\mid y\in\mathbb{R}^p,\ y\neq 0 \text{ and } n_i^Ty=0 \text{ if } \alpha_i>0,\ n_i^Ty\geqq 0 \text{ if } \alpha_i=0 \text{ for } i=1(1)q\}$. (2.9)

In (2.7)

$$H_j=H_j(g):=-(I-BN_j(N_j^TBN_j)^{-1}N_j^T)B,\ j=1(1)q,\quad H_0:=-B, \qquad (2.10)$$

where $B=LL^T$, possibly dependent on $g\in\mathcal{G}$, is uniformly Lipschitz-continuous and uniformly positive definite on $\mathcal{G}_0$. By $A^{\dagger}$ we denote the pseudoinverse of a matrix A,

$$A^{\dagger}=(A^TA)^{-1}A^T \text{ if } A^TA \text{ is nonsingular.}$$

Remark 1: H_j is symmetric negative semidefinite and nonorthogonally projects any $x\in\mathbb{R}^p$ onto a flat parallel to the supporting hyperplane of $\mathcal{G}$ in g which is orthogonal to $\operatorname{span}(n_1,\dots,n_j)$.

Remark 2: In the case of only finitely many linear constraints condition (2.8) is a well known second order sufficiency condition, [7].

proof of theorem 1: a) q=0. This case is trivial since from (2.9) $\mathcal{R}(g)=\mathbb{R}^p\setminus\{0\}$.

b) $q\neq 0$.

b1) Sufficiency (observe that if h=0 then $c\in\mathcal{G}$, $c\neq g$ implies $c-g\in\mathcal{R}(g)$).

$H_qh=0\Rightarrow \exists b_q\in\mathbb{R}^q$: $h=N_qb_q$. Since L^TN_q is of rank q, $b_q=(L^TN_q)^{\dagger}L^Th=a_q$. Let $c\in\mathcal{G}$, $c\neq g$, $z:=c-g$. Then $g+\tau z\in\mathcal{G}$ for $\tau\in[0,1]$ and $N_q^Tz\geqq 0$. From (2.8) and the continuity of the spectrum of J with respect to g we have for sufficiently small τ_0, $1\geqq\tau_0>0$,

$$y^TJ(g+\tau z)y>0 \text{ for } y\in\mathcal{R} \text{ and } 0\leqq\tau\leqq\tau_0. \qquad (2.11)$$

Clearly, for $0\leqq\tau\leqq\tau_0$ and some θ, $0<\theta<1$,

$$E(g+\tau z)-E(g)=\tau z^Th+\tau^2z^TJ(g+\tau\theta z)z=\tau\sum_{i=1}^{q}\alpha_i(z^Tn_i)+\tau^2z^TJ(g+\tau\theta z)z, \qquad (2.12)$$

and therefore $E(g+\tau z)>E(g)$ for $z\in\mathcal{R}$, $0<\tau\leqq\tau_0$. If $z\notin\mathcal{R}$ then $\alpha_{i_0}n_{i_0}^Tz>0$ for at least one i_0, $1\leqq i_0\leqq q$. Since $\|J\|$ and $\|z\|$ are uniformly bounded on $\mathcal{G}$, we again conclude $E(g+\tau z)>E(g)$ for $z\notin\mathcal{R}$ and $0<\tau\leqq\tau_0$.

Remark 3: By replacing ">" in (2.8) by "$\geqq$" we obviously obtain a necessary condition.

b2) Necessity (observe that $\neg(A)\Rightarrow h\neq 0$). We have $\neg(A)\Rightarrow(B)\dot\vee(C)$, where

$$(B):\ -h^TH_qh>\delta$$

$$(C):\ -h^TH_qh\leqq\delta \text{ and } \alpha_k<0$$

with $\delta=\max(0,-\operatorname{sign}(\alpha_k)\alpha_k^2\beta_k^2)$ and k: $\alpha_k\beta_k=\min\{\alpha_i\beta_i\mid i=1(1)q\}$. Without loss of generality we may assume k=q. The weights β_i are specified below

$$\beta_i:=(e_i^T(N_q^TBN_q)^{-1}e_i)^{-1/2},\ i=1(1)q,\ e_i=(0,\dots,0,\underset{\uparrow_i}{1},0,\dots,0)^T\in\mathbb{R}^q. \qquad (2.13)$$

Now let

$$z=z(g):=H_{q'}h,\quad q':=\begin{cases} q & \text{if (B)} \\ q-1 & \text{if (C)} \end{cases} . \qquad (2.14)$$

Then

$$-z^T h \geqq \delta \geqq \max(\|H_q h\|^2/\|B\|,\ \alpha_q^2\beta_q^2\ ), \qquad (2.15)$$

i.e. z is a downhill-direction, $z\neq 0$. (2.15) is easily shown if one observes that

$$H_{q'}=-L(I-UU^T)L^T,$$
$$h^T H_q h = h^T H_{q-1} h + \alpha_q^2\beta_q^2\ ,$$

where $U=(u_1,\dots,u_{q'})$ and the u_i form an orthogonal basis of $\mathrm{span}(L^T n_1,\dots,L^T n_{q'})$. The second relation can be proved using a QR-decomposition of the partitioned matrix $(L^T N_{q-1}, L^T n_q)$. Let

$$\omega := \frac{1}{2}\min\{\ \|c_1-c_2\|\ |c_1\in\partial\mathfrak{D},\ c_2\in\partial\mathfrak{D}(0)\}. \qquad (2.16)$$

Clearly $\omega>0$ and $J\in C^\infty\{c\ |c\in\mathbb{R}^p,\ \|c-g\|\leqq\frac{3}{2}\omega\}$. Let

$$\kappa_0:=\max\{\|J(c)\|\ |\ \|c-g\|\leqq\frac{3}{2}\omega\}. \qquad (2.17)$$

Then $E(g)-E(g+\lambda z)\geqq\frac{1}{2}\lambda|h^T z|$ for

$$0<\lambda\leqq\lambda_0:=\min\{\omega/\|z\|,\ |z^T h|/(2\kappa_0\|z\|^2)\}.$$

We now consider the case $q'\geqq 1$ first. Let $d\in\mathring{\mathfrak{Q}}$ (e.g. $d=(d_D^T,d_N^T)^T$, $d_D=\frac{1-\varepsilon}{2m}(1,\dots,1)^T$). Clearly $Q_m(\xi;d_D)\geqq\varepsilon_d>\varepsilon$ for $\xi\in[-1,1]$. Since $g_D+\lambda z_D$ lies in a supporting hyperplane of the convex compact set $\mathfrak{D}$, for any $\lambda\in\mathbb{R}$ there exists an unique $g_D(\lambda)\in\partial\mathfrak{D}$ on the ray $\{c|\ c=d_D+\psi(g_D+\lambda z_D-d_D),\psi\geqq 0\}$. Obviously we may write

$$g_D(\lambda)=g_D+\lambda\psi(g_D,\lambda)z_D+(1-\psi(g_D,\lambda))(d_D-g_D), \qquad (2.18)$$

where $\psi(g_D,\lambda):\partial\mathfrak{D}\times\mathbb{R}\to\mathbb{R}$ is well defined for every $\lambda\in\mathbb{R}$ and $g_D\in\partial\mathfrak{D}$. If $z_D=0$ we may put $\psi(g_D,\lambda)\equiv 1$ for $\lambda\in\mathbb{R}$. Moreover, $\psi(g_D,\lambda)$ is continuous with respect to λ and $\psi(g_D,0)\equiv 1$ on $\partial\mathfrak{D}$. Since $d_D\in\mathring{\mathfrak{D}}$ and $\mathfrak{D}$ is compact, for every $\lambda^*>0$ there exists a $\psi_0>0$ such that

$$0<\psi_0\leqq\psi(g_D,\lambda)\leqq 1\quad\text{uniformly on } \partial\mathfrak{D}\times[-\lambda^*,\lambda^*]. \qquad (2.19)$$

In order to prove a sufficient decrease of E along the arc

$$\{(\lambda,g(\lambda))\ |\lambda\geqq 0,\ g(\lambda)=g+\psi(g_D,\lambda)\lambda z+(1-\psi(g_D,\lambda))(d-g)\}$$

we have to prove that $1-\psi$ decreases superlinearly in λ for λ sufficiently small, uniformly on $\partial\mathfrak{D}$.

Proposition:

$$\exists\bar\lambda>0,\ \kappa_1>0,\ \kappa_2>0:\ 1-\psi(g_D,\lambda)\leqq\kappa_1\lambda^{1+\kappa_2}\ \text{for}\ 0\leqq\lambda\leqq\bar\lambda,\ g_D\in\partial\mathfrak{D}. \qquad (2.20)$$

Proof: Suppose

$$\forall\bar\lambda>0,\ \kappa_1>0,\ \kappa_2>0\quad\exists\bar g_D\in\partial\mathfrak{D},\ \exists\lambda'\in[0,\bar\lambda]:\ 1-\psi(\bar g_D,\lambda')\geqq\kappa_1(\lambda')^{1+\kappa_2}. \qquad (2.21)$$

Since clearly we must have $\lambda'>0$ and since $\sup\{\lambda^{\kappa_2}|\kappa_2>0\}=1$ for $0<\lambda\leqq 1$ we get

$$\forall\bar\lambda\in]0,1],\ \forall\kappa_1>0\ \exists\bar g_D\in\partial\mathfrak{D},\ \exists\lambda'\in]0,\bar\lambda]:\ 1-\psi(\bar g_D,\lambda')\geqq\kappa_1\lambda'. \qquad (2.22)$$

Now from (2.18), (2.19), (2.22) for $\xi\in[-1,1]$

$$Q_m(\xi;\bar g_D(\lambda'))=Q_m(\xi;\bar g_D+(1-\psi(\bar g_D,\lambda'))(d_D-\bar g_D))+\psi(\bar g_D,\lambda')\lambda'(T_1(\xi),\dots,T_m(\xi))\bar z_D$$

and therefore

$$Q_m(\xi;\bar{g}_D(\lambda'))\geqq\varepsilon+(1-\psi(\bar{g}_D,\lambda'))(\varepsilon_d-\varepsilon)-\lambda'\sqrt{m}\|\bar{z}_D\|\geqq\varepsilon+\kappa_1\lambda'(\varepsilon_d-\varepsilon)-\lambda'\sqrt{m}\,\bar{\kappa},$$

where $\bar{\kappa}:=\max\{\|B\|\|h\| \quad | g\in\mathcal{G}_{0,r}\}$. Clearly

$Q_m(\xi;\bar{g}_D(\lambda'))\geqq\varepsilon+\lambda'\sqrt{m}\,\bar{\kappa}>\varepsilon$ if $\kappa_1\geqq 2\sqrt{m}\bar{\kappa}/(\varepsilon_d-\varepsilon)>0$, which contradicts $\bar{g}_D\in\partial\mathcal{D}$. This proves (2.20). Now for $0\leqq\lambda\leqq\min(\bar{\lambda},\lambda^*)$

$$h^T(g(\lambda)-g)=\psi(g_D,\lambda)\lambda h^Tz+(1-\psi(g_D,\lambda))h^T(d-g)\leqq\psi_0\lambda h^Tz+(1-\psi(g_D,\lambda))\kappa_3\kappa_4, \tag{2.23}$$

where $\kappa_3=\max\{\|h\| \quad |g\in\mathcal{G}_{0,r}\}$, $\kappa_4=\max\{\|d-g\| \quad |g\in\mathcal{G}_{0,r}\}$, and

$$1-\psi(g_D,\lambda)\leqq\lambda\psi_0|h^Tz|/(2\kappa_3\kappa_4) \tag{2.24}$$

for $\lambda\in[0,\tilde{\lambda}]$, $\tilde{\lambda}:=(\psi_0|h^Tz|/(2\kappa_3\kappa_4))^{1/\kappa_2}$. Therefore

$$h^T(g(\lambda)-g)\leqq\lambda\psi_0 h^Tz/2 \text{ for } 0\leqq\lambda\leqq\phi(|h^Tz|):=\min(\lambda^*,\bar{\lambda},\tilde{\lambda}). \tag{2.25}$$

Clearly ϕ is a F-function of $|h^Tz|$. Furthermore from (2.19), (2.24) for $\lambda\in[0,\phi(|h^Tz|)]$:

$$\|g(\lambda)-g\|\leqq\lambda\|z\|+\lambda\psi_0\|h\|\|z\|\|d-g\|/(2\kappa_3\kappa_4)\leqq\tfrac{3}{2}\lambda\|z\|.$$

Therefore

$$E(g(\lambda))-E(g)<\lambda\psi_0 h^Tz/2+9\lambda^2\|z\|^2\kappa_0/4\leqq\lambda\psi_0 h^Tz/4 \tag{2.26}$$

if $\lambda\in]0,\phi^*(|h^Tz|)]$, where

$$\phi^*(|h^Tz|):=\min\{\bar{\lambda},\lambda^*,\tilde{\lambda},\omega/\|z\|,|z^Th|/(9\|z\|^2\kappa_0)\}. \tag{2.27}$$

Since $\omega/\|z\|\geqq\omega/\bar{\kappa}>0$, $|z^Th|/\|z\|^2\geqq 1/\max\{\|B\| \quad |g\in\mathcal{G}_0\}>0$, ϕ^* is a F-function of $|z^Th|$ and therefore of $\max\{\|H_qh\|, -\mathrm{sign}(\alpha_q)\alpha_q^2\beta_q^2\}$. Clearly in this case g is not a constrained minimum of E on $\mathcal{G}$. Moreover, on the arc $\{(\lambda,g(\lambda))\}$ the decrease of E can be bounded from below by a F-function of $|z^Th|$. It remains to consider the case q=1, q'=0. Since $n_1^Tz=-\alpha_1\beta_1^2>0$, we have

$$\hat{\lambda}:=\max\{\lambda| \ \lambda\geqq 0,\ g_D+\lambda z_D\in\mathcal{D}\}>0. \tag{2.28}$$

Since z=-Bh, for $\lambda\in[0,\min(\lambda_0,\hat{\lambda})]$:

$$E(g)-E(g+\lambda z)\geqq\tfrac{1}{2}\lambda\ \|h\|^2/\max\{\|B^{-1}\| \quad |g\in\mathcal{G}_0\}. \tag{2.29}$$

Clearly g is not a constrained minimum. This completes the proof of theorem 1 ∎

<u>Remark 4:</u> g+z is the minimum of the quadratic

$$E(g)+(c-g)^Th+\tfrac{1}{2}(c-g)^TB^{-1}(c-g)$$

with respect to $c\in\mathbb{R}^p$ under the constraints $N_{q'}^T(c-g)=0$. In an algorithmic approach therefore $B:=\frac{1}{2}J^{-1}$ should be best. Unfortunately J will not be positive definite in general. Moreover this choice will involve numerical problems since often J is extremely illconditioned. Now

$$J=A^TA-S,\quad \text{where } A=A(g):=\frac{\partial}{\partial g^T}f(g),\quad S=\mathcal{O}(\|y-f(g)\|).$$

Therefore for sufficiently good fits $B:=\frac{1}{2}(A^TA)^{-1}$ will be a reasonable choice. Under weak assumptions it can be shown that it fulfilles the requirements of theorem 1. In addition it makes possible the computation of z, a_q, .. by stable numerical methods, using the results in [2], [3], [18].

In the following we will have to deal with sequences $(g_i)\in\mathcal{G}_0$. Using the variables introduced in this section, e.g. h, J, B, q, z, ,... the notation h_i, J_i, B_i, and so on will imply the validity of $h_i=h(g_i)$, $J_i=J(g_i)$, $B_i=B(g_i)$ and so on.

To establish the proof of convergence for the algorithm to be described in the following section it remains to be shown that for a sequence $(g_i)_{i=1}^{\infty}\in\mathcal{G}_{0,r}$ such that $q_i=1$, $q_i'=0$ hold throughout, the decrease of E can be bounded below by a F-function of $\max\{\|H_q h\|,-\mathrm{sign}(\alpha_q)\alpha_q^2\beta_q^2\}$ as has been done in theorem 1 in the case $q'\geq 1$.
Let

$$\mathcal{C}_0:=\{g \mid g\in\mathcal{G}_{0,r},\ q=q(g)=1,\ -h^T H_q h\leq\alpha_1^2\beta_1^2,\ \alpha_1<0\}.$$

$\mathcal{C}_0$ is the subset of boundary points of $\mathcal{G}_0$ where the case q=1, q'=0 applies in theorem 1. For $\rho>0$ let

$$\mathcal{C}_{0,\rho}:=\{g \mid g\in\mathcal{C}_0,\ \alpha_1^2\beta_1^2\geq\rho\ \}.$$

<u>Theorem 2:</u> If $(g_i)_{i=1}^{\infty}\in\mathcal{C}_0$ and $\hat{\lambda}_i\to 0$ as $i\to\infty$, then $|\alpha_{1,i}\beta_{1,i}|\to 0$ as $i\to\infty$, i.e. $E(g_i)-E(g_{i+1})\geq\bar{\phi}(\max\{\|H_{1,i}h_i\|,\ -\mathrm{sign}(\alpha_{1,i})\alpha_{1,i}^2\beta_{1,i}^2\ )$, where $\bar{\phi}$ is a F-function independent of i. (Compare (2.29) and, for the definition of α, β, $\hat{\lambda}$:(2.7), (2.13), (2.28))

<u>Proof:</u> Assume that for $\rho>0$ given there exists a sequence $(g_i)_{i=1}^{\infty}\in\mathcal{C}_{0,\rho}$ such that

as $i\to\infty$. Clearly $g_{i+1}=g_i-\hat{\lambda}_i B_i h_i$. Since $q_i=1$ there exists a unique inward normal

for any i, i=1,2,... . Since necessarily $m\geq 2$ in this case and $z_i=-B_i h_i$,

$\ldots_{,i}B_i n_{1,i})^{1/2}$, $\alpha_{1,i}=-n_{1,i}^T z_i/\beta_{1,i}^2$ for i=1,2,.. we get

$$z_i^T n_{1,i}=|\alpha_{1,i}\beta_{1,i}^2|\geq(\rho\times\min\{\|(T_1(\xi),\ldots,T_m(\xi))^T\|\ \xi\in[-1,1]\}/\max\{\|B^{-1}\|\ |g\in\mathcal{G}_{0,r}\})^{1/2}$$

and therefore

$$z_i^T n_{1,i}\geq\bar{\rho}>0 \text{ for } i=1,2,\ldots..$$

Since $\mathcal{G}$ is convex

$$z_i^T n_{1,i+1}<0 \quad \text{for } i=1,2,\ldots..$$

Since h and B are uniformly Lipschitz-continuous on the compact set $\mathcal{G}_0$

$$\|z_i-z_{i+1}\|\leq\kappa\hat{\lambda}_i,\quad i=1,2,\ldots..$$

Now by assumption $0\leq\hat{\lambda}_i\leq\bar{\rho}/(2\kappa)$ for $i\geq i_0\geq 1$, which yields the contradiction

$$0>z_i^T n_{1,i+1}=(z_{i+1}^T+(z_i-z_{i+1})^T)n_{1,i+1}\geq\bar{\rho}/2>0 \text{ for } i\geq i_0.$$

This proves theorem 2.■

<u>3. Computational algorithm</u>

The method to be given in this section follows directly the necessity part of the proof of theorem 1. Each step may be characterized by a triple

$$\{g,q,\mathcal{X}^*\} \tag{3.1}$$

where $g\in\mathcal{G}$, $q=q(g)\in\{0,1,\ldots,[m/2]+1\}$ and $\mathcal{X}^*=\{\xi_i^*\mid i=1(1)q\}$, where ξ_i^* are the zeroes of $Q_m(\xi;g_D)-\varepsilon$ on $[-1,1]$. The algorithm constructs a finite or infinite sequence of triples $\{g_i,q_i,\mathcal{X}_i^*\}$ such that $E(g_{i+1})<E(g_i)$ and $E(\lambda g_i+(1-\lambda)g_{i+1})\leq E(g_i)$ for $0\leq\lambda\leq 1$, i=1,2,.. . Each step $g_i\mapsto g_{i+1}$ consists of three basic parts which will be described separately. The index i is dropped in the following to simplify the notation.
The choice of d and B=B(g) is specified first. d is held fixed throughout:

$$d_D=\frac{1-\varepsilon}{2m}(1,\ldots,1)^T,\quad d_N=(\Delta(d_D)\Pi_P)^{+}y,\quad d=(d_D^T,d_N^T)^T. \tag{3.2}$$

In (3.2)

$$\Delta(c)=\mathrm{diag}^{-1}(Q_m(\xi_i;c)) \quad \text{and} \quad (\Pi_1)_{i,j}=T_j(\xi_i),\ i=1(1)n,\ j=0(1)l. \tag{3.3}$$

d_N is chosen optimal in the least squares sense. B is chosen according remark 4 in 2:

$$B=\frac{1}{2}(A^TA)^{-1}, \text{ where } A=\frac{\partial}{\partial g^T}f(g). \tag{3.4}$$

Step 1 . For $\{g,q,\mathfrak{X}\}$ given compute H_qh and a_q. This amounts in solving a linear least squares problem with q equality-constraints. If (A) is not satisfied, compute q' and z. (If q=0 then q'=0 of course.) If q'<q z may be obtained from H_qh by updating ([18]). The weights β_i, i=1(1)q, may be obtained as a byproduct of the matrix-decomposition used for computing H_qh and a_q.

Step 2 . case 1 : q'=0. Let

$$g(\lambda):=g+\lambda z. \tag{3.5}$$

Minimize $E(g(\lambda))$ with respect to λ on $[0,\hat{\lambda}]$. Any of the stepsize-algorithms mentioned in [12] may be applied. Since $E(g(\lambda))$ is a fairly simple function whose higher derivatives can easily be computed analytically, we may do even better and apply a three-term Taylor-expansion of E' with a bound for the remainder obtained by means of simple interval arithmetic to compute the smallest positive zero of odd order of E' to high accuracy. Thereby instabilities inherent in many stepsize-algorithms, see [5], can be avoided. A large amount of numerical experience indicates the use of sophisticated methods worthwhile in this application.

case 2 : q'>0. Let

$$g(\lambda):=g+z\lambda\psi(g_D,\lambda)+(1-\psi(g_D,\lambda))(d-g). \tag{3.6}$$

Minimize $E(g(\lambda))$ with respect to λ on $[0,\lambda^*]$ with side-condition $\mu g(\lambda)+(1-\mu)g\in\mathcal{L}(g)\cap\mathcal{G}$ for $0\leq\mu\leq 1$. A value of $\lambda^*=10$ has been found reasonable in practice. Zettl's method [19], originally designed for the localization of the minimum of a function along a line, may be used to bracket the minimizing value λ. One should be aware that this part of the algorithm is the most shaky one since it relies on the use of function values alone. Further study is necessary to increase both its efficiency and numerical reliability. The main obstacle here is the fact, that $E(g(\lambda))$ is only piecewise differentiable with respect to λ, where the discontinuities are not known in advance. -

Let λ_{opt} be the minimizing value found either in case 1 or case 2 and $\tilde{g}:=g(\lambda_{opt})$. The corresponding values $\tilde{q}$ and $\tilde{\mathfrak{X}}^*$ are obtained simultaneously of course by evaluation either of the restriction $\lambda\leq\hat{\lambda}$ or of $\psi(g_D,\lambda)$.

Step 3. Replace $\tilde{g}$ by

$$\begin{pmatrix} \tilde{g}_D \\ (\Delta(\tilde{g}_D)\Pi_1)^+y \end{pmatrix}. \tag{3.7}$$

This step, the recomputation of the coefficients of the numerator polynomial optimally in the least squares sense for the denominators coefficients $\tilde{g}_D$ given, has proven very effective in practice although it is superfluous from a theoretical standpoint.

This completes the determination of the new triple $\{\tilde{g},\tilde{q},\tilde{\mathfrak{X}}^*\}$.

It remains to describe the computation of $\psi(g_D,\lambda)$ according (2.18). This problem may be reformulated as follows: given $g_1\in\mathring{\mathcal{G}}$, $g_2\notin\mathring{\mathcal{G}}$, (d and $g+\lambda z$ in our case), compute $\alpha\in[0,1]$ such that $g_1+\alpha(g_2-g_1)\in\partial\mathcal{G}$. From the convexity of $\mathcal{G}$, α is unique. It may be found by the following algorithm.

Let $\xi_0\in[-1,1]$ be given and a corresponding value α_0 which is maximal in [0,1], such that

$$Q_m(\xi;g_D(\alpha_0))\geqq\varepsilon \text{ for } \xi\in[-1,\xi_0],$$
$$Q_m(\xi_i^*;g_D(\alpha_0))=\varepsilon \text{ for } i=1(1)k,\ k\geqq 0,\ \xi_1^*<\ldots<\xi_k^*\in[-1,\xi_0], \tag{3.9}$$

where

$$g(\alpha_0)=g_1+\alpha_0(g_2-g_1)\ .$$

For k=0 $\{\xi_1^*,\ldots,\xi_k^*\}$ is understood to be empty. Now a second order Taylor-expansion of $Q_m^{(j)}(\xi;g_D(\alpha_0))$ with respect to ξ about ξ_0, where $Q_m^{(j)}$ is the lowest nonvanishing derivative of Q_m at $\xi=\xi_0$ and $g_D(\alpha_0)$ and where a reasonable upper bound for $|Q_m^{(j+2)}(\zeta;g_D)|$ for $g_D:=g_D(\alpha_0)$ and $\zeta\in[\xi_0,\xi_0+\delta_\xi]$ is known, makes possible the computation of ξ_1, $\xi_0<\xi_1\leqq 1$, such that

$$Q_m(\xi;g_D(\alpha_0))\geqq\varepsilon \quad \text{for } \xi\in[-1,\xi_1], \tag{3.10}$$

where the set of ξ_j^* is possibly increased by one, ore

$$Q_m^{(1)}(\xi;g_D(\alpha_0))<0 \text{ for } \xi\in[\xi_0,\xi_1]. \tag{3.11}$$

If only the second case applies, $Q_m(\xi_1;g_D(\alpha_0))<\varepsilon$ indicates a decrease of α_0 necessary which in turn leads to $k:=1$, $\xi_1^*:=\xi_1$. ξ_1 and the possibly modified k and $\{\xi_1^*,\ldots,\xi_k^*\}$ then replace ξ_0 and the corresponding values in (3.9). - This algorithm can be initialized trivially at $\xi_0=-1$ and completes if $\xi_1=1$. Combined with some safeguards against roundoff effects it has proved very fast and safe, the effort becoming increasingly large only with increasing number or illconditioning of the ξ_i^*, which in practice will not occur. $\delta_\xi>0$ may be fixed in advance.

Remark 1: As a first triple one may choose, with no better information available,

$$\{d,0,\emptyset\}. \tag{3.12}$$

Remark 2: Since ε is a auxialary parameter only, it may be desirable or even necessary to find a final approximation with a very small value for it, e.g. a bound for the roundoff-error in evaluating Q_m. In this case one should start the minimization with a rather large value for ε, e.g. .1, and decrease this value, by *.1 e.g., every time the solution settles down on the boundary of $\mathcal{G}$ as long as the final level for ε is reached. If this is done carefully numerical problems, which unevitably occur on the boundary for ε small, especially in step 2 of the algorithm given above, can be decreased.

Remark 3: Updating techniques [8], [13], [14], [18] or the Quasi-Newton methods [8], [19], common tools in nonlinear optimization, have not been taken into account because of the following reasons:

a) We will have $\mathcal{X}_{j+1}^*\cap\mathcal{X}_j^*=\emptyset$ in general.

b) The effort necessary to compute A(g) is very small, if one computes

Π_k for k=max(1,m) once for all. Since E(g) has to be computed anyway, A(g) then can be obtained with at most 2kn additional multiplications.

c) The numerical stability of the present method, if implemented reasonably, is much better than that of Quasi-Newton methods. The situation here is the same as with orthogonalization and iterative refinement [2], [3] compared with the direct solution of the normal equations in the linear least squares case.

Remark 4: The algorithm given for the computation of $\psi(g_D,\lambda)$ equally applies for the evaluation of the restriction $\lambda\leq\hat{\lambda}$.

Remark 5: The minimization algorithm given in this section may be considered as a method of feasible directions in the sense of Zoutendijk [20]. The construction of $g(\lambda)$ (3.6) corresponds to his antizigzagging requirement AZ3.

4. Convergence of the algorithm.

In order to prove the convergence of the method described we need some regularity condition which will be specified first.

Definition: $g^*\in\mathcal{G}$ is said to be stationary ("S") iff $H^*_{q^*}h^*=0$.

Definition: $g^*\in\mathcal{G}$ is said to be stationary with nonnegative multipliers ("SNNM") iff

$$h^*=\begin{cases}0 \text{ if } q^*=0\\ N_{q^*}a^*_{q^*},\ a^*_{q^*}\geq 0 \text{ for } q^*>0\ .\end{cases}$$

Definition: $(E,\mathcal{G})$ is said to be regular iff

$\Omega=\{g\,|\,g\in\mathcal{G}$ is "S"$\}$ is finite,

$z^TJ^*z>0$ for $z\in\tilde{\mathcal{R}}(g^*)$ for any g^* which is "SNNM".

Lemma 1: $g^*\in\mathcal{G}$ is "SNNM" iff $\sigma(g^*)=0$, where

$$\sigma(g)=\min(\|h\|,\max(\|H_qh\|,-\text{sign}(\alpha_q)\alpha_q^2\beta_q^2)). \tag{4.1}$$

Lemma 2: Let $(E,\mathcal{G})$ be regular. $g^*\in\mathcal{G}$ is a strong local minimum of E on $\mathcal{G}$ iff g^* is "SNNM".

Proofs: For the case $q^*>0$ Lemmata 1,2 directly follow from theorem 1. The case $q^*=0$ is trivial.∎

Lemma 3: Let g_0 be given such that for $g\in\mathcal{G}_0$:

a) $f(\xi;g)$ irreducible

b) $\partial P_1=1$, $\partial Q_m=m$. (4.2)

Then $B=(A^TA)^{-1}/2$ is uniformly Lipschitz-continuous and uniformly positive definite on $\mathcal{G}_0$.

Proof: From (4.2) and (1.2) it follows that A is of rank p (uniqueness of rational interpolation over $\mathcal{G}_0$). Therefore A^TA is positive definite on $\mathcal{G}_0$. Since A is differentiable on $\mathcal{G}_0$ and $\mathcal{G}_0$ is compact the proposition follows.∎

Remark: (4.2) constitutes no essential restriction, since a degenerate $R_{1,m}$ is a best approximation only to itself in our case, as shown by Dunham [6].

Lemma 4: Let g_0 be given such that for $g\in\mathcal{G}_0$ (4.2) is valid. For a sequence $(g_i)_{i=1}^{\infty}\subset\mathcal{G}_0$

such that

$$\text{a) } g_{i+1}=g_i-\lambda_i B_i h_i \ , \ B_i=\frac{1}{2}(A_i^T A_i)^{-1} \qquad\qquad (4.3)$$

$$\text{b) } g_i\in \mathring{\mathcal{G}}_0 \ , \ i=1,2,..,$$

where λ_i has been chosen by one of the stepsize-algorithms mentioned in [12]. Then

$$E(g_i)-E(g_{i+1})\geq\bar{\phi}\ (\|h_i\|),$$

where $\bar{\phi}$ is a F-function independent of i.

Proof: This follows immediately from the results in [12], observing that by condition b) the unconstrained case applies.■

Theorem 3: Let $(E,\mathcal{G})$ be regular and let $g_0\in\mathcal{G}$ be given such that for $g\in\mathcal{G}_0$ (4.2) is valid and $E(g_0)\leq E(0)$. Then the algorithm given in 3. converges globally in $\mathring{\mathcal{G}}_0$ in the sense that for every triple $\{g_1,q_1,\mathcal{X}_1^*\}$, where $g_1\in\mathcal{G}_0$, it constructs a sequence of triples $\{g_i,q_i,\mathcal{X}_i^*\}$ such that either $g_i=\bar{g}$ for $i=i_0$, for some $i_0\geq 1$, or $g_i\to\bar{g}$ for $i\to\infty$, where $\bar{g}$ is a strong local minimum of E on $\mathcal{G}$.

Proof: From Lemma 3 it follows that $B=\frac{1}{2}(A^TA)^{-1}$ is uniformly Lipschitz-continuous and uniformly positive definite on $\mathcal{G}_0$. Now either (A) is satisfied after a finite number of i_0 steps or a infinite sequence of triples $\{g_i,q_i,\mathcal{X}_i^*\}_1^\infty$ arises, where by construction

$$E(g_{i+1})<E(g_i)$$
$$E(\lambda g_{i+1}+(1-\lambda)g_i)\leq E(g_i),0\leq\lambda\leq 1. \qquad i=1,2,...$$

Because Lemmata 1,2 and theorem 1 apply it remains to be proofed that for a infinite subsequence of steps $g_{i_j}\mapsto g_{i_j+1}$

$$E(g_{i_j})-E(g_{i_j+1})\geq\mu(\sigma(g_{i_j})),\ j=1,2,...$$

where μ is a F-function independent of j. Now either there exists a infinite subsequence of steps according (3.6) or (4.3), or theorem 2 applies. Now (2.26), (2.27), the results of theorem 2 and Lemma 4 together with the comparison-principle [12], applied to the stepsize-algorithm given, complete the proof.

Remark: One should observe that the algorithm described uses local information only, which indeed can easily be obtained by numerical computation.

5. Numerical results

In order to demonstrate the main features of the present method we give the results obtained for the following fairly simple problem:

$$l=0,\ m=2,\ n=5,$$
$$\{(\xi_i,\eta_i)\}=\{(-1,\tfrac{1}{50}),(-\tfrac{2}{5},-1),(0,\tfrac{1}{25}),(\tfrac{1}{2},\tfrac{1}{125}),(1,\tfrac{1}{300})\}\ .$$

Here $\eta_i=\hat{R}_{0,2}(\xi_i)=\frac{1}{25}\ \frac{1}{6\xi_i^2+5\xi_i+1}$. $\hat{R}_{0,2}$ has poles at $\xi=-\frac{1}{2}$ and $\xi=-\frac{1}{3}$. Clearly the best fit $f(\xi;\bar{g})$ in $\mathcal{G}(\varepsilon)$ must be of the form

$$-\frac{\bar{\alpha}}{(\xi-\bar{\xi})^2+\bar{\beta}}\ , \qquad (5.1)$$

where $\bar{\alpha}>0$, $\bar{\beta}>0$, $\bar{\alpha}\to 0$, $\bar{\beta}\to 0$, $\bar{\xi}\to -2/5$ and $|\bar{\alpha}-\bar{\beta}|\to 0$ as $\varepsilon\to 0$.

$E(g)>.2075\bar{1}-2$ [1]) for $g\in\mathcal{G}(\varepsilon)$ and $\varepsilon>0$.

Problem (1.9) has been solved for a decreasing finite sequence of ε-values $\varepsilon_i \geqq \varepsilon_{min}$, where $\varepsilon_{min}=.5-7$ has been chosen so that

$$Q_2(\xi;g)-|fl(Q_2(\xi;g))-Q_2(\xi;g)| \geq \bar{\varepsilon}>0$$

holds for all $\xi\in[-1,1]$, $g\in\dot{\mathcal{G}}(\varepsilon)$ and $\varepsilon\geqq\varepsilon_{min}$. Here fl(.) denotes the result of computer arithmetic.

d and g_1 purposely have been chosen very poor, namely d=0 and $g_{D,1}$ in the (unique) corner of $\mathcal{D}(\varepsilon_1)$, almost opposite the final solution. Details of the results are given in table 1. Only the last three columns of table 1 need further comment. Since any linear system arising in step 1 of the algorithm has been solved by Björck's and Golub's method [3], from their results a roundoff estimate for the computed direction z_i is available if one assumes the precision in the evaluation of A(g) and N_q given. This estimate ($\|\Delta z_i\|$) has been incorporated into table 1 assuming that A(g) and N_q are computed to machine precision. ζ_i is the absolute value of the absolutely smallest component of z_i. Since the absolute values of the absolutely largest components of z_i are smaller than $60\zeta_i$ with the exception of steps 5, 21, 22 and 24, there the corresponding factors are 120, 100, 136 and 685 respectively, it can be assumed that the computed direction will be perturbed severely if ζ_i and $\|\Delta z_i\|$ have the same order of magnitude. In this roundoff analysis as a lower bound for $cond_2(A_i)$

$$\kappa_i := \max_{1\leqq k,j\leqq p} \left| \frac{\rho_{kk}^{(i)}}{\rho_{jj}^{(i)}} \right|$$

has been used, where $A_i=Q_iR_i$ is a QR-decomposition with $Q_i^TQ_i=I$, $R=(\rho_{jk}^{(i)})$. A large amount of test results indicates that κ_i gives a quite realistic estimate for $cond_2(A_i)$ in this application.

Obviously the progress of minimization is fairly good for $\varepsilon\geqq.5-2$. In the last four steps however the unimodal minimization fails due to large errors in the computed direction. Therefore a decrease of ε is enforced followed by an ordinary Gauß-Newton-step (3.5). Of course these can bring no much help here because of the small $\hat{\lambda}_i$. This illustrates our warning to avoid steps of type (3.6) for very small ε. Nevertheless the final accuracy must be regarded very good taking in mind that our problem is not well posed for $\varepsilon=0$. We have

$E(\bar{g})=.2147-2$,

$f(\xi;\bar{g})=(-.1088-3)/((\xi-\bar{\xi})^2+(.2901-7))$,

$\bar{\xi}=-.3896$,

ξ	-.5	-.4	$\bar{\xi}$	-.3
f	-.89-2	-.10+1	-.37+4	-.14-1

Zeroes of $Q_2(\xi;\bar{g})$: $\bar{\xi}\pm i\times.1703-3$.

This computation took a total of 15.6 sec. 451 boundary points have been computed in the course of unimodal minimization on $\partial\mathcal{G}$.

[1]) denotes $2.075\bar{1}\times10^{-3}$

computer: TR 440, RHR Kaiserslautern

macheps: 2^{-38}

arithmetic: floating point with postnormalization and rounding

base of mantissa: 16

Table 1

i	E_i	$\gamma_{1,i}$	$\gamma_{2,i}$	$\gamma_{3,i}$	ε_i, q_i, q_i'	$\mathfrak{X}_i^*$	$z_i^T h_i$	λ_i	$\hat{\lambda}_i, \psi_i$	κ_i	ζ_i	$\Vert \Delta z_i \Vert$	
1	.1271+1	o	-.5ooo	-.25oo	.5, 2,2	-1,1	-.631	.1oo+1	.1oo+1	.424+1	.24-2	.46-8	1)
2	.956o	o	-.5ooo	-.6913-1	.5, 2,1	-1,1	-.137	.153+1	.753	.153+2	.65-1	.43-7	
3	.64o4	.6866	.3o98	-.1961	.5, 1,1	-.554o	-.2o2	.256	.966	.1o1+2	.37-2	.24-7	
4	.6147	.6o83	.3775	-.2o38	.5, 1,1	-.4o28	-.138-2	.2o2	.1oo+1	.961+1	.32-3	.21-7	
5	.6146	.6ooo	.3823	-.2o39	.5, 1,1	-.3923	-.128-8	.326	.1oo+1	.961+1	.32-6	.21-7	R, 2)
6	.6146	.6ooo	.3823	-.2o39	.5-1,o,o	---	---	.292	---	.961+1	.33	.18-7	
7	.1988	.1o93+1	.6382	-.1o55	.5-1,o,o	---	---	.418	.359	.124+2	.12	.35-8	
8	.5972-1	.12o9+1	.6822	-.4995-1	.5-1,1,1	-.443o	-.217-1	.128	.999	.426+2	.16-1	.7o-8	
9	.5817-1	.1185+1	.6986	-.4841-1	.5-1,1,1	-.4242	-.628-4	.1o1	.1oo+1	.471+2	.19-2	.72-8	
1o	.5814-1	.1181+1	.7o12	-.4823-1	.5-1,1,1	-.4212	-.138-7	.29o	.1oo+1	.477+2	.79-5	.72-8	R
11	.5814-1	.1181+1	.7o13	-.4823-1	.5-2,o,o	---	---	.519	.519	.477+2	.36-1	.5o-8	
12	.1o61-1	.1242+1	.72oo	-.1357-1	.5-2,o,o	---	---	.392+1	.268	.545+3	.97-3	.24-7	
13	.5638-2	.1258+1	.72o2	-.693o-2	.5-2,1,1	-.4367	-.1o3-1	.355-1	.1oo+1	.2o1+4	.18-1	.17-6	
14	.53o9-2	.125o+1	.7257	-.6357-2	.5-2,1,1	-.43o7	-.966-2	.272-1	.1oo+1	.245+4	.17-1	.18-6	
15	.5o74-2	.1243+1	.73o6	-.5931-2	.5-2,1,1	-.4254	-.91o-2	.2o6-1	.1oo+1	.288+4	.17-1	.18-6	
16	.49o8-2	.1237+1	.7349	-.5617-2	.5-2,1,1	-.42o6	-.856-2	.152-1	.1oo+1	.327+4	.17-1	.16-6	
17	.4794-2	.1231+1	.7388	-.539o-2	.5-2,1,1	-.4164	-.79o-2	.1o9-1	.1oo+1	.361+4	.16-1	.14-6	
18	.472o-2	.1225+1	.7421	-.5232-2	.5-2,1,1	-.4127	-.691-2	.774-2	.1oo+1	.389+4	.15-1	.12-6	
19	.4676-2	.1221+1	.7451	-.5128-2	.5-2,1,1	-.4o96	-.527-2	.552-2	.1oo+1	.41o+4	.13-1	.11-6	
2o	.4653-2	.1217+1	.7474	-.5o64-2	.5-2,1,1	-.4o69	-.284-2	.438-2	.1oo+1	.424+4	.93-2	.11-6	
21	.4644-2	.1214+1	.7492	-.5o29-2	.5-2,1,1	-.4o5o	-.7o5-3	.445-2	.1oo+1	.433+4	.39-2	.12-6	
22	.4642-2	.1212+1	.75o3	-.5o14-2	.5-2,1,1	-.4o38	-.319-4	.5o3-2	.1oo+1	.438+4	.66-3	.13-6	R
23	.4642-2	.1212+1	.75o1	-.5o11-2	.5-3,o,o	---	---	.928	.291	.439+4	.4o-2	.1o-6	
24	.229o-2	.121o+1	.7579	-.5oo8-3	.5-3,1,1	-.3992	-.369-2	.12o-3	.1oo+1	.445+6	.56-2	.24-5	R
25	.229o-2	.1211+1	.7576	-.5oo2-3	.5-4,o,o	---	---	.46o-1	.3oo-1	.445+6	.11-1	.13-5	
26	.2152-2	.1198+1	.7659	-.1772-3	.5-4,1,1	-.39o9	-.427-2	---	---	.336+7	.97-2	.89-2	F 3)
					.5-5,o,o	---	---	.168-1	.9o3-3	.336+7	.1o-1	.48-2	
27	.2148-2	.1196+1	.767o	-.1678-3	.5-5,1,1	-.3897	-.426-2	---	---	.41o+7	.96-2	.11-1	F
					.5-6,o,o	---	---	.159-1	.787-4	.41o+7	.1o-1	.6o-2	
28	.2147-2	.1195+1	.7671	-.167o-3	.5-6,1,1	-.3896	-.426-2	---	---	.414+7	.96-2	.11-1	F
					.5-7,o,o	---	---	.158-1	.777-5	.414+7	.1o-1	.61-2	
29	.2147-2	.1195+1	.7671	-.167o-3	.5-7,1,1	-.3896	-.426-2	---	---	.415+7	.96-2	.11-1	F, 4)

1) replacement of numerator coefficients, 2) reduction of ε because solution settles down (after computation of g_{i+1})
3) reduction of ε because of failure in unimodal minimization (before computation of g_{i+1}), 4) $\varepsilon_{29}=\varepsilon_{min}$: stop.

Remark: From the analysis in 2. and 4. it follows that we may use a variable d=d(g) too, if only it is assured that

$$Q_m(\xi;d(g)) \geqq \varepsilon(1+\theta),\ 0<\theta\quad,\ \text{for } \xi\in[-1,1] \text{ and } g\in\mathcal{G}_0. \tag{5.2}$$

Clearly a "good" choice would be d(g)-g in the cone of directions of decrease of E at g. However such a d(g) could be found only in the course of the computation of g(λ) itself, see (2.18) and (2.25). A much simpler strategy (where $-(d_i-g_i)$ will be "nearly" in this cone at least for sufficiently large i) would be

$$\begin{aligned} &d_0 := d_1 := g_1 \in \mathring{\mathcal{G}},\\ &d_{i+1} := (\theta_i d_i + (1-\theta_i) g_{i+1})(1-\theta) + \theta d_0,\\ &0 \leqq \theta_i \leqq 1,\ 0<\theta<<1. \end{aligned} \tag{5.3}$$

It serves the purpose "not too much" to disturb the downhill property of g(λ)-g for reasonable large λ. This strategy has been used for the example above with $\theta_i=1/2$ and θ=0 (!), giving after only 23 steps with ε_{23}=.5-7 a $\hat{g}$ where

$E(\hat{g})$=.2089-2,

$f(\xi;\hat{g})=(-.2174-4)/((\xi-\hat{\xi})^2+(.3277-7))$,

$\hat{\xi}$=-.3953,

ξ	-.5	-.4	$\hat{\xi}$	-.3
f	-.24-2	-.10+1	-.66+3	-.20-2

,

and the zeroes of $Q_2(\xi;\hat{g})$: $\hat{\xi}\pm i\times$.181-3 .

Of course neither $Q_2(\bar{\xi};\bar{g})$=.5-7 nor $Q_2(\hat{\xi};\hat{g})$=.5-7 due to roundoff effects.

A large set of further examples has been run successfully, with n up to 166, l up to 8 and m up to 5. For increasing n and m however the problem of occurence of several local minima of E on $\partial\mathcal{G}$ becomes more and more pronounced. The problem of finding the global minimum of E, i. e. the global minimum of a general nonlinear nonconvex function, has not yet been solved satisfactory.

References

1. Barrodale, I. B. : Best Rational Approximation And Strict Quasi-Convexity, SIAM J. Numer. Anal. 10, (1973), 8-12.
2. Björck, Å. : Iterative Refinement Of Linear Least Squares Solutions II, BIT 8, (1968), 8-30.
3. Björck, Å., Golub, G. H. : Iterative Refinement Of Linear Least Squares Solutions By Householder Transformation, BIT 7, (1967), 322-337.
4. Braun, B.: Nichtlineare Gauß-Approximation, Lösungsverfahren und Kondition, mit Anwendung auf Exponentialsummenapproximation, Thesis, Mainz 1967.
5. Dietze, S., Schwetlick, H.: Über die Schrittweitenwahl bei Abstiegsverfahren zur Minimierung konvexer Funktionen, ZAMM 51, (1971), 451-454.
6. Dunham, C. B.: Best Discrete Mean Rational Approximation, Aeq. Math. 11, (1974), 8-12.
7. Fiacco, A. V., McCormick, G. P.: Nonlinear Programming (Sequential Unconstrained Minimization Techniques), New York, London, Toronto, Sydney: J. Wiley (1968)

8. Goldfarb, D.: Extension Of Davidon's Variable Metric Method To Maximization Under Linear Inequality And Equality Constraints, SIAM J. Appl. Math. 17, (1969), 739-764.
9. Golub, G. H., Pereyra, V.: The Differentiation Of Pseudo-Inverses And Nonlinear Least Squares Problems Whose Variables Separate, SIAM J. Numer. Anal. 10,(1973), 413-432.
10. Hartley, H. O.: The Modified Gauß-Newton Method For The Fitting Of Nonlinear Regression Functions By Least Squares, Technometrics 3, (1961), 269-280.
11. Pomentale, T.: On Rational Least Squares Approximation, Num. Math. 12, (1968), 40-46.
12. Rauch, S. W.: A Convergence Theory For A Class Of Nonlinear Programming Problems, SIAM J. Numer. Anal. 10, (1973), 207-228.
13. Rosen, J. B.: The Gradient Projection Method For Nonlinear Programming, Part I: Linear Constraints. SIAM J. Appl. Math. 8,(1960), 181-217.
14. Rosen, J. B.: The Gradient Projection Method For Nonlinear Programming, Part II: Nonlinear Constraints. SIAM J. Appl. Math. 9,(1961), 514-532.
15. Spellucci, P.: Algorithms For Rational Discrete Least Squares Approximation, Part I: Unconstrained Optimization. To be published.
16. Spellucci, P.: Über den Koeffizientenbereich gewisser positiver Polynome. To be published.
17. Spellucci, P.: Einige neue Ergebnisse auf dem Gebiet der diskreten rationalen Approximation, Thesis, Ulm 1972.
18. Stoer, J.: On The Numerical Solution Of Constrained Least Squares Problems, SIAM J. Numer. Anal. 8,(1971), 382-411.
19. Zettl, G.: Ein Verfahren zum Minimieren einer Funktion bei eingeschränktem Variationsbereich der Parameter, Num. Math. 15, (1970), 415-432.
20. Zoutendijk, G.: Methods Of Feasible Directions: A Study In Linear And Nonlinear Programming. Amsterdam: Elsevier (1960).

Dr. Peter Spellucci
Johannes Gutenberg-Universität
Fachbereich Mathematik
6500 Mainz
Saarstr. 21

Über das Prinzip der Eindeutigen Fortsetzbarkeit in der Kontrolltheorie

Norbert Weck

In der Kontrolltheorie ergeben sich einige neue Fragestellungen, wenn die betrachteten Systeme nicht mehr durch gewöhnliche, sondern durch partielle Differentialgleichungen beschrieben werden. So tritt bei der Auswertung des Pontryaginschen Maximumprinzips zur Charakterisierung der optimalen Steuerung ("bang-bang-Sätze") das Problem auf, zu zeigen, daß für die betrachteten Differentialgleichungen ein Prinzip der eindeutigen Fortsetzbarkeit gilt. Dieses besagt, daß Lösungen, die lokal sehr stark gegen Null gehen, identisch verschwinden müssen. Solche Ergebnisse gibt es in zwei im wesentlichen äquivalenten Versionen:

(a) Lösungen, die in einer nichtleeren offenen Menge verschwinden, verschwinden überall.

(b) Lösungen, deren Cauchydaten (Ableitungen bis zu einer bestimmten Ordnung) auf einem Flächenstück S verschwinden, verschwinden überall.

Sätze dieser Art sind für große Klassen von Differentialgleichungen bekannt (vgl. z.B. [3], [6]). Für die Zwecke der Kontrolltheorie muß man sie jedoch verschärfen und die folgende Version beweisen:

(b') Lösungen, von denen ein Teil der Cauchydaten auf einem Flächenstück S verschwindet, während der Rest auf einer Teilmenge $\tilde{S} \subset S$ von positivem Maß gleich Null ist, verschwinden überall.

Wir wollen im folgenden über zwei Kontrollprobleme berichten, bei denen ein Ergebnis (b') wichtig ist. Beim ersten betrachten wir die Wärmeleitungsgleichung. Hier läßt sich (b') bisher nicht direkt beweisen, sondern man muß analytische Daten voraussetzen, um dann über die Analytizität der Lösung von (b) zu (b') zu gelangen. Dies ist eine bekannte und in der Kontrolltheorie übliche Vorgehensweise (vgl. etwa [8]). Die Besonderheit an unserem Wärmeleitungsproblem ist, daß wir mit Hilfe einer L_∞-Randsteuerung ein Minimum-Norm-Problem bezüglich der Supremumsnorm untersuchen. Es handelt sich um die in [11] nur skizzierte Verallgemeinerung von [11] auf n Dimensionen. Das Problem, einen Fortsetzungssatz (b') direkt zu zeigen, besteht aber auch für andere Versionen des Wärmeleitungsproblems und ist noch offen.

Im zweiten Teil untersuchen wir ein Kontrollproblem im Zusammenhang mit einer elliptischen Differentialgleichung zweiter Ordnung. Hier folgt die Kontrolltheorie den bekannten Wegen. Wir können aber hier ein Ergebnis (b') direkt beweisen, ohne Analytizität voraussetzen zu müssen. Dies läßt sich durch Zurückführung auf eine schärfere Version von (a) bewerkstelligen, die in [2] gezeigt wurde.

Wir wollen schon hier die Generalvoraussetzung machen, daß alle im folgenden auftretenden Koeffizienten und Gebietsränder zur Klasse C_∞ gehören. Dies ist an einer Stelle wesentlich, wo wir es noch einmal besonders anmerken. Ansonsten würden wir auch mit genügend hohen endlichen Differenzierbarkeitsordnungen auskommen.

1. Ein Kontrollproblem aus der Wärmeleitung

Wir führen die folgenden Bezeichnungen ein

$$Lw(x) := \sum_{i,j=1}^{n} \partial_i (a_{ij}(x) \partial_j w(x)) + a(x) w(x)$$

$\partial_i := \frac{\partial}{\partial x_i}$; (a_{ij}) sei symmetrisch und gleichmäßig positiv definit.

$\Omega \subset \mathbb{R}^n$ sei ein Gebiet; $T \in \mathbb{R}^+$

$G := (o,T) \times \Omega \subset \mathbb{R}^{n+1}$; $\Gamma := (o,T) \times \partial\Omega$

$\dot{\partial} w(\xi) := \sum_{i,j} n_i(\xi) a_{ij}(\xi) \partial_j w(\xi)$; $\xi \in \partial\Omega$; $n(\xi)$: Normale

Wir untersuchen das folgende Kontrollproblem:

$$(1) \qquad \begin{aligned} -\partial_t y(t,x) + L_x y(t,x) &= o & (t,x) &\in G \\ \dot{\partial} y(t,\xi) + b y(t,\xi) &= b\, u(t,\xi) & (t,\xi) &\in \Gamma \\ y(o,x) &= o & x &\in \Omega \end{aligned}$$

Die Steuerung u variiert in

$$U := \{ u \in L_\infty(\Gamma) \mid |u| \le 1 \}$$

Für die zu u gehörende Lösung schreiben wir $y(u;t,x)$.

Gegeben ist $z \in C_o(\bar{\Omega})$. Optimalitätskriterium ist

$$(*) \qquad \| y(u;T,\cdot) - z \|_{C_o(\bar{\Omega})} \overset{!}{=} \min$$

Die verallgemeinerte Lösung von (1) kann mit Hilfe einer Greenschen Funktion g definiert werden.

$$y(u;t,x) = \int_\Gamma g(t,x;\tau,\xi)\, u(\tau,\xi)\, d\tau\, d\xi$$

Es seien v_k die Eigenlösungen zu

$$Lv_k(x) + \lambda_k v_k(x) = o \qquad x \in \Omega$$

$$\partial v_k(\xi) + b v_k(\xi) = o \qquad \xi \in \Omega$$

Dann gilt

$$g(t,x;\tau,\xi) = b \sum_{k=1}^{\infty} e^{-\lambda_k(t-\tau)} v_k(x)\, v_k(\xi)$$

Für jedes $\varepsilon > o$ konvergiert diese Reihe mitsamt allen ihren Ableitungen gleichmäßig in $[o,t-\varepsilon] \times \bar{\Omega}$. Dies ergibt sich aus den folgenden Eigenschaften der λ_k und v_k, die aus [1] (theorem 14.6) und bekannten Regularitätssätzen (vgl. auch [1]) folgen:

(2) $\qquad \lambda_k = c\, k^{2/n} + o(k^{2/n}) \;,\; k \to \infty$

(3) $\qquad |v_k(x)| \leq C \cdot k^{\beta} \qquad \beta \in \mathbb{R}^+ \;,\quad x \in \bar{\Omega}.$

Entsprechendes gilt für alle Ableitungen von v_k.
Aus dem Maximumprinzip (vgl. [10], p. 174, theorem 6) folgt weiter für g :

(4) $\qquad g$ ist stetig und positiv für $o \leq \tau < t \;,\; x \in \bar{\Omega} \;,\; \xi \in \partial\Omega$

(5) $\qquad g(T,x;\cdot,\cdot) \in L_1(\Gamma)$

<u>Lemma 1:</u> Es gibt eine Funktion ε mit $\varepsilon(\delta) \to o$ für $\delta \to o$, so daß für $u \in U$ gilt:

$$u(\tau,\xi) = o \text{ für } \tau \leq T-\delta \quad \Rightarrow \quad \| y(u;T,\cdot) \|_{C_o(\bar{\Omega})} \leq \varepsilon(\delta) .$$

Beweis:

$$\chi_\delta(\tau,\xi) := \begin{cases} 1 & \text{für } o \leq \tau \leq T-\delta \\ o & \text{sonst} \end{cases}$$

Dann folgt aus (4)

$$y(\chi_\delta;T,\cdot) \in C_o(\bar{\Omega})$$

$$y(\chi_{\delta_1};T,x) \leq y(\chi_{\delta_2};T,x) \leq y(1;T,x) \qquad \text{für } \delta_1 < \delta_2$$

Aus (5) folgt mit Hilfe des Satzes von Lebesgue

$$\lim_{\delta \to o} y(\chi_\delta;T,x) = y(1;T,x)$$

Wegen bekannter Regularitätssätze gilt $y(1;T,\cdot) \in C_o(\bar{\Omega})$. Deshalb können wir den Satz von Dini anwenden und erhalten:

$$\varepsilon(\delta) := \| y(1;T,\cdot) - y(\chi_\delta;T,\cdot) \|_{C_o(\bar{\Omega})} \to o$$

Für $u \in U$ mit $u(\tau,\xi) = o$ für $\tau \leq T-\delta$ gilt aber wegen (4):

$$|y(u;T,x)| \leq y(1-\chi_\delta;T,x) \qquad \text{also}$$

$$\| y(u;T,\cdot) \|_{C_o(\bar{\Omega})} \leq \varepsilon(\delta) \to o$$

<u>Korollar:</u> $u \in U \;\Rightarrow\; y(u;T,\cdot) \in C_o(\bar{\Omega})$.

Lemma 2: Die lineare Hülle V von $\{v_k \mid k \in \mathbb{N}\}$ liegt dicht in $C_o(\bar{\Omega})$.

Beweis: Jedes $f \in C_o(\bar{\Omega})$ läßt sich als Fourierreihe in den v_k schreiben; die Konvergenz braucht aber nicht gleichmäßig zu sein. Lemma 2 ist bewiesen, wenn es gelingt, eine dichte Teilmenge $F \subset C_o(\bar{\Omega})$ anzugeben, derart, daß für $w \in F$ die Fourierreihe gleichmäßig konvergiert. Dies ist der Fall für hinreichend großes $r \in \mathbb{N}$ und

$$F := \{ w \in C_\infty(\bar{\Omega}) \mid (\dot{\partial}+b)w = (\dot{\partial}+b)Lw = \dots = (\dot{\partial}+b)L^r w = o \}$$

Für $w \in F$ gilt nämlich

$$|(w,v_k)| = |\lambda_k^{-r-1} (L^{r+1}w,v_k)| \leq K \cdot |\lambda_k|^{-r-1}$$

und die gleichmäßige Konvergenz folgt aus (2) und (3).

Den Lösungsoperator $u \longmapsto y(u;,T,\cdot)$ bezeichnen wir mit S.

Im folgenden denken wir uns Funktionen aus $L_2((o,T-\delta) \times \partial\Omega)$ durch Null auf Γ fortgesetzt. Wegen (4) sind dann die Abbildungen

$$\begin{aligned} S_\delta : L_2((o,T-\delta) \times \partial\Omega) &\longrightarrow C_o(\bar{\Omega}) \\ u &\longmapsto Su \end{aligned}$$

kompakt. Da die Mengen

$$U_\delta := \{ \chi_\delta \cdot u \mid u \in U \}$$

in $L_2((o,T-\delta) \times \partial\Omega)$ schwach kompakt sind, sind die SU_δ kompakt in $C_o(\bar{\Omega})$, und wir erhalten aus Lemma 1:

Satz 1: SU ist kompakt in $C_o(\bar{\Omega})$. Das Kontrollproblem $(*)$ besitzt eine Lösung.

Im folgenden wollen wir nur noch den Fall $z \notin SU$ betrachten. Es sei $\hat{u}$ eine Lösung zu z und $\hat{y} := S\hat{u}$ sowie $\rho := \| z - \hat{y} \|$. Dann sind SU und $K := \{w \in C_o(\bar{\Omega}) \mid \|w-z\| < \rho\}$ disjunkte konvexe Mengen und K enthält innere Punkte. Wir können daher einen Trennungssatz ([7], vgl. auch [5]) anwenden und erhalten die Existenz eines stetigen linearen Funktionals $\alpha \in C_o(\bar{\Omega})'$ mit $\alpha K < o$ und $\alpha(SU) \geq o$. Wegen $\hat{y} \in SU \cap \bar{K}$ gilt $\langle \alpha,\hat{y} \rangle = o$ und es folgt:

$$(6) \qquad \begin{aligned} \langle \alpha , S\hat{u}-Su \rangle &\leq o \qquad \text{für } u \in U \\ \langle S'\alpha, \hat{u}-u \rangle &\leq o \qquad \text{für } u \in U \end{aligned}$$

Wir wollen noch $S'\alpha$ bestimmen. Hierzu bilden wir

$$w(\alpha;\tau,x) := \sum_k \langle \alpha,v_k \rangle \; e^{-(T-\tau)\lambda_k} v_k(x)$$

Wegen (2) und (3) konvergiert diese Reihe gleichmäßig mit allen ihren Ableitungen in $[o,T-\delta] \times \bar{\Omega}$. Hieraus folgt

$$(7) \qquad \begin{aligned} (-\partial_t - L_x)\, w(\alpha;t,x) &= o \qquad t < T , \quad x \in \Omega \\ (\dot{\partial}+b)w(\alpha;t,\xi) &= o \qquad t < T , \quad \xi \in \partial\Omega \end{aligned}$$

Für $v \in U$ gilt:

$$\langle S'\alpha, v\rangle = \langle \alpha, Sv\rangle = \lim_{\delta \to o} \langle \alpha, S(\chi_\delta v)\rangle =$$

$$= \lim_{\delta \to o} \sum_k \langle \alpha, v_k\rangle \int_\Gamma e^{-(T-\tau)\lambda_k} v_k(\xi)\ b\ v(\tau,\xi)\ \chi_\delta(\tau)\ d\tau\ d\xi$$

$$= \lim_{\delta \to o} \int_\Gamma w(\alpha;\tau,\xi)\ b\ \chi_\delta(\tau)\ v(\tau,\xi)\ d\tau\ d\xi$$

Für $v(\tau,\xi) = \operatorname{sgn} w(\alpha;\tau,\xi)$ folgt hieraus

$$w(\alpha;\cdot,\cdot)|_\Gamma \in L_1(\Gamma)$$

und für beliebiges v gilt:

$$\text{(8)} \qquad \langle S'\alpha, v\rangle = \int_\Gamma w(\alpha;\tau,\xi)\ v(\tau,\xi)\ d\tau\ d\xi$$

Sei $N := \{ (\tau,\xi) \in \Gamma \mid w(\alpha;\tau,\xi) = o \}$. Dann folgt aus (8) und (6):

Lemma 3: $|\hat{u}(\tau,\xi)| = 1$ fast überall in $\Gamma - N$.

Es ist unser Ziel, diese "bang-bang-Eigenschaft" für eine möglichst große Menge zu beweisen. Hierzu beachten wir zunächst

Lemma 4: $\alpha \neq o \Rightarrow w(\alpha) \neq o$.

Beweis: $w(\alpha) = o \Rightarrow \int_\Omega w(\alpha;t,x)\ v_k(x)\ dx = o$

$\Rightarrow \langle \alpha, v_k\rangle = o \Rightarrow \alpha = o$ (wegen Lemma 2).

Weiterhin berücksichtigen den folgenden Fortsetzungssatz vom Typ (b):

Lemma 5 (Mizohota [9]): Enthält N eine offene Menge, so gilt $w(\alpha) = o$

Aus den Lemmas 3-5 folgt:

Satz 2: Die Menge der Punkte $(\tau,\xi) \in \Gamma$ mit $|\hat{u}(\tau,\xi)| \neq 1$ ist (von einer Nullmenge abgesehen) in Γ nirgends dicht.

Dies sagt noch nicht viel über $\hat{u}$ aus; das Maß der Menge mit $|\hat{u}(\tau,\xi)| = 1$ kann trotzdem beliebig klein sein. Ein Fortsetzungssatz vom Typ (b') wäre nötig. Dies läßt sich bisher nur unter Analytizitätsvoraussetzungen zeigen (vgl. [4]).

Lemma 5': Sind $\partial\Omega$ und alle Koeffizienten von L analytisch und hat N positives Maß, so gilt $w = o$.

Aus den Lemmas 3,4 und 5' folgt:

Satz 2': Sind $\partial\Omega$ und alle Koeffizienten von L analytisch, so gilt fast überall $|\hat{u}(\tau,\xi)| = 1$.

Satz 3: Sind $\partial\Omega$ und alle Koeffizienten von L analytisch, so gibt es genau eine optimale Steuerung $\hat{u}$.

Aus dem obigen Beweisgang ist ersichtlich, daß es interessant wäre, zu untersuchen, ob ein Ergebnis vom Typ (b') auch ohne Analytizitätsvoraussetzungen gilt.

2. Ein Kontrollproblem für einen elliptischen Differentialoperator zweiter Ordnung

Wie in 1. sei $\Omega \subset \mathbb{R}^n$ ein Gebiet; sein Rand $\partial\Omega$ bestehe aus zwei Komponenten Γ_1 und Γ_2 . L und $\dot{\partial}$ seien wie bisher definiert. Um Existenzproblemen aus dem Wege zu gehen, setzen wir $a(x) \leq o$ voraus. Wir untersuchen das folgende Kontrollproblem:

$$\text{(9)} \quad \begin{array}{lll} Ly(x) & = o & x \in \Omega \\ y(\xi) & = o & \xi \in \Gamma_1 \\ y(\xi) & = u(\xi) & \xi \in \Gamma_2 \end{array}$$

Die Steuerung u variiert in

$$U := \{\, u \in L_\infty(\Gamma_2) \mid |u| \leq 1 \,\}$$

Gegeben ist $z \in L_2(\Gamma_1)$. Für die $u \in U$ gehörende Lösung schreiben wir $y(u;x)$.
Optimalitätskriterium ist

$$(**) \quad \| \dot{\partial} y(u;\cdot) - z \| L_2(\Gamma_1) \overset{!}{=} \min$$

Dieses Problem wird z. B. in [8] (p. 78) diskutiert.
Aufgrund von Regularitätssätzen ist klar, daß die folgende Abbildung kompakt ist:

$$\begin{array}{rcl} S : L_2(\Gamma_2) & \longrightarrow & L_2(\Gamma_1) \\ u & \longmapsto & \partial y(u;\cdot) \end{array}$$

Da U in $L_2(\Gamma_2)$ schwach kompakt ist, folgt

Satz 4: SU ist kompakt. Das Kontrollproblem (**) besitzt eine Lösung.

Wir betrachten im folgenden stets den Fall $z \notin SU$. Das Problem (**) kann man dann als die Aufgabe betrachten, zu $z \in L_2$ ein Element bester Approximation in der konvexen und abgeschlossenen Menge SU zu finden. Sei $\hat{u}$ eine optimale Steuerung und $\hat{y} = S\hat{u}$. Dann ist $\hat{y}$ (aber nicht notwendig $\hat{u}$) eindeutig bestimmt und wird charakterisiert durch

$$(z-\hat{y},\hat{y}-y)L_2(\Gamma_1) \geq o \qquad y \in SU$$

$$\Rightarrow \quad (S^*\hat{z},\hat{u}-u)L_2(\Gamma_2) \geq o \qquad u \in U \quad ; \quad \hat{z} := z-\hat{y}$$

Wir definieren $w(\hat{z})$ als Lösung von

$$Lw = o \quad , \quad w|\Gamma_1 = \hat{z} \quad , \quad w|\Gamma_2 = o$$

Da $w(z)$ und $y(v)$ niemals in gleichen Punkten nicht regulär sind, können wir die Greensche Formel anwenden, und wir erhalten

$$(S^*\hat{z},v)L_2(\Gamma_2) = (-\mathring{\partial}w(\hat{z}),v)L_2(\Gamma_2)$$

Somit erhalten wir aus (10) als notwendiges Optimalitätskriterium:

$$(10') \quad \int_{\Gamma_2} \mathring{\partial}w(\hat{z};\xi)\ (\hat{u}(\xi)-u(\xi))\ d\xi \geq o.$$

Sei $N := \{\xi \in \Gamma_2 |\ \mathring{\partial}w(z;\xi) = o\}$. Wir erhalten

Lemma 6: $|\hat{u}(\xi)| = 1$ fast überall in Γ_2-N

Lemma 7: $\hat{z} \neq o \implies w(\hat{z}) \neq o$.

(Folgt aus $\hat{z} = w|_{\Gamma_1}$) . Aus bekannten Fortsetzungssätzen vom Typ (a) (z. B. [2]) folgt

Lemma 8: Enthält N eine offene Menge, so ist $w(\hat{z}) = o$.

Satz 5: Die Menge der Punkte $\xi \in \Gamma_2$ mit $|\hat{u}(\xi)| \neq 1$ ist (von einer Nullmenge abgesehen) nirgends dicht.

Bei diesem Beispiel können wir dies verschärfen:

Lemma 8': Hat N positives Maß, so gilt $w = o$.

Dies wird im folgenden Abschnitt gezeigt. Es folgen

Satz 5': Es gilt $|\hat{u}(\xi)| = 1$ fast überall auf Γ_2.

Satz 6: Das Kontrollproblem (**) besitzt genau eine Lösung.

3. Ein Fortsetzungssatz für Lösungen von $Lu = o$

In diesem Abschnitt beweisen wir Lemma 8' und betrachten Lösungen von $Lu = o$ in einem Gebiet $\Omega \subset \mathbb{R}^n$. Diese sollen die folgenden Voraussetzungen erfüllen:

(11) $\quad u|_\Gamma = o \qquad \Gamma$: offene Teilmenge von $\partial\Omega$.

(12) $\quad \mathring{\partial}u|_N = o \qquad N$: Teilmenge von Γ mit positivem Maß.

Da L elliptisch ist, kann (12) (wegen (11)) auch wie folgt formuliert werden:

(12') $\quad \nabla u|_N = o$

Weiterhin können wir o. B. d. A. annehmen, daß Γ eben ist, d. h.

$$\Gamma \subset \{\ x \in \mathbb{R}^n |\ x_n = o\ \}$$

Mit der gleichen Technik wie in [4] folgert man aus (11) und (12) die Existenz eines Punktes $\xi_o \varepsilon \Gamma$ mit

(13) $|u(x)| \leq C(n)|x-\xi_o|^n$

(Nur an dieser Stelle benötigen wir, daß alle Daten und damit u zu C_∞ gehören.) Für $\Gamma_1 := \{ \xi \varepsilon \Gamma \,|\, |\xi-\xi_o| < \delta \}$ betrachten wir die Abbildung

$$\begin{aligned} \sigma : \Gamma_1 \times (-\varepsilon,\varepsilon) &\longrightarrow \mathbb{R}^n \\ \xi \;,\quad s &\longmapsto \xi - s \cdot N(\xi) \end{aligned}$$

mit $N_i(\xi) := a_{in}(\xi)$.

Sind ε und δ hinreichend klein, so ist σ bijektiv und C_∞ , und es gilt $\sigma(\Gamma_1 \times [o,\varepsilon)) \subset \bar{\Omega}$. Wir können daher $v(\xi,s) := u(\sigma(\xi,s))$ für $(\xi,s) \varepsilon \Gamma_1 \times [o,\varepsilon)$ definieren und erhalten

(14) $L_1 v = o$

(11') $v(\xi,o) = o$

(13') $|v(\xi,s)| \leq \tilde{C}(n)\ (s^2+|\xi-\xi_o|^2)^{n/2}$

Dabei ist

$$L_1 = \sum_{i,k} b_{ik}\, \partial_i \partial_k + \ldots \qquad (b_{ik} = b_{ki})$$

$$\partial_i := \frac{\partial}{\partial \xi_i} \quad (i=1,\ldots,n-1) \; ; \quad \partial_n := \frac{\partial}{\partial s}$$

wieder ein gleichmäßig elliptischer Differentialoperator mit C_∞-Koeffizienten. Außerdem zeigt die Kettenregel

(15) $b_{in}(\xi,o) = o \qquad i = 1, \ldots, n-1$

Wir setzen nun v durch

$$\tilde{v}(\xi,-s) := -v(\xi,s) \qquad s > o$$

auf $\Gamma_1 \times (-\varepsilon,\varepsilon)$ fort. Dann gilt in $-\varepsilon < s \leq o$ für $\tilde{v}$ die folgende Differentialgleichung:

$$\sum_{i,k=1}^{n-1} b_{ik}(\xi,-s)\ \partial_i \partial_k\ \tilde{v}(\xi,\, s)$$

$$+ \sum_{i=1}^{n-1} -b_{in}(\xi,-s)\ \partial_i \partial_n\ \tilde{v}(\xi,\, s)$$

$$+ b_{nn}(\xi,-s)\ \partial_n^2\ \tilde{v}(\xi,\, s) + \ldots = o$$

Wegen (15) genügt $\tilde{v}$ in $\Gamma_1 \times (-\varepsilon,\varepsilon)$ einer elliptischen Differentialgleichung, deren Hauptteilkoeffizienten lipschitzstetig sind. Wegen (13') folgt aus einem bekannten Fortsetzungssatz [2] $\tilde{v} = o$. Hieraus folgt $u = o$ in einer Umgebung von ξ_o . Erneute Anwendung von [2] liefert $u = o$ in Ω.

Literatur:

[1] AGMON, S.: Lectures on elliptic boundary value problems.
Van Nostrand, Princeton 1965.

[2] ARONSZAJN, N.; KRZYWICKI, A.; SZARSKI, J.: A unique continuation theorem for exterior differential forms on Riemannian manifolds.
Ark. Mat. 4, 417-453 (1962).

[3] CALDERON, A. P.: Uniqueness in the Cauchy problem for partial differential equations.
Amer. J. Math. 80. 16-36 (1958).

[4] FRIEDMAN, A.: Optimal control for parabolic equations.
J. Math. Anal. Appl. 18, 479-491 (1967).

[5] GLASHOFF, K.: Optimal control of one-dimensional linear parabolic differential equations.
Erscheint in diesem Symposiumsband.

[6] HÖRMANDER, L.: On the uniqueness of the Cauchy problem.
Math. Scand. 6, 213-225 (1958).

[7] KÖTHE, G.: Topologische lineare Räume.
Springer-Verlag. Grundlehren der mathematischen Wissenschaften, Band 107.

[8] LIONS, J. L.: Optimal control of systems governed by partial differential equations.
Springer-Verlag, Grundlehren der mathematischen Wissenschaften, Band 170.

[9] MIZOHATA, S.: Unicité du prolongement des solutions pour quelques opérateurs différentiels paraboliques.
Mém. Coll. Sc. Univ. Kyoto, A 31, 219-239 (1958).

[10] PROTTER, F.; WEINBERGER, H.: Maximum principles in differential equations.
Prentice-Hall, 1967.

[11] WECK, N.: Über Existenz, Eindeutigkeit und das "bang-bang-Prinzip" bei Kontrollproblemen aus der Wärmeleitung.
Erscheint demnächst in "Bonner Mathematische Schriften".

Der Sattelpunktsatz von Kuhn und Tucker in Geordneten Vektorräumen

Jochem Zowe
Inst.f.Angew.Mathematik d.Univ.Würzburg

Zusammenfassung: Es wird ein konvexes Programmierungsproblem über geordneten Vektorräumen betrachtet. Die Zielfunktion ist eine konvexe Abbildung von einem reellen Vektorraum in einen geordneten Vektorraum; der zulässige Bereich wird durch endlich viele Ungleichungen in geordneten Vektorräumen beschrieben. Notwendige und hinreichende Optimalitätskriterien in Form einer Sattelpunktaussage für eine dem Problem zugeordnete Lagrangefunktion werden gegeben. Unsere Ergebnisse sind eine direkte Erweiterung des Sattelpunktsatzes von Kuhn und Tucker.

1. DEFINITIONEN UND PROBLEMSTELLUNG

Es sei im folgenden X ein reeller Vektorraum und $Z_1,\dots,Z_m$ geordnete Vektorräume, d.h., Z_i, $i=1,\dots,m$, ist ein reeller Vektorraum versehen mit einer binären reflexiven und transitiven Relation '$\geqq$' (wir benutzen dasselbe Symbol für alle Vektorraumordnungen), die, wie folgt, mit der linearen Struktur von Z_i verträglich ist:

$z_i \geqq 0$ impliziert $\lambda z_i \geqq 0$ für alle $z_i \in Z_i$ und reellen $\lambda \geqslant 0$,

$z_i \geqq z_i'$ impliziert $z_i+z_i'' \geqq z_i'+z_i''$ für alle $z_i, z_i', z_i'' \in Z_i$.

Gleichbedeutend damit ist, daß die Menge der positiven Elemente von Z_i

$$Z_{i+} := \{z_i \in Z_i \mid z_i \geqq 0\}$$

ein konvexer Kegel mit Scheitel in 0 ist

$$\alpha Z_{i+} + \beta Z_{i+} \subset Z_{i+} \text{ für alle reellen } \alpha,\beta \geqslant 0.$$

Weiter sei Y ein geordneter Vektorraum, dessen Ordnung überdies antisymmetrisch ist

$$y \geqq y' \geqq y \text{ impliziert } y=y' \text{ für alle } y,y' \in Y.$$

Für den positiven Kegel $Y_+ := \{y \in Y \mid y \geqq 0\}$ bedeutet das

(1.1) $$Y_+ \cap -Y_+ = \{0\}.$$

Diese Bedingung garantiert, daß eine Teilmenge B von Y genau eine größte untere Schranke (infB) besitzt, falls überhaupt eine größte untere Schranke zu B existiert. Analoges gilt für das Minimum.

Gegeben seien ferner konvexe Funktionen $f: D(f) \to Y$ und $g_i: D(g_i) \to Z_i$, $i=1,\dots,m$, mit $D(f)$ und $D(g_i)$ enthalten in X. Hierbei wird z.B. f <u>konvex</u> genannt, falls D(f) eine nichtleere konvexe Teilmenge von X ist und für alle $x_1, x_2 \in D(f)$ und reellen λ, $0 \leqslant \lambda \leqslant 1$,

$$f(\lambda x_1+(1-\lambda)x_2) \leq \lambda f(x_1) + (1-\lambda)f(x_2).$$

Den gemeinsamen Definitionsbereich von f und $g_1,\ldots,g_m$ bezeichnen wir mit D,

$$D = D(f) \cap D(g_1) \cap \ldots \cap D(g_m) .$$

Die Funktionen f und g_i definieren ein <u>konvexes Programmierungsproblem</u> in dem geordneten Vektorraum Y mit Nebenbedingungen in den geordneten Räumen Z_i:

(MP) <u>minimiere</u> f(x) <u>unter der Nebenbedingung</u>
$x \in D$ <u>und</u> $g_i(x) \leq 0$ <u>für</u> $i=1,\ldots,m$.

Ein Punkt $\overline{x} \in D$ mit $g_i(\overline{x}) \leq 0$ für $i=1,..,m$ wird <u>optimale Lösung</u> des Minimierungsproblems (MP) genannt, falls

$$f(\overline{x}) = \min\{f(x) \mid x \in D,\ g_i(x) \leq 0 \text{ für } i=1,..,m\}.$$

Dem Problem (MP) wollen wir eine Lagrangefunktion zuordnen. Dazu bezeichnen wir mit L_i, $i=1,\ldots,m$, den Raum der linearen Abbildungen von Z_i nach Y. Ein $T_i \in L_i$ wird <u>positiv</u> genannt, falls $T_i z_i \geq 0$ für alle $z_i \geq 0$. Als <u>Lagrangefunktion</u> Φ des Problems (MP) definieren wir auf der Teilmenge

$$D(\Phi) := \{(x,T_1,\ldots,T_m) \mid x \in D, T_i \in L_i \text{ und } T_i \text{ positiv für } i=1,..,m\}$$

des Produktraumes $X \times L_1 \times \ldots \times L_m$ die Funktion

$$\Phi(x,T_1,..,T_m) := f(x) + \sum_{i=1}^{m} T_i \circ g_i(x) . \tag{1.2}$$

Φ bildet $D(\Phi)$ in Y ab. Wir sagen, daß $(\overline{x},\overline{T}_1,\ldots,\overline{T}_m) \in D(\Phi)$ ein <u>Sattelpunkt</u> von Φ ist, falls

$$\Phi(\overline{x},T_1,..,T_m) \leq \Phi(\overline{x},\overline{T}_1,..,\overline{T}_m) \leq \Phi(x,\overline{T}_1,..,\overline{T}_m) \tag{1.3}$$

für alle $x \in D$ und positiven $T_i \in L_i$, $i=1,..,m$. In dem Spezialfall $Y=Z_1=\ldots=Z_m=\mathbb{R}$ liefern (1.2) und (1.3) gerade die übliche Definition der Lagrangefunktion, die (MP) zugeordnet wird bzw. die Definition eines Sattelpunktes von Φ (siehe z.B.[9]).

In Abschnitt 2 und 3 erörtern wir den Zusammenhang zwischen den optimalen Lösungen des Minimierungsproblemes (MP) und den Sattelpunkten von Φ. Unsere Hauptergebnisse, Satz 2.3 und Satz 3.3, enthalten den Satz von Kuhn und Tucker [6] als Spezialfall. Als unmittelbare Folgerung aus unserem Sattelpunktsatz geben wir in Abschnitt 4 eine Erweiterung des Lemma von Farkas an und einen Fortsetzungssatz für monotone lineare Operatoren.

Ähnliche Ergebnisse wurden von Ritter [7] erzielt unter der Voraussetzung, daß X und Z_i geordnete Banachräume sind, daß Y ein geordneter reflexiver Banachraum ist und daß die betrachteten Funktionen stetig differenzierbar sind. Unser Zugang vermeidet diese topologischen Voraussetzungen und benutzt statt dessen allein die Ordnungsstruktur der zugrunde liegenden Räume.

Es sollen einige weitere im folgenden gebrauchte Definitionen zusammengestellt werden. Wir nennen einen geordneten Vektorraum F <u>ordnungsvollständig</u>, falls jede nichtleere nach unten beschränkte Teilmenge B von F eine größte untere Schranke (inf B) besitzt.(Vgl. [5]. In unserer Definition wird nicht verlangt, daß F ein Vektorverband ist. Andere Sprechweisen sind: "F besitzt die least upper bound property" [8] oder "F_+ ist fully minihedral" [3].)

Der Punkt x einer Teilmenge B eines reellen Vektorraumes heißt <u>relativ innerer</u> Punkt von B, falls auf jeder in der affinen Hülle von B liegenden Geraden durch x eine x als inneren Punkt enthaltende Strecke in B liegt. Die Gesamtheit der relativ inneren Punkte von B bezeichnen wir mit B^I. Weiter nennen wir B <u>linear abgeschlossen</u>, wenn der Schnitt einer jeden Geraden mit B eine abgeschlossene Teilmenge der Geraden ist.

2. EIN SATTELPUNKT IST HINREICHEND

Der folgende Trennungssatz wurde für endlichdimensionale Räume von Fenchel [4] bewiesen. Seine Verallgemeinerung auf beliebige reelle Vektorräume stammt von Bair und Jongmanns [2].

SATZ 2.1. <u>Seien</u> A <u>und</u> B <u>konvexe Teilmengen eines reellen Vektorraumes</u> E <u>und</u> $A^I \neq \emptyset$, $B^I \neq \emptyset$. A <u>und</u> B <u>können genau dann durch eine Hyperebene</u> H <u>mit</u> $H \not\supset A \cup B$ <u>getrennt werden</u>, <u>wenn</u> $A^I \cap B^I = \emptyset$.

Mit Hilfe von 2.1 beweisen wir

SATZ 2.2. <u>Sei</u> E <u>ein reeller Vektorraum und</u> $A \subset E$ <u>ein linear abgeschlossener konvexer Kegel mit Scheitel in</u> 0 <u>und nichtleerem relativ Inneren</u>. <u>Sei weiter</u> $w \in E$ <u>und</u> $w \notin A$. <u>Dann gibt es eine Linearform</u> t <u>auf</u> E <u>mit</u> $t(A) \geq 0$ <u>und</u> $t(w) < 0$.

BEWEIS. Sei $u \in A^I$ und v der Punkt der Strecke von u nach w, für den das abgeschlossene Intervall [u,v] in A liegt, aber der Schnitt von A mit dem halboffenen Intervall (v,w] leer ist; ein solcher Punkt existiert, da A konvex und linear abgeschlossen ist. Es ist $A^I \neq \emptyset$, $(v,w]^I \neq \emptyset$ und $A^I \cap (v,w]^I = \emptyset$. Somit gibt es nach Satz 2.1 eine

Linearform t auf E und ein reelles α mit

$$t(A) \geqslant \alpha \geqslant t((v,w])$$

aber

$$H := \{x \in E \mid t(x) = \alpha\} \not\supset A \cup (v,w] .$$

Da A ein Kegel ist, folgt $t(A) \geqslant 0 \geqslant \alpha$. Wir nehmen $t(w) = 0$ an. Dies impliziert $\alpha = 0$ und $t(v) = 0$, d.h. $H \supset (v,w]$. Da u auf der Geraden durch v und w liegt, ist auch $t(u) = 0$, und H trennt somit A und $\{u\}$. Wegen $A^I \cap \{u\}^I \neq \emptyset$ zeigt Satz 2.1, $H \supset A$ und somit $H \supset A \cup (v,w]$ im Widerspruch zur Wahl von H.

Die folgende Aussage liefert eine Richtung unseres Sattelpunktsatzes. Die Konvexität der Funktionen f und g_i wird hierzu nicht gebraucht.

SATZ 2.3. <u>Betrachtet werde das konvexe Programmierungsproblem</u> (MP). <u>Es seien die positiven Kegel</u> $Z_{1+},\dots,Z_{m+}$ <u>linear abgeschlossen und</u> $Z^I_{1+},\dots,Z^I_{m+}$ <u>nichtleer</u>. <u>Ist dann</u> $(\bar{x},\bar{T}_1,\dots,\bar{T}_m)$ <u>ein Sattelpunkt der Lagrangefunktion</u> Φ, <u>so ist</u> $\bar{x}$ <u>eine optimale Lösung von</u> (MP).

BEWEIS. Für alle $x \in D$ und positiven $T_i \in L_i$, $i=1,\dots,m$, gilt

$$(2.1) \quad f(\bar{x})+\sum_{i=1}^{m} T_i\circ g_i(\bar{x}) \leqq f(\bar{x})+\sum_{i=1}^{m} \bar{T}_i\circ g_i(\bar{x}) \leqq f(x)+\sum_{i=1}^{m} \bar{T}_i\circ g_i(x) .$$

Sei $i \in \{1,\dots,m\}$ fest. Mit $T_j := \bar{T}_j$ für $j \neq i$ wird aus der ersten Ungleichung in (2.1)

$$(2.2) \quad \bar{T}_i\circ g_i(\bar{x}) \geqq T_i\circ g_i(\bar{x}) \text{ für alle positiven } T_i \in L_i$$

und somit

$$(2.3) \quad \bar{T}_i\circ g_i(\bar{x}) \geqq 0.$$

Wir nehmen für den Augenblick an, daß

$$(2.4) \quad g_i(\bar{x}) \leqq 0 .$$

Dann ist $\bar{T}_i\circ g_i(\bar{x}) \leqq 0$ und wegen der Antisymmetrie der Ordnung von Y folgt hieraus und aus (2.3), daß $\bar{T}_i\circ g_i(\bar{x}) = 0$. Dies gilt für alle i, und aus der zweiten Ungleichung in (2.1) wird

$$f(\bar{x}) \leqq f(x)+\sum_{i=1}^{m} \bar{T}_i\circ g_i(x) \text{ für alle } x \in D.$$

Die Behauptung ergibt sich unmittelbar.

Es bleibt (2.4) zu zeigen. Dazu nehmen wir an, daß (2.4) nicht richtig ist, d.h. $-g_i(\bar{x}) \notin Z_{i+}$. Nach Satz 2.2 gibt es dann eine Linearform t auf Z_i mit

$$(2.5) \quad t(Z_{i+}) \geqslant 0 \quad \text{und} \quad t(g_i(\bar{x})) > 0.$$

Wir wählen ein $y_o \in Y_+$ aber $y_o \neq 0$ (falls $Y_+ = \{0\}$ ist die Behauptung trivialerweise richtig) und definieren ein $T_i^* \in L_i$ durch

$$T_i^* z_i := (t(g_i(\bar{x})))^{-1}\ t(z_i) y_o \text{ für alle } z_i \in Z_i \ .$$

Wegen (2.5) ist T_i^* positiv und (2.2) zusammen mit der Definition von T_i^* liefert

(2.6) $$\bar{T}_i \circ g_i(\bar{x}) \geqq T_i^* \circ g_i(\bar{x}) = y_o \ .$$

Nach Wahl von y_o und wegen der Transitivität und der Antisymmetrie der Ordnung von Y erhält man

(2.7) $$\bar{T}_i \circ g_i(\bar{x}) \geqq 0 \quad \text{und} \quad \bar{T}_i \circ g_i(\bar{x}) \neq 0 \ .$$

Folglich gilt (2.6) insbesondere für $y_o := 2\bar{T}_i \circ g_i(\bar{x})$. Mit diesem speziellen y_o erhält man aus (2.6)

$$\bar{T}_i \circ g_i(\bar{x}) \leqq 0$$

und zusammen mit (2.3), $\bar{T}_i \circ g_i(\bar{x}) = 0$ im Widerspruch zu (2.7).

3. EIN SATTELPUNKT IST NOTWENDIG

Für den Beweis der (teilweisen) Umkehrung von Satz 2.3 benötigen wir die folgende Verallgemeinerung der analytischen Form des Satzes von Hahn-Banach (siehe z.B.[3,chapter VI, §3, Theorem 1]). Der übliche Beweis des reellen Hahn-Banach Satzes benutzt von $\mathbb{R}$ nur die Ordnungsvollständigkeit und 3.1 läßt sich ganz analog zeigen.

SATZ 3.1. Es sei E ein reeller Vektorraum, F ein ordnungsvollständiger Vektorraum und p eine sublineare Abbildung von E in F, d.h.

$$p(x_1+x_2) \leqq p(x_1)+p(x_2), \quad p(\lambda x_1) = \lambda p(x_1)$$

für alle $x_1, x_2 \in E$ und reellen $\lambda \geqslant 0$. Ist T_M eine lineare Abbildung von einem linearen Teilraum M von E in F und $T_M x \leqq p(x)$ für alle $x \in M$, so läßt sich T_M linear fortsetzen zu einem T von E in F mit $Tx \leqq p(x)$ für alle $x \in E$.

Wir führen folgende vereinfachende Schreibweise ein. Es bezeichne Z den Produktraum $Z_1 \times \cdots \times Z_m$. Z sei komponentenweise geordnet, d.h.

$$Z_+ = \{z \in Z \mid z \geqq 0\} = Z_{1+} \times \cdots \times Z_{m+} .$$

Dann ist

$$g(x) := (g_1(x), \ldots, g_m(x))$$

eine konvexe Funktion definiert auf $\bigcap_{i=1}^{m} D(g_i)$ mit Werten in Z.

Weiter setzen wir

(3.1) $$N := \{\lambda g(x)+z \mid x \in D,\ z \geqq 0,\ \lambda \text{ reell} \geqslant 0\} \ .$$

In Satz 3.3 werden wir verlangen, daß N ein linearer Teilraum von Z ist. Es gilt

SATZ 3.2. *Jede der beiden folgenden Bedingungen garantiert, daß N ein linearer Teilraum von Z ist.*

(a) *Es gibt ein* $x_o \in D$, *so daß* $-g(x_o)$ *Ordnungseinheit bezüglich* N *ist* (d.h., *für alle* $z \in N$ *gibt es ein reelles* $\lambda \geq 0$ *mit* $\lambda(-g(x_o)) \geqq z$).

(b) *Es ist* D *ein linearer Teilraum von* X, g *linear und* $g(D)+Z_+ = g(D)-Z_+$.

BEWEIS. (a) Offensichtlich ist N ein Kegel mit Scheitel in 0. Sei $n_i = \lambda_i g(x_i)+z_i$ mit $\lambda_i > 0$, $x_i \in D$ und $z_i \in Z_+$ für i=1,2. Dann ist

$$n_1+n_2 = (\lambda_1+\lambda_2)\left(\frac{\lambda_1}{\lambda_1+\lambda_2}\, g(x_1) + \frac{\lambda_2}{\lambda_1+\lambda_2}\, g(x_2)\right) + z_1+z_2$$

und wegen der Konvexität von g hat man mit einem geeigneten $z \geqq 0$

$$n_1+n_2 = (\lambda_1+\lambda_2)g\left(\frac{\lambda_1}{\lambda_1+\lambda_2}\, x_1 + \frac{\lambda_2}{\lambda_1+\lambda_2}\, x_2\right) + ((\lambda_1+\lambda_2)z+z_1+z_2),$$

d.h. $n_1+n_2 \in N$. Somit ist N ein konvexer Kegel mit Scheitel in 0, und es bleibt zu zeigen, daß mit n auch -n zu N gehört. Sei also $n \in N$. Nach Voraussetzung gibt es dann ein reelles $\lambda \geq 0$, so daß $n \leqq \lambda(-g(x_o))$, d.h. $n = \lambda(-g(x_o))-z$ mit einem geeigneten $z \geqq 0$. Offensichtlich gehört dann auch $-n = \lambda g(x_o)+z$ zu N.

(b) Trivial.

Die Voraussetzung (a) des obigen Satzes ist z.B. erfüllt, falls es ein $x_o \in D$ gibt, so daß $-g_i(x_o)$ Ordnungseinheit in Z_i ist für i=1,..,m. Im Spezialfall $Z_1 = \ldots = Z_m = \mathbb{R}$ ist dies gerade die "constraint qualification" von Slater (siehe z.B. [1,S.34]). Man beachte, daß im linearen Fall (a) immer (b) impliziert, daß aber (b) gelten kann, ohne daß (a) gilt (z.B. falls g(D) den positiven Kegel Z_+ enthält und es in g(D) keine Ordnungseinheit gibt).

Als Gegenstück zu Satz 2.3 beweisen wir nun

SATZ 3.3. *Betrachtet werde das konvexe Programmierungsproblem* (MP). *Es sei* N (*siehe* (3.1)) *ein linearer Teilraum von* Z *und* Y *ordnungsvollständig. Dann gibt es zu jeder optimalen Lösung* $\bar{x}$ *von* (MP) *einen Sattelpunkt der Gestalt* $(\bar{x},\bar{T}_1,\ldots,\bar{T}_m)$ *von* Φ.

BEWEIS. Wir betrachten in $Z \times Y$ die Menge

$$B := \{\lambda(g(x),f(x)-f(\bar{x})) + (z,y) \mid \lambda \text{ reell} \geq 0,\ x \in D,\ z \geqq 0 \text{ und } y \geqq 0\}.$$

Man verifiziert leicht, daß B ein konvexer Kegel mit Scheitel in (0,0) ist. Sei weiter für $z \in Z$

$$S_z := \{y \in Y \mid (z,y) \in B\}\ .$$

Offensichtlich ist

(3.2) $S_z \neq \emptyset$ für alle $z \in N$.

Ferner gilt, da B ein konvexer Kegel ist,

(3.3) $S_{z_1} + S_{z_2} \subset S_{z_1+z_2}$ für alle $z_1, z_2 \in N$.

Sei nun $z=0$ und $y \in S_0$. Nach Definition von S_z und B hat man mit einem geeigneten $\lambda \geq 0$ und $x \in D$

$$\lambda g(x) \leqq 0 \text{ und } \lambda(f(x)-f(\overline{x})) \leqq y$$

und, da $\overline{x}$ optimal sein soll, $y \geqq \lambda(f(x)-f(\overline{x})) \geqq 0$. Wir notieren

(3.4) $S_0 \subset Y_+$.

Über die Mengen S_z wollen wir eine sublineare Abbildung p von N in Y definieren. Sei dazu $z \in N$ fest. Nach Voraussetzung über N und nach (3.2) ist $S_{-z} \neq \emptyset$. Mit einem beliebigen $\overline{y} \in S_{-z}$ ergibt sich aus (3.3) und (3.4)

$$S_z + \{\overline{y}\} \subset S_z + S_{-z} \subset S_0 \subset Y_+,$$

d.h., $-\overline{y}$ ist eine untere Schranke von S_z. Da Y ordnungsvollständig ist und die Ordnung antisymmetrisch ist, existiert eine eindeutig bestimmte größte untere Schranke, inf S_z. Somit ist durch

$$p(z) := \inf S_z \text{ für } z \in N$$

eine Abbildung von N nach Y definiert.

Aus $\lambda B = B$ für reelles $\lambda > 0$ folgt $\lambda S_z = S_{\lambda z}$ und damit $\lambda p(z) = p(\lambda z)$ für $\lambda > 0$. Da weiter $p(0) = 0$ (S_0 enthält $f(\overline{x})-f(\overline{x})=0$) hat man $p(\lambda z) = \lambda p(z)$ für alle $z \in N$ und $\lambda \geq 0$. Weiter zeigt (3.3) für $z_1, z_2 \in N$

$$p(z_1+z_2) \leqq y_1+y_2 \text{ für alle } y_1 \in S_{z_1} \text{ und } y_2 \in S_{z_2}$$

und damit $p(z_1+z_2) \leqq y_1+p(z_2)$ für alle $y_1 \in S_{z_1}$ und schließlich

$$p(z_1+z_2) \leqq p(z_1)+p(z_2) .$$

Somit ist p sublinear und nach 3.1 gibt es eine lineare Abbildung T_N von dem reellen Vektorraum N in den ordnungsvollständigen Vektorraum Y mit $T_N z \leqq p(z)$ für alle $z \in N$

(setze dazu in Satz 3.1 $M := \{0\}$). Da $f(x)-f(\overline{x}) \in S_{g(x)+z}$ für alle $x \in D$ und $z \geqq 0$ folgt

(3.5) $T_N(g(x)+z) \leqq f(x)-f(\overline{x})$ für alle $x \in D$, $z \geqq 0$.

Sei $\overline{T}$ eine lineare Fortsetzung von $-T_N$ auf ganz Z und $(\overline{T}_1,..,\overline{T}_m)$ die kanonische Zerlegung von $\overline{T}$ in Elemente $\overline{T}_i \in L_i$. Aus (3.5) wird dann

(3.6) $f(\overline{x}) \leqq f(x) + \sum_{i=1}^{m} \overline{T}_i \cdot g_i(x) + \sum_{i=1}^{m} \overline{T}_i z_i$ für alle $x \in D$ und $z_i \geqq 0$.

Insbesondere erhält man aus (3.6)

$$\sum_{i=1}^{m} \overline{T}_i \circ g_i(\overline{x}) \geq 0.$$

Wir nehmen für den Augenblick an, daß $\overline{T}_i$ positiv ist für alle i. Wegen $g_i(\overline{x}) \leq 0$ folgt

$$\sum_{i=1}^{m} \overline{T}_i \circ g_i(\overline{x}) \leq 0$$

und somit $\sum_{i=1}^{m} \overline{T}_i \circ g_i(\overline{x})=0$. Dies zusammen mit (3.6) liefert eine der beiden einen Sattelpunkt definierenden Ungleichungen

$$f(\overline{x})+\sum_{i=1}^{m} \overline{T}_i \circ g_i(\overline{x}) \leq f(x)+\sum_{i=1}^{m} \overline{T}_i \circ g_i(x) \text{ für alle } x \in D.$$

Die andere Ungleichung ist trivialerweise erfüllt, da $\sum_{i=1}^{m} T_i \circ g_i(\overline{x}) \leq 0$ für alle positiven $T_i \in L_i$.

Es bleibt zu zeigen, daß $\overline{T}_i$ positiv ist für i=1,...,m. Sei i fest. Für festes $z_i \geq 0$ gewinnt man aus (3.6) die Ungleichung

$$\lambda(\overline{T}_i z_i)+\sum_{j=1}^{m} \overline{T}_j \circ g_j(\overline{x}) = \overline{T}_i(\lambda z_i)+\sum_{j=1}^{m} \overline{T}_j \circ g_j(\overline{x}) \geq 0 \text{ für } \lambda \geqslant 0$$

und damit

$$\overline{T}_i z_i+\frac{1}{\lambda}\left(\sum_{j=1}^{m} \overline{T}_j \circ g_j(\overline{x})\right) \in Y_+ \quad \text{für} \quad \lambda > 0.$$

Da der positive Kegel eines ordnungsvollständigen Vektorraumes linear abgeschlossen ist (siehe z.B. [8,Theorem 1]), folgt $\overline{T}_i z_i \geq 0$, d.h. $\overline{T}_i$ ist positiv.

Ist $Z_1 = \ldots = Z_m = Y = \mathbb{R}$, so fallen Satz 2.3 und 3.3 mit dem Satz von Kuhn und Tucker [6] zusammen.

BEMERKUNG. Satz 2.3 und 3.3 lassen sich auf den Fall verallgemeinern, daß in (MP) zusätzlich lineare Restriktionen als Gleichungen $g_{m+j}(x)=0$, j=1,.,n, gegeben sind. Die constraint qualification von Satz 3.3 "N ist ein linearer Teilraum von Z" ist dann zu ersetzen durch "es gibt ein $x_o \in D$, eine Menge $U \subset X$ und Elemente $\overline{z}_i \in Z_i$, i=1,.,m, so daß $x_o + U \subset D$, U absorbiert jedes x aus der linearen Hülle von D, $(\overline{z}_1,..,\overline{z}_m)$ ist eine Ordnungseinheit in N, $g_i(x_o+x) \leq -\overline{z}_i$ für alle $x \in U$ und i=1,...,m, und $g_{m+j}(x_o) = 0$ für j=1,...,n".

4. FOLGERUNGEN

Wie vorher sei X ein reeller Vektorraum, $Y, Z_1, \ldots, Z_m$ geordnete Vektorräume und die Ordnung von Y antisymmetrisch.

Das folgende Resultat ist eine Verallgemeinerung des Lemma von Farkas.

SATZ 4.1. *Sei* F *eine lineare Abbildung von* X *in* Y *und* G_i *eine lineare Abbildung von* Z_i *in* Y, $i=1,..,m$. *Sei weiter* Y *ordnungsvollständig und*

$$\{(G_1x,..,G_mx)+(z_1,..,z_m) \mid x \in X, z_i \geqq 0\} = \{(G_1x,..,G_mx)-(z_1..,z_m) \mid x \in X, z_i \geqq 0\}.$$

Dann sind die beiden folgenden Aussagen äquivalent:

(a) $G_i(x) \leqq 0$ *für* $i=1,..,m$ *impliziert* $Fx \geqq 0$ *für alle* $x \in X$.

(b) *Es gibt positive lineare Abbildungen* T_i *von* Z_i *in* Y, $i=1,...,m$, *so daß* $F + \sum_{i=1}^{m} T_i \circ G_i = 0$.

BEWEIS. Trivialerweise zieht (b) die Aussage (a) nach sich. Gelte nun (a). Dann ist $x=0$ eine optimale Lösung des Minimierungsproblems (MP) mit den speziellen Funktionen $f = F$ und $g_i = G_i$. Nach Satz 3.3 und 3.2(b) besitzt die zugehörige Lagrangefunktion einen Sattelpunkt $(0,T_1,..,T_m)$ mit $T_i \in L_i$ und T_i positiv. Es folgt leicht, daß damit

$$Fx + \sum_{i=1}^{m} T_i \circ G_i x \geqq 0 \quad \text{für alle} \quad x \in X.$$

Die Aussage (b) ergibt sich unmittelbar hieraus.

Als weitere Folgerung aus 3.3 notieren wir einen Fortsetzungssatz für monotone lineare Operatoren.

SATZ 4.2. *Sei* T_M *eine lineare Abbildung von einem linearen Teilraum* M *von* X *in* Y. *Weiter sei* $K \subset X$ *ein konvexer Kegel mit Scheitel in* 0 *und* $T_M(M \cap K) \geqq 0$. *Falls* Y *ordnungsvollständig ist und* $M+K = M-K$, *so existiert eine lineare Fortsetzung* T *von* T_M *auf ganz* X *mit* $T(K) \geqq 0$.

BEWEIS. Man setze in (MP)

$Z := X$, $Z_+ := K$, $f := T_M$ mit $D(f) := M$ und $g := -I$ (I Einheitsoperator auf X). Dann ist

$$\min\{f(x) \mid x \in D,\ g(x) \leqq 0\} = f(0) = 0$$

und nach 3.2 und 3.3 gibt es ein $T : Z \to Y$ mit $T(Z_+) \geqq 0$ und

$$0 \leqq f(x)+T \circ g(x) = T_M x - Tx \text{ für alle } x \in D = M.$$

Offensichtlich hat T die gewünschten Eigenschaften.

BEMERKUNG. Ist Y ein geordneter Vektorraum und gilt die Aussage von Satz 4.2 bei festem Y für alle X, M, T_M und K (die den Voraussetzungen des Satzes genügen), dann muß Y ordnungsvollständig sein (siehe [3, chapter VI, § 3, Theorem 1]). Es folgt, daß Satz 3.3 für einen geordneten Vektorraum Y genau dann in Allgemeinheit gilt, wenn Y ordnungsvollständig ist.

LITERATUR

1 Arrow, K.J., Hurwicz, L., Uzawa, H.: Studies in linear and non-linear programming, Stanford Univ.Press, Stanford, 1958.

2 Bair, Jongmanns: Séparation franche dans un espace vectoriel, Bull. Soc. Roy. Sci. Liège 39, 474-477 (1970).

3 Day, M.M.: Normed linear spaces, 3rd ed., Springer-Verlag, Berlin-Heidelberg-New York, 1973.

4 Fenchel, W.: Convex cones, sets and functions, Lecture notes, Princeton University, 1953.

5 Jameson, G.: Ordered linear spaces, Lecture notes in Mathematics 141, Springer-Verlag, Berlin-Heidelberg-New York, 1970.

6 Kuhn, H., Tucker, A.W.: Nonlinear programming, in Proceedings of the Second Berkeley Symposium on Mathematical Statistical Problems, University of California Press, 481-492, 1951.

7 Ritter, K.: Optimization theory in linear spaces, Math.Ann. 182, 189-206 (1969), 183, 169-180 (1969) und 184, 133-154 (1970).

8 Silverman, R.J., Yen, T.: The Hahn-Banach theorem and the least upper bound property, Trans. Amer. Math. Soc. 90, 523-526 (1959).

9 Stoer, J., Witzgall, C.W.: Convexity and optimization in finite dimensions I, Springer-Verlag, Berlin-Heidelberg-New York, 1970.

Vol. 309: D. H. Sattinger, Topics in Stability and Bifurcation Theory. VI, 190 pages. 1973. DM 20,–

Vol. 310: B. Iversen, Generic Local Structure of the Morphisms in Commutative Algebra. IV, 108 pages. 1973. DM 18,–

Vol. 311: Conference on Commutative Algebra. Edited by J. W. Brewer and E. A. Rutter. VII, 251 pages. 1973. DM 24,–

Vol. 312: Symposium on Ordinary Differential Equations. Edited by W. A. Harris, Jr. and Y. Sibuya. VIII, 204 pages. 1973. DM 22,–

Vol. 313: K. Jörgens and J. Weidmann, Spectral Properties of Hamiltonian Operators. III, 140 pages. 1973. DM 18,–

Vol. 314: M. Deuring, Lectures on the Theory of Algebraic Functions of One Variable. VI, 151 pages. 1973. DM 18,–

Vol. 315: K. Bichteler, Integration Theory (with Special Attention to Vector Measures). VI, 357 pages. 1973. DM 29,–

Vol. 316: Symposium on Non-Well-Posed Problems and Logarithmic Convexity. Edited by R. J. Knops. V, 176 pages. 1973. DM 20,–

Vol. 317: Séminaire Bourbaki – vol. 1971/72. Exposés 400–417. IV, 361 pages. 1973. DM 29,–

Vol. 318: Recent Advances in Topological Dynamics. Edited by A. Beck. VIII, 285 pages. 1973. DM 27,–

Vol. 319: Conference on Group Theory. Edited by R. W. Gatterdam and K. W. Weston. V, 188 pages. 1973. DM 20,–

Vol. 320: Modular Functions of One Variable I. Edited by W. Kuyk. V, 195 pages. 1973. DM 20,–

Vol. 321: Séminaire de Probabilités VII. Edité par P. A. Meyer. VI, 322 pages. 1973. DM 29,–

Vol. 322: Nonlinear Problems in the Physical Sciences and Biology. Edited by I. Stakgold, D. D. Joseph and D. H. Sattinger. VIII, 357 pages. 1973. DM 29,–

Vol. 323: J. L. Lions, Perturbations Singulières dans les Problèmes aux Limites et en Contrôle Optimal. XII, 645 pages. 1973. DM 46,–

Vol. 324: K. Kreith, Oscillation Theory. VI, 109 pages. 1973. DM 18,–

Vol. 325: C.-C. Chou, La Transformation de Fourier Complexe et L'Equation de Convolution. IX, 137 pages. 1973. DM 18,–

Vol. 326: A. Robert, Elliptic Curves. VIII, 264 pages. 1973. DM 24,–

Vol. 327: E. Matlis, One-Dimensional Cohen-Macaulay Rings. XII, 157 pages. 1973. DM 20,–

Vol. 328: J. R. Büchi and D. Siefkes, The Monadic Second Order Theory of All Countable Ordinals. VI, 217 pages. 1973. DM 22,–

Vol. 329: W. Trebels, Multipliers for (C, α)-Bounded Fourier Expansions in Banach Spaces and Approximation Theory. VII, 103 pages. 1973. DM 18,–

Vol. 330: Proceedings of the Second Japan-USSR Symposium on Probability Theory. Edited by G. Maruyama and Yu. V. Prokhorov. VI, 550 pages. 1973. DM 40,–

Vol. 331: Summer School on Topological Vector Spaces. Edited by L. Waelbroeck. VI, 226 pages. 1973. DM 22,–

Vol. 332: Séminaire Pierre Lelong (Analyse) Année 1971-1972. V, 131 pages. 1973. DM 18,–

Vol. 333: Numerische, insbesondere approximationstheoretische Behandlung von Funktionalgleichungen. Herausgegeben von R. Ansorge und W. Törnig. VI, 296 Seiten. 1973. DM 27,–

Vol. 334: F. Schweiger, The Metrical Theory of Jacobi-Perron Algorithm. V, 111 pages. 1973. DM 18,–

Vol. 335: H. Huck, R. Roitzsch, U. Simon, W. Vortisch, R. Walden, B. Wegner und W. Wendland, Beweismethoden der Differentialgeometrie im Großen. IX, 159 Seiten. 1973. DM 20,–

Vol. 336: L'Analyse Harmonique dans le Domaine Complexe. Edité par E. J. Akutowicz. VIII, 169 pages. 1973. DM 20,–

Vol. 337: Cambridge Summer School in Mathematical Logic. Edited by A. R. D. Mathias and H. Rogers. IX, 660 pages. 1973. DM 46,–

Vol: 338: J. Lindenstrauss and L. Tzafriri, Classical Banach Spaces. IX, 243 pages. 1973. DM 24,–

Vol. 339: G. Kempf, F. Knudsen, D. Mumford and B. Saint-Donat, Toroidal Embeddings I. VIII, 209 pages. 1973. DM 22,–

Vol. 340: Groupes de Monodromie en Géométrie Algébrique. (SGA 7 II). Par P. Deligne et N. Katz. X, 438 pages. 1973. DM 44,–

Vol. 341: Algebraic K-Theory I, Higher K-Theories. Edited by H. Bass. XV, 335 pages. 1973. DM 29,–

Vol. 342: Algebraic K-Theory II, "Classical" Algebraic K-Theory, and Connections with Arithmetic. Edited by H. Bass. XV, 527 pages. 1973. DM 40,–

Vol. 343: Algebraic K-Theory III, Hermitian K-Theory and Geometric Applications. Edited by H. Bass. XV, 572 pages. 1973. DM 40,–

Vol. 344: A. S. Troelstra (Editor), Metamathematical Investigation of Intuitionistic Arithmetic and Analysis. XVII, 485 pages. 1973. DM 38,–

Vol. 345: Proceedings of a Conference on Operator Theory. Edited by P. A. Fillmore. VI, 228 pages. 1973. DM 22,–

Vol. 346: Fučík et al., Spectral Analysis of Nonlinear Operators. II, 287 pages. 1973. DM 26,–

Vol. 347: J. M. Boardman and R. M. Vogt, Homotopy Invariant Algebraic Structures on Topological Spaces. X, 257 pages. 1973. DM 24,–

Vol. 348: A. M. Mathai and R. K. Saxena, Generalized Hypergeometric Functions with Applications in Statistics and Physical Sciences. VII, 314 pages. 1973. DM 26,–

Vol. 349: Modular Functions of One Variable II. Edited by W. Kuyk and P. Deligne. V, 598 pages. 1973. DM 38,–

Vol. 350: Modular Functions of One Variable III. Edited by W. Kuyk and J.-P. Serre. V, 350 pages. 1973. DM 26,–

Vol. 351: H. Tachikawa, Quasi-Frobenius Rings and Generalizations. XI, 172 pages. 1973. DM 20,–

Vol. 352: J. D. Fay, Theta Functions on Riemann Surfaces. V, 137 pages. 1973. DM 18,–

Vol. 353: Proceedings of the Conference on Orders, Group Rings and Related Topics. Organized by J. S. Hsia, M. L. Madan and T. G. Ralley. X, 224 pages. 1973. DM 22,–

Vol. 354: K. J. Devlin, Aspects of Constructibility. XII, 240 pages. 1973. DM 24,–

Vol. 355: M. Sion, A Theory of Semigroup Valued Measures. V, 140 pages. 1973. DM 18,–

Vol. 356: W. L. J. van der Kallen, Infinitesimally Central-Extensions of Chevalley Groups. VII, 147 pages. 1973. DM 18,–

Vol. 357: W. Borho, P. Gabriel und R. Rentschler, Primideale in Einhüllenden auflösbarer Lie-Algebren. V, 182 Seiten. 1973. DM 20,–

Vol. 358: F. L. Williams, Tensor Products of Principal Series Representations. VI, 132 pages. 1973. DM 18,–

Vol. 359: U. Stammbach, Homology in Group Theory. VIII, 183 pages. 1973. DM 20,–

Vol. 360: W. J. Padgett and R. L. Taylor, Laws of Large Numbers for Normed Linear Spaces and Certain Fréchet Spaces. VI, 111 pages. 1973. DM 18,–

Vol. 361: J. W. Schutz, Foundations of Special Relativity: Kinematic Axioms for Minkowski Space Time. XX, 314 pages. 1973. DM 26,–

Vol. 362: Proceedings of the Conference on Numerical Solution of Ordinary Differential Equations. Edited by D. Bettis. VIII, 490 pages. 1974. DM 34,–

Vol. 363: Conference on the Numerical Solution of Differential Equations. Edited by G. A. Watson. IX, 221 pages. 1974. DM 20,–

Vol. 364: Proceedings on Infinite Dimensional Holomorphy. Edited by T. L. Hayden and T. J. Suffridge. VII, 212 pages. 1974. DM 20,–

Vol. 365: R. P. Gilbert, Constructive Methods for Elliptic Equations. VII, 397 pages. 1974. DM 26,–

Vol. 366: R. Steinberg, Conjugacy Classes in Algebraic Groups (Notes by V. V. Deodhar). VI, 159 pages. 1974. DM 18,–

Vol. 367: K. Langmann und W. Lütkebohmert, Cousinverteilungen und Fortsetzungssätze. VI, 151 Seiten. 1974. DM 16,–

Vol. 368: R. J. Milgram, Unstable Homotopy from the Stable Point of View. V, 109 pages. 1974. DM 16,–

Vol. 369: Victoria Symposium on Nonstandard Analysis. Edited by A. Hurd and P. Loeb. XVIII, 339 pages. 1974. DM 26,–

Vol. 370: B. Mazur and W. Messing, Universal Extensions and One Dimensional Crystalline Cohomology. VII, 134 pages. 1974. DM 16,-

Vol. 371: V. Poenaru, Analyse Différentielle. V, 228 pages. 1974. DM 20,-

Vol. 372: Proceedings of the Second International Conference on the Theory of Groups 1973. Edited by M. F. Newman. VII, 740 pages. 1974. DM 48,-

Vol. 373: A. E. R. Woodcock and T. Poston, A Geometrical Study of the Elementary Catastrophes. V, 257 pages. 1974. DM 22,-

Vol. 374: S. Yamamuro, Differential Calculus in Topological Linear Spaces. IV, 179 pages. 1974. DM 18,-

Vol. 375: Topology Conference 1973. Edited by R. F. Dickman Jr. and P. Fletcher. X, 283 pages. 1974. DM 24,-

Vol. 376: D. B. Osteyee and I. J. Good, Information, Weight of Evidence, the Singularity between Probability Measures and Signal Detection. XI, 156 pages. 1974. DM 16.-

Vol. 377: A. M. Fink, Almost Periodic Differential Equations. VIII, 336 pages. 1974. DM 26,-

Vol. 378: TOPO 72 - General Topology and its Applications. Proceedings 1972. Edited by R. Alò, R. W. Heath and J. Nagata. XIV, 651 pages. 1974. DM 50,-

Vol. 379: A. Badrikian et S. Chevet, Mesures Cylindriques, Espaces de Wiener et Fonctions Aléatoires Gaussiennes. X, 383 pages. 1974. DM 32,-

Vol. 380: M. Petrich, Rings and Semigroups. VIII, 182 pages. 1974. DM 18,-

Vol. 381: Séminaire de Probabilités VIII. Edité par P. A. Meyer. IX, 354 pages. 1974. DM 32,-

Vol. 382: J. H. van Lint, Combinatorial Theory Seminar Eindhoven University of Technology. VI, 131 pages. 1974. DM 18,-

Vol. 383: Séminaire Bourbaki - vol. 1972/73. Exposés 418-435 IV, 334 pages. 1974. DM 30,-

Vol. 384: Functional Analysis and Applications, Proceedings 1972. Edited by L. Nachbin. V, 270 pages. 1974. DM 22,-

Vol. 385: J. Douglas Jr. and T. Dupont, Collocation Methods for Parabolic Equations in a Single Space Variable (Based on C^1-Piecewise-Polynomial Spaces). V, 147 pages. 1974. DM 16,-

Vol. 386: J. Tits, Buildings of Spherical Type and Finite BN-Pairs. IX, 299 pages. 1974. DM 24,-

Vol. 387: C. P. Bruter, Eléments de la Théorie des Matroïdes. V, 138 pages. 1974. DM 18,-

Vol. 388: R. L. Lipsman, Group Representations. X, 166 pages. 1974. DM 20,-

Vol. 389: M.-A. Knus et M. Ojanguren, Théorie de la Descente et Algèbres d' Azumaya. IV, 163 pages. 1974. DM 20,-

Vol. 390: P. A. Meyer, P. Priouret et F. Spitzer, Ecole d'Eté de Probabilités de Saint-Flour III - 1973. Edité par A. Badrikian et P.-L. Hennequin. VIII, 189 pages. 1974. DM 20,-

Vol. 391: J. Gray, Formal Category Theory: Adjointness for 2-Categories. XII, 282 pages. 1974. DM 24,-

Vol. 392: Géométrie Différentielle, Colloque, Santiago de Compostela, Espagne 1972. Edité par E. Vidal. VI, 225 pages. 1974. DM 20,-

Vol. 393: G. Wassermann, Stability of Unfoldings. IX, 164 pages. 1974. DM 20,-

Vol. 394: W. M. Patterson 3rd, Iterative Methods for the Solution of a Linear Operator Equation in Hilbert Space - A Survey. III, 183 pages. 1974. DM 20,-

Vol. 395: Numerische Behandlung nichtlinearer Integrodifferential- und Differentialgleichungen. Tagung 1973. Herausgegeben von R. Ansorge und W. Törnig. VII, 313 Seiten. 1974. DM 28,-

Vol. 396: K. H. Hofmann, M. Mislove and A. Stralka, The Pontryagin Duality of Compact O-Dimensional Semilattices and its Applications. XVI, 122 pages. 1974. DM 18,-

Vol. 397: T. Yamada, The Schur Subgroup of the Brauer Group. V, 159 pages. 1974. DM 18,-

Vol. 398: Théories de l'Information, Actes des Rencontres de Marseille-Luminy, 1973. Edité par J. Kampé de Fériet et C. Picard. XII, 201 pages. 1974. DM 23,-

Vol. 399: Functional Analysis and its Applications, Proceedings 1973. Edited by H. G. Garnir, K. R. Unni and J. H. Williamson. XVII, 569 pages. 1974. DM 44,-

Vol. 400: A Crash Course on Kleinian Groups - San Francisco 1974. Edited by L. Bers and I. Kra. VII, 130 pages. 1974. DM 18,-

Vol. 401: F. Atiyah, Elliptic Operators and Compact Groups. V, 93 pages. 1974. DM 18,-

Vol. 402: M. Waldschmidt, Nombres Transcendants. VIII, 277 pages. 1974. DM 25,-

Vol. 403: Combinatorial Mathematics - Proceedings 1972. Edited by D. A. Holton. VIII, 148 pages. 1974. DM 18,-

Vol. 404: Théorie du Potentiel et Analyse Harmonique. Edité par J. Faraut. V, 245 pages. 1974. DM 25,-

Vol. 405: K. Devlin and H. Johnsbråten, The Souslin Problem. VIII, 132 pages. 1974. DM 18,-

Vol. 406: Graphs and Combinatorics - Proceedings 1973. Edited by R. A. Bari and F. Harary. VIII, 355 pages. 1974. DM 30,-

Vol. 407: P. Berthelot, Cohomologie Cristalline des Schémas de Caracteristique $p > o$. VIII, 598 pages. 1974. DM 44,-

Vol. 408: J. Wermer, Potential Theory. VIII, 146 pages. 1974. DM 18,-

Vol. 409: Fonctions de Plusieurs Variables Complexes, Séminaire François Norguet 1970-1973. XIII, 612 pages. 1974. DM 47,-

Vol. 410: Séminaire Pierre Lelong (Analyse) Année 1972-1973. VI, 181 pages. 1974. DM 18,-

Vol. 411: Hypergraph Seminar. Ohio State University, 1972. Edited by C. Berge and D. Ray-Chaudhuri. IX, 287 pages. 1974. DM 28,-

Vol. 412: Classification of Algebraic Varieties and Compact Complex Manifolds. Proceedings 1974. Edited by H. Popp. V, 333 pages. 1974. DM 30,-

Vol. 413: M. Bruneau, Variation Totale d'une Fonction. XIV, 332 pages. 1974. DM 30,-

Vol. 414: T. Kambayashi, M. Miyanishi and M. Takeuchi, Unipotent Algebraic Groups. VI, 165 pages. 1974. DM 20,-

Vol. 415: Ordinary and Partial Differential Equations, Proceedings of the Conference held at Dundee, 1974. XVII, 447 pages. 1974. DM 37,-

Vol. 416: M. E. Taylor, Pseudo Differential Operators. IV, 155 pages. 1974. DM 18,-

Vol. 417: H. H. Keller, Differential Calculus in Locally Convex Spaces. XVI, 131 pages. 1974. DM 18,-

Vol. 418: Localization in Group Theory and Homotopy Theory and Related Topics Battelle Seattle 1974 Seminar. Edited by P. J. Hilton. VI, 171 pages. 1974. DM 20,-

Vol. 419: Topics in Analysis - Proceedings 1970. Edited by O. E. Lehto, I. S. Louhivaara, and R. H. Nevanlinna. XIII, 391 pages. 1974. DM 35,-

Vol. 420: Category Seminar. Proceedings, Sydney Category Theory Seminar 1972/73. Edited by G. M. Kelly. VI, 375 pages. 1974. DM 32,-

Vol. 421: V. Poénaru, Groupes Discrets. VI, 216 pages. 1974. DM 23,-

Vol. 422: J.-M. Lemaire, Algèbres Connexes et Homologie des Espaces de Lacets. XIV, 133 pages. 1974. DM 23,-

Vol. 423: S. S. Abhyankar and A. M. Sathaye, Geometric Theory of Algebraic Space Curves. XIV, 302 pages. 1974. DM 28,-

Vol. 424: L. Weiss and J. Wolfowitz, Maximum Probability Estimators and Related Topics. V, 106 pages. 1974. DM 18,-

Vol. 425: P. R. Chernoff and J. E. Marsden, Properties of Infinite Dimensional Hamiltonian Systems. IV, 160 pages. 1974. DM 20,-

Vol. 426: M. L. Silverstein, Symmetric Markov Processes. IX, 287 pages. 1974. DM 28,-

Vol. 427: H. Omori, Infinite Dimensional Lie Transformation Groups. XII, 149 pages. 1974. DM 18,-

Vol. 428: Algebraic and Geometrical Methods in Topology, Proceedings 1973. Edited by L. F. McAuley. XI, 280 pages. 1974. DM 28,-